communities and conservation

United Nations Educational, Scientific and Cultural Organization

communities and conservation

natural resource management in South and Central Asia

Editors
ASHISH KOTHARI
NEEMA PATHAK
R.V. ANURADHA
BANSURI TANEJA

Sage Publications
NEW DELHI • THOUSAND OAKS • LONDON

First published in 1998 by

Sage Publications India Pvt Ltd
M–32 Market, Greater Kailash–I
New Delhi–110 048

Sage Publications Inc.
2455 Teller Road
Thousand Oaks, California 91320

Sage Publications Ltd
6 Bonhill Street
London EC2A 4PU

Published by Tejeshwar Singh for Sage Publications India Pvt Ltd, lasertypeset by Line Arts, Pondicherry and printed at Chaman Enterprises, Delhi.

Library of Congress Cataloging-in-Publication Data

Communities and conservation: natural resource management in South and Central Asia / editors, Ashish Kothari ... [et al.].

p. cm. (cloth: alk. paper) (pbk.: alk. paper)

Papers first presented at a regional workshop on Community-Based Conservation: Policy and Practice, sponsored by UNESCO and organized by the Indian Institute of Public Administration in New Delhi, February 9–11, 1997—CIP galley.

Includes bibliographical references.

1. Conservation of natural resources—South Asia—Citizen participation—Congresses. 2. Conservation of natural resources—Asia, Central—Citizen participation—Congresses. 3. Natural resources—South Asia—Management—Citizen participation—Congresses. 4. Natural resources—Asia, Central—Management—Citizen participation—Congresses. I. Kothari, Ashish, 1961–

S934.S64C65 333.7'2'0954—dc21 1998 98–24648

ISBN: 0–7619–9279–0 (US-hb) 81–7036–739–5 (India-hb)
0–7619–9280–4 (US-pb) 81–7036–740–9 (India-pb)

Sage Production Team: Jaya Chowdhury, N.K. Negi and Santosh Rawat

To the hundreds of communities in South and Central Asia who, against all odds, are conserving their natural resources…and to those government officials and non-governmental people who are supporting them.

MAP OF SOUTH AND CENTRAL ASIA (WITH CASE STUDY SITES)

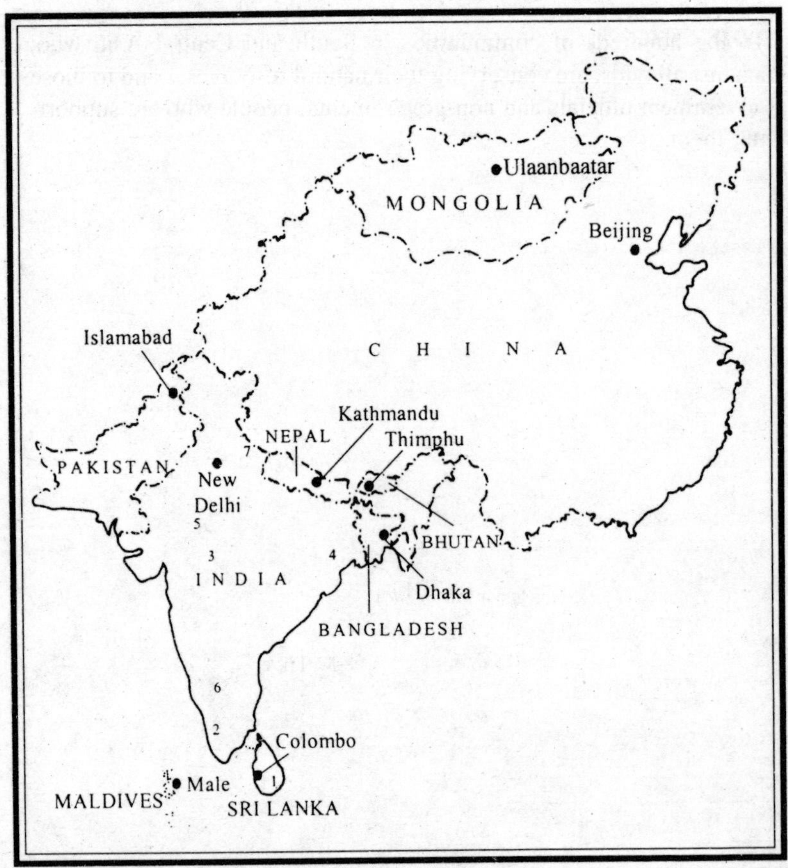

CASE STUDY SITES

1. Rekawa Lagoon, Sri Lanka
2. Chinnar Wildlife Sanctuary, Kerala
3. Kailadevi Wildlife Sanctuary, Rajasthan
4. Dalma Wildlife Sanctuary, Bihar
5. Alwar District, Rajasthan
6. Biligiri Rangaswamy Temple Sanctuary, Karnataka
7. Nanda Devi Biosphere Reserve, U.P.

Contents

PART 4: Case Studies

List of tables

List of maps and figures

Maps

Figures

List of abbreviations

ACAP	Annapurna Conservation Area Project
ACF	Assistant Conservator of Forests
ADB	Asian Development Bank
ADR	Assistant District Registrar
AIADMA	All India Drug Manufacturers Association
AIR	All India Reporter
AKRSP	Aga Khan Rural Support Programme, Ahmedabad
APNE	Agency for Protection of Nature and Environment
BAP	Biodiversity Action Plan
BCN	Biodiversity Conservation Network
BHCP	Biodiversity Hotspots Conservation Programme
BOSTID	Board on Science and Technology for International Development
BRSP	Baluchistan Rural Support Programme
BR	Biosphere Reserve
BRT	Biligiri Rangaswamy Temple
BSP	Biodiversity Support Programme
CA	Conservation Area
CAMPFIRE	Communal Areas Management Plan for Indigenous Resources
CAPART	Council for the Advancement of People's Actions and Rural Technology
CBC	Community-Based Conservation
CBCM	Community Biodiversity Conservation Movement
CBO	Community-Based Organization
CBRM	Community-Based Resource Management
CBS	Central Bureau of Statistics
CCD	Coast Conservation Department
CCR	Community-Controlled Research
CCs	Cluster Committees
CD	Cooperatives Department
CES	Centre for Ecological Sciences
CFDP	Community Forestry Development Project
CFM	Community Forest Management
CFR	Common Food Resource
CH	Chancery Division
CIDA	Canadian International Development Agency
CITES	Convention on International Trade in Endangered Species

CLEO	Crown Lands Encroachment Ordinance
CPT	Cattle-proof Trenches
CR	Critically Endangered
CRMP	Coastal Resources Management Project
CSIR	Council of Scientific and Industrial Research
CSW/CSWI	Committee on the Status of Women in India
CWS	Chinnar Wildlife Sanctuary
CWDS	Centre for Women's Development Studies
CZMP	Coastal Zone Management Plan
DD	Data Deficient
DDC	District Development Committee
DFO	Divisional Forest Officer
DNPWC	Department of National Parks and Wildlife Conservation
DOB	Department of Botany
DOF	Department of Forests
DOT	Department of Tourism
DRDA	District Rural Development Agency
DS	Divisional Secretary
DSD	Divisional Secretary Division
DWCRA	Development of Women and Children in Rural Areas
DWLC	Department of Wildlife Conservation
DWS	Dalma Wildlife Sanctuary
EC	European Commission
EC*	Executive Committee
EEZ	Exclusive Economic Zone
EFYP	Eighth Five Year Plan
EIA	Environmental Impact Assessment
EN	Endangered
EX	Extinct
EW	Extinct in Wild
FAO	Food and Agriculture Organization
FD	Forest Department
FECOFUN	Federation of Community Forestry Users in Nepal
FFO/FFPO	Fauna and Flora Protection Ordinance
FO	Farmers' Organization
FPC	Forest Protection Committee
FPO	Food Products Order
FPS	Forest Policy Statement
FPU	Food Processing Unit
FRLHT	Foundation for Revitalization of Local Health Traditions
FSI	Forest Survey of India
FSMP	Forestry Sector Master Plan
FUG	Forest Users' Group

GATT	General Agreement on Tariffs and Trade
GB	General Body
GBPIHED	G.B. Pant Institute of Himalayan Environment and Development
GDP	Gross Domestic Product
GEF	Global Environment Facility
GFD	Gujarat Forest Department
GKMSS	Ghad Kshetra Mazdoor Sangharsh Samiti
GO	Government Order
GOI	Government of India
GONWFP	Government of North-West Frontier Province
GOP	Government of Pakistan
GOSL	Government of Sri Lanka
GOWB	Government of West Bengal
GRAIN	Genetic Resources Action International
GTZ	German Agency for Technical Development
HARC	Himalayan Action Research Centre
HEC	Human–Elephant Conflict
HFD	Haryana Forest Department
HMG	His Majesty's Government
HMG/N	His Majesty's Government of Nepal
HP	Himachal Pradesh
HPU	Honey Processing Unit
HR	Hunting Reserve
HRMS	Hill Resource Management Society
ICAR	Indian Council of Agricultural Research
ICDP	Integrated Conservation Development Project
ID	Irrigation Department
IDS	India Development Service
IDA	International Development Agency
IFA	Indian Forest Act
IG	Implementation Grant
IIED	International Institute for Environment and Development
IIMI	International Irrigation Management Institute
IIPA	Indian Institute of Public Administration
ILO	International Labour Office
IMD	Irrigation Management Division
INR	Indian Rupees
IPR	Intellectual Property Rights
ISM	Indigenous System of Medicine
IUCN	International Union for Conservation of Nature and Natural Resources (now World Conservation Union)
JFM	Joint Forest Management
JPAM	Joint Protected Area Management

KFRI	Kerala Forest Research Institute
KFW	Kreditan stalt for wiederaufbau
KWS	Kailadevi Wildlife Sanctuary
LAMPS	Large-scale Adivasi Multi-purpose Societies
LCK	Local Community Knowledge
LDO	Land Development Ordinance
LI	Local Institution
LR	Low Risk
LSO	Land Settlement Ordinance
MAB	Man and Biosphere
MACNE	Mongolian Association for Conservation of Nature and the Environment
M/ALF	Ministry of Agriculture, Lands and Forestry
MBCP	Makalu–Barun Conservation Project
MC	Managing Committee
MBCA	Makalu-Barun Conservation Area
MEP	Mahaweli Environmental Project
MFP	Minor Forest Produce
MFSC	Ministry of Forest and Soil Conservation
MNE	Ministry of Nature and Environment
MOA	Ministry of Agriculture
MOEF	Ministry of Environment and Forests
MOFA	Ministry of Fisheries and Agriculture
MPAs	Marine Protected Areas
MPCA	Medicinal Plant Conservation Area
MPCP	Medicinal Plant Conservation Project
MPDA	Medicinal Plant Development Area
MPFS	Master Plan for the Forestry Sector
MPHRE	Ministry of Planning, Human Resources and Environment
MRS	Marine Research Section
MS	Multiple Shoot
MSL	Metres above Sea Level
MTCA	Ministry of Tourism and Civil Aviation
M/TEWA	Ministry of Transport, Environment and Women's Affairs
MWR	Ministry of Water Resources
NARA	National Aquatic Resources Agency
NARC	National Agricultural Research Council
NAREPP	Natural Resources Environment and Policy Project
NARESA	Natural Resources Energy and Science Authority of Sri Lanka
NAS	National Academy of Sciences
NBS	Nari Bikas Sangh
NCPE	National Commission for the Protection of Environment
NCS	National Conservation Strategy

NDBR	Nanda Devi Biosphere Reserve
NDNP	Nanda Devi National Park
NEA	National Environmental Act
NEA	Nepal Electricity Authority
NEPAP	Nepal Environmental Policy and Action Plan
NFP	National Forest Policy
NGO	Non-governmental Organization
NHM	Natural and Historical Monuments
NP	National Park
NPC	National Planning Commission
NR	Nature Reserve
NRM	Natural Resource Management
NRSP	National Rural Support Programme
NTFP	Non-timber Forest Produce
NWFP	North-West Frontier Province
NWP	North-Western Province
OBC	Other Backward Classes
ODA	Overseas Development Administration
OGDs	Other Government Departments
PA	Protected Area
PC	Privy Council/Provincial Council
PF	Panchayat Forest
PG	Planning Grant
PI	People's Institutions
PL	Post-larvae
PMS	Pragatisheel Mahila Sabha
PPF	Panchayat Protected Forest
PRA	Participatory Rural Appraisal
PS	Pradeshiya Sabha
RA	Regional Assessment
R&D	Research and Development
RAFI	Rural Advancement Foundation International
RCF	Reserve Civil Forest
RF	Reserve Forest
RLFCS	Rekawa Lagoon Fishermen's Cooperative Society
RITICOE	Ritigala Community-Based Development and Environment Management Foundation
RNA	Royal Nepal Army
S	Sanctuary
SAARC	South Asian Association for Regional Cooperation
SAM	Special Area Management
SARTHI	Social Action for Rural and Tribal Inhabitants of India
SC	Supreme Court

SCC	Supreme Court Cases, Volume 1
SCOR	Shared Control of Natural Resources
SDA	Southern Development Authority
SIFP	Self-initiated Forest Protection
SIFPG	Self-initiated Forest Protection Group
SLO	State Lands Ordinance
SNC	Save Nilgiri Campaign
SNR	Strict Nature Reserve
SOE	State of the Environment
SPA	Strictly Protected Area
SRSC	Sarhad Rural Support Corporation
SSD	Society for Sustainable Development
ST	Scheduled Tribe
SWRC	Social Work and Research Centre
TAAS	Travel Agents Association of Sikkim
TAF	The Asia Foundation
TBGRI	Tropical Botanical Garden Research Institute
TBS	Tarun Bharat Sangh
TERI	Tata Energy Research Institute
TDCC	Tribal Development Cooperative Corporation
TGCS	Tree Growers' Cooperative Society
TRIPS	Trade Related Intellectual Property Rights
TRR	Traditional Resource Rights
TU	Tribhuvan University
UMB	University of Massachusetts/Boston
UN	United Nations
UNCED	United Nations Conference on Environment and Development
UNDP	United Nations Development Programme
UNESCO	United Nations Educational, Scientific and Cultural Organization
UPOV	Union for the Protection of New Varieties of Plants
URI	University of Rhode Island
USAEP	United States Asia Environmental Partnership
USAID	United States Agency for International Development
VDC	Village Development Committees
VFMPSs	Village Forest Management and Protection Samitis
VFPMC	Village Forest Planning and Management Committee
VGKK	Vivekananda Girijana Kalayana Kendra
VIKSAT	Vikram Sarabhai Centre for Development Interaction
VPF	Van Panchayat Forest
VSS	Van Suraksha Samiti/Van Samrakshana Samiti
VU	Vulnerable
WB	World Bank

WBFD	West Bengal Forest Department
WC	Watershed Committee
WCMC	World Conservation Monitoring Centre
WCPA	World Commission on Protected Areas
WEC	Water and Energy Commission
WGSP	Wayamba Govi Sanwardana Padanama
WHO	World Health Organization
WJMS	Wana Jana Mithuro Sanvidanaya
WLO	Waste Lands Ordinance
WLPA	Wild Life Protection Act
wrt	with respect to
WTO	World Trade Organization
WWF	World Wide Fund for Nature

Foreword

Improving scientific understanding of natural and social processes relating to humanity's interactions with its environment, providing information useful to decision-making on resource use, promoting the conservation of genetic diversity as an integral part of land management, enjoining the efforts of scientists, policy-makers and local people in problem-solving ventures, mobilizing resources for field activities, strengthening of regional cooperative frameworks—these are some of the generic characteristics of UNESCO's Man and Biosphere (MAB) Programme.

Community-based conservation and management strategies of biological resources are amongst the important elements under the MAB Programme. The present volume, *Communities and Conservation*, has its origins in a regional workshop on Community-Based Conservation: Policy and Practice, held in New Delhi, India, from 9 to 11 February 1997. This workshop was organized by the Indian Institute of Public Administration in New Delhi and was sponsored by UNESCO under its MAB Programme. The publication contains a wealth of information—it includes presentations made during this workshop which provide an overview of community-based conservation efforts which are going on regionally and worldwide—and discusses their strengths and their weaknesses.

Two main products have been an outcome of this workshop—a detailed report with recommendations, and this publication. We hope that this comparative assessment of ongoing experiences in community-based conservation, both inside and outside of protected areas/biosphere reserves, and its implications for policy, will be of interest to national governments particularly in South and Central Asia, regional and institutional forums formulating and implementing legislations, donor agencies, scientists, researchers, NGOs, students, and, importantly, planners and decision-makers who shape, move and implement policies related to conservation and resource management.

It is a pleasure for UNESCO to thank Mr M. Wadhwani, then Director, Indian Institute of Public Administration, New Delhi, for hosting this workshop. Our very special thanks are due to Mr Ashish Kothari, Ms Neema Pathak, Ms R.V. Anuradha and to my colleague, Ms Sudha Mehndiratta, who were responsible for the implementation of this workshop, and to all those who prepared the synthesis papers. We thank the IIPA team, for their valuable support, diligence and perseverance in seeing the present volume into print.

Professor Moegiadi
Director

Preface

This volume is a compilation of papers first presented at a regional workshop on Community-Based Conservation: Policy and Practice, sponsored by UNESCO and organized by the Indian Institute of Public Administration in New Delhi from 9–11 February 1997. The workshop focused on countries of South and Central Asia. The papers were subsequently revised by the authors, and edited to complement each other in analyzing the main subject: the factors behind the success and failure of community participation in natural resource conservation.

Countries in the region, as elsewhere in the world, have increasingly been exploring decentralized, community-based conservation (CBC) approaches. What is perhaps most remarkable about these approaches is their range and diversity: from community-initiated efforts to state-initiated ones, from those in which communities are only receiving some benefits to those in which they are the sole decision-making and benefiting party, from those which depend on external inputs like NGOs and donor agencies to those which are running entirely on local steam.

But the move towards CBC, and the resulting processes, is not without hurdles and problems. The major challenges have been to build on local knowledge and practices, safeguard livelihoods based on sustainable use of resources, provide incentives for continued conservation, devise alternatives for unsustainable resource extraction, tackle inequities in local decision-making and benefit-sharing, and enhance the management capacity of community-based institutions.

While considerable experience on CBC has by now accumulated in the region, there have not been many attempts at comparing and analyzing the lessons being learnt. The above-mentioned workshop was one such attempt, and this book hopes to fill a part of the gap in analytical and process documentation on the subject. It covers country overviews, issue papers and case studies from India, Maldives, Mongolia, Nepal, Pakistan and Sri Lanka. Papers from Bhutan and Iran, though presented at the workshop, could not be included for various reasons.

The papers in this book are arranged in the following way. Section I provides an overview of issues relating to CBC in general, and with reference to the South and Central Asian region. Section II provides country profiles of CBC in the region. Section III includes papers with an in-depth focus on specific issues related to CBC, mainly from the Indian experience. Section IV contains case studies from specific areas where communities are involved with the management of natural resources in various capacities.

As is the case with any edited volume, several people and organizations contributed to make this possible. UNESCO is gratefully acknowledged for sponsoring the workshop and subsidizing this book. In particular, Sudha Mehndiratta of UNESCO has been extremely supportive. The facilities of the Indian Institute of Public Administration were also critical for the workshop and the editorial work on the book. Warm thanks are due to Sangeeta Kaintura, Vishal Thakre, Virender Anand and Kheema Rawat, for smilingly undertaking the arduous task of keying in and making unending corrections to the manuscript. Assistance in editing and related work was also given by Farhad Vania, Swati Sreshtha, Anuprita Patel and Sunita Rao, the latter two in desperate last-minute circumstances! Thanks are also due to the anonymous reviewer of the manuscript, for very useful comments, and to Omita Goyal, Sumitra Srinivasan, Jaya Chowdhury and others of Sage Publications for all their help. Heartfelt thanks go out to all the contributors, for the revisions they made on their papers, and for tolerating our persistent queries and our butcher-like editing. All the participants of the workshop, and some people whose papers could not be included here, also deserve our gratitude, for their inputs went into the making of the overview paper and the editorial process for other papers.

PART 1

introduction

1

Community-based conservation: Issues and prospects

Ashish Kothari, R.V. Anuradha and Neema Pathak

A sea change is taking place in conservation across the world. From standardized policies and programmes initiated by centralized and urban-based agencies, a slow but definite shift is taking place towards decentralized, site-specific, community-based activities.

What is community-based conservation?

The move towards community-based conservation (CBC) is occurring in highly diverse forms, and therefore any single definition of the phenomenon would be highly simplistic. Broadly, and for the purposes of this book, CBC could be described as conservation of biological diversity[1] (or wildlife) based on the involvement of local communities, in decision-making. This (*a*) *excludes* conservation attempts by official or private agencies which either have no participation of local people, or have participation only in the form of labour; and (*b*) *includes* a whole range of situations from one extreme in which official/private agencies predominantly retain control but consult with local communities in planning or implementation; to the other extreme in which communities are completely in control.

Definitions of the above nature inevitably raise a whole host of questions. This paper attempts to deal with some of these by presenting an introductory overview of CBC in general and in the South and Central Asia region in particular, picking from our own experiences, available literature, and the contribution of other authors in this book.

Why community-based conservation?

There are a variety of reasons why a worldwide shift to CBC is taking place:

1. In virtually all 'developing' countries, local communities continue a day-to-day interaction with the areas and species sought to be conserved; even if not *de jure*, there is *de facto* use. This is the case even with strictly protected areas (PAs). Across the world, perhaps half the PAs are inhabited (Borrini–Feyerabend 1996). In India, over two-thirds of PAs have human habitation; and a larger proportion have some form of human activity apart from tourism (Kothari et al. 1989). Exclusion is not always desirable or possible.

2. Attempts at exclusion have created severe conflicts, often violent. A mid-1980s survey of 222 PAs in India revealed that at least 47 (21 per cent) had had physical clashes between people and forest officials (Kothari et al. 1989). In some areas, people are demanding dereservation of the PA status, since they legitimately perceive their alienation from the resources they need for survival, and in some cases from the habitats they hold culturally valuable, as being unjust. Resolving such conflicts requires that communities be treated as equal partners in conservation attempts.

3. Considerable wildlife still exists outside of PAs, in lands/wetlands which are owned by or are under control of local communities. CBC is possibly the only option in these areas.

4. All over the world, it is being realized that central agencies are simply not able to carry out the task of conservation, being understaffed, underfunded, ill-trained and ill-equipped to handle the myriad threats that habitats and species face. Public support for conservation therefore becomes a necessity. Indeed, local people because of their day-to-day interaction and dependence on these areas are often at the forefront of protests against the degradation caused by outside commercial interests. Mining, which was destroying the forests of the Sariska Tiger Reserve in western India, was fought against and stopped by the local people through both legal action and on-the-ground agitations.

5. Political support for conservation is declining in many countries, especially where it is seen (even if mistakenly) to be a hindrance to poverty alleviation or to developmental aspirations, or where it hampers the activities of powerful vested interests. Without a substantial public base, such political support is unlikely to be forthcoming.

6. The assumption that conservation is only possible through the exclusion of human activities, is under serious challenge, as researchers show that there can be many (though certainly not all) situations in which human activities and desirable levels of biodiversity can co-exist (Arhem 1985; Michon and Foresta 1990; Posey 1985; Saberwal 1996; Western 1989). This possibility of co-existence provides the basis for a more inclusionary approach to conservation, though conservationists may still be justified in keeping at least some areas as 'inviolate'.

7. Experience suggests that the costs involved in conservation may go down once CBC is in place, as the community shares in responsibilities like patrolling, fire-fighting and regenerative and protective measures.

8. Local communities, especially those with a long tradition of resource use in a particular area, hold in-depth knowledge and experience of wildlife and habitats, which can be invaluable for conservation efforts.

9. Not least, people everywhere are demanding a greater voice in decision-making, and aspiring to re-establish some control over the resources which sustain their lives and livelihoods. The move towards CBC is therefore as much a matter of fundamental human rights and social justice, as of necessity. It is both a result of a societal move towards democratic functioning, and a potential stimulant to such functioning.

The above issues are dealt with in detail in a number of recent publications, which also bring out the worldwide move towards participatory conservation models (Kemf 1993; Kothari et al. 1996; West and Brechin 1991; Western and Wright 1994b).

Issues for resolution

Who is the local community?

One fundamental question which is often sidestepped in the advocacy of CBC is: Who is the local community? Lele et al. and Sarin et al. point out in this book, that the use of the term '*the* community' could sometimes be an over-simplification as it assumes that there is one single homogeneous entity. DeCosse and Jayawickrama (in this book) add that the term 'community' presupposes the potential of a group of stakeholders to come together to manage a common resource, even though the conditions for this may not exist.

Various definitions and criteria have been used to identify local communities. One is the group of people who have a *historic relationship* with the land or resources, and therefore a claim to them. However, by accepting traditional use patterns as the primary determinant for the selection of decision-makers and right-holders, other criteria, like *resource-dependence* or management capacity, tend to get sidelined (Sarin et al. in this book). Sometimes, traditional users or right-holders may no longer be dependent on adjoining forests, whereas a more recently settled group may be in a more disadvantaged position. Can the rights which are recognized for the traditional users be transferred to the more dependent, and therefore more needy group?

In the Shiwalik hills of Haryana, northern India, there is a conflict between Jholuwal Jats (traditional users), who are only marginally dependent on forests to supplement their fodder requirements, and Moinawali Banjaras (more recent settlers) who are primarily dependent on grass from the forests to make ropes, their primary livelihood source (Sarin et al. 1996). Who has the primary or prior right in such instances? Such problems are very often associated with displacement of people resulting from 'development' projects or migration resulting from population increase; groups settling themselves, or being settled by official agencies in areas where resources are already being used by local communities. The point is, for a law to be effective, it should have a clear sense of who it is addressing, and may have to be flexible enough to accommodate at least *bona fide* requirements of even relative newcomers. However, there also needs to be some cut-off date for the acceptance of people's claims, as otherwise the claimants can increase indefinitely to the detriment of the conservation status of the area.

Physical proximity of the persons/community to the resource may also be a useful criterion to apply. The Aranya Panchayat Samithi in Uttara Kannada, southern India, referred to later, tackles this problem by stipulating that the right to elect members of the Panchayat Samithi lies only with resident landowners and cattle owners. Another example is that of the Hill Resource Management Societies (HRMSs) of Sukhomarji and Nada villages in Haryana, northern India, where communities decided that permanent residence in the villages next to the forest was a non-negotiable condition for HRMS membership (Sarin et al. 1996). This helped to ensure that non-forest-dependent individuals living in urban or distant settlements could not usurp control over forest resources on which local residents were highly dependent.

However, the proximity rule can have its own problems: it can leave out highly resource-dependent people, including both settled populations at a distance as also nomadic populations, while involving those who might be staying in the vicinity but who are not dependent. In Attidiya Sanctuary in Sri Lanka, the local inhabitants do not depend on the sanctuary for livelihood, but rather on wages from the nearby cities and towns (DeCosse and Jayawickrama in this book). It is important to note that pastoralist communities' traditional special arrangements with villages en route, such as the use of their fallow lands for livestock grazing, which in turn helped to enrich the land with manure, are fast breaking down because of various reasons (Agarwal and Narain 1989). Handling the claims of nomads in such a situation is critical from both the social justice and conservation points of view.

For the above and other reasons, it is important to define primary, secondary and tertiary stakeholders. This, however, should be flexible, based on subsistence, resource-dependency, proximity, antiquity, capacity and willingness to participate in conservation and other factors. DeCosse and Jayawickrama (in this book), present a detailed analysis on this issue.

Role of local community knowledge/practices

Wildlife conservation programmes in most countries have emphasized the use of 'modern' wildlife/ecological science and practices, and the involvement of 'experts' trained in these. The knowledge and practices of local communities are only now being acknowledged by conservationists, as being of equal importance and relevance. Such knowledge/practices comprise a complex and dynamic mix of the age-old with the new, the theoretical with the practical. Papers by Gadgil and Ramakrishnan in this book deal extensively with this aspect.

Characteristics of local community knowledge/practice

Local community knowledge/practice (LCK) has characteristics which provide a host of advantages for attempts at CBC. Rules regarding resource use are embedded in cultural and religious systems which give them a strong legitimacy, going beyond scientific/ecological arguments. For example, sacred groves often protect important catchments or resource banks (an ecological function), but are protected as the homes of ancestors or deities (a cultural function or justification). Building conservation strategies upon LCK therefore ensures far better acceptance of the strategies.

Second, the range of LCK is often astounding; it incorporates information, attitudes, values, skills and practices concerning a high diversity of biological resources. There is also considerable dynamism and innovation in LCK, especially in the forms of resource use. Typically though, change is gradual, which is why LCK appears to be static. Finally, a substantial part of LCK has some kind of communal or shared ownership and use, though individual innovation and privately held knowledge/practices are also common.

There are, however, also weaknesses, e.g., a general inability to cope with sudden or large-scale changes. Since all parts of the society are intricately linked with LCK, much like a rainforest, a change in one part can have a chain effect on others. The introduction of market mechanisms, or government-controlled institutions, will affect local traditional knowledge, which in turn will affect the way resources are managed/used. In addition, there are often major gaps in local knowledge, relating especially to species which are not in use or not in some way impinging on the lives of the villagers, such as small fauna or micro-organisms. Finally, some misconceptions about species lead to outright elimination, e.g., in many parts of India, snakes are killed regardless of whether they are poisonous or not.

Erosion of LCK

Across the world, LCK systems have been severely eroded by a variety of factors including the displacement/devaluation of traditional knowledge and practice by the modern 'scientific' knowledge system. Local traditions of medicinal plant use, or of conservation, have been devalued and replaced by the modern allopathic system, and state-sponsored practices of conservation, respectively. In more recent times, local knowledge has even been appropriated by the state and private sectors in the form of intellectual property rights (IPRs, such as patents) on products and processes derived from LCK. The institutional takeover of resources by the state/private sector, including the displacement or infiltration of local village institutions by political parties, government agencies and market forces, further erodes LCK systems. These factors have been brought out by a number of contributors to this book, including Choudhury and Gadgil.

The erosion of LCK directly leads to negative consequences on biodiversity, as the basis for well-tried systems of management, use and conservation break down. This can be seen especially in the case of common property resources; instead of a tightly regulated system of use by the community, the system tends towards free-for-all use, turning the resource into an open access property. Forests, wetlands, pastures, all have suffered serious consequences. Habitats and species once protected for their sacred qualities, are no longer revered and have been over-exploited. Incentives and systems of innovation, e.g., in agriculture, disappear with the increasing dependence on government agencies. As knowledge of the intricacies of habitats and species disappears, opportunities for their conservation also diminish.

Revival and protection of LCK

In the face of the above erosion, a number of efforts are being made to protect and revive LCK, either in the traditional form, or in new innovative forms. And a number of interesting initiatives at empowering local communities with appropriate information, skills and technologies, suggest that there is substantial meeting ground between LCK and modern knowledge systems. For instance, Rajendra Singh shows in his case study for this book, that a revival of the local traditional knowledge of water harvesting was used in conjunction with modern engineering science to convert drought-hit areas into water-surplus ones. Participatory mapping of resources is another field where integration has been successfully attempted (Poole 1995); yet another is that of medicinal plants, as hinted at by Darshan Shankar's article (in this book).

Not much has been done, so far, to expose students and experts of formal modern systems to the LCK systems. The formal/modern sector needs to learn the underlying premises, methods and other aspects of LCK, recognizing it as a valid form of knowledge. There is a long way to go before the local community forest expert can work with the modern forestry scientist (or the farmer with the formal agricultural scientist, the fisherperson with the modern fisheries expert), without the latter assuming a position of superiority over the former. Also, farmers, forest dwellers and fisherfolk need to be equal partners in formal R&D programmes, the two sectors learning from each other. The innovative use of a village study group in central India, described later in the section on institutional structures, is an example of such mutual learning.

There is then the additional issue of the increasing tendency of outside agencies and individuals to appropriate LCK for their own purposes, and even apply monopolistic IPRs on them or their derivatives without providing any returns. In the face of this, communities need to be empowered with community intellectual rights, which safeguard their knowledge and the values they attach to it, and ensure that they are fairly benefited by the use of this knowledge by the outside world (benefit-sharing is discussed in the next section). Several models of such community IPRs are being suggested all over the world (Nijar 1995; Posey and Dutfield 1996); some have suggested the use of appropriate individual IPRs also for rural innovators (Gupta 1996); in India, an innovative strategy of documenting LCK to benefit from its use is the preparation of Community Biodiversity Registers.

Creating a stake, sharing benefits

The current conservation scenario in India and many other developing countries is such that while the local communities pay the major cost of conservation (loss of access to resources, etc.), the benefits are mainly enjoyed by the non-local national/international communities (so-called 'national or global benefit'). This feeling is heightened when tangible economic benefits being derived from conservation areas are cornered by outsiders. For instance, it is estimated that at Keoladeo (Bharatpur) National Park in Rajasthan, India, almost all the substantial tourism revenue generated goes to private tourist agencies or the government, while the local communities suffer a loss of about INR 20 million (US$ 60,000) per year on account of agricultural losses due to waterlogging related to the wetland management in the Park, restrictions on fuel and fodder, etc. (Murty 1996).

This scenario does not provide a conducive environment for any step towards CBC. While local communities can be the best protectors of their natural surroundings, they can also be the worst destroyers if there are no stakes involved. The present alienation and indifference of the local communities towards the 'government forests' can be dissolved if they themselves become the prime recipients of the benefits accruing from conservation. Studies show that forest regeneration and protection measures in eastern India were initiated in most cases by the local people, when the natural resources on which their livelihood depended went down to critically low levels, and the government's Joint Forest Management (JFM) orders (see Box 1.1 on p. 44) appeared to them to offer substantial benefits (Poffenberger 1994).

Can conservation programmes actually be reoriented to benefit local people, and are there situations in which all those who have a stake in conserving a particular resource can share its benefits? This question is especially pertinent in the case of wildlife PAs, which conventionally have been viewed as non-exploitable, and therefore with little or no 'benefits' to 'share'. However, conservation agencies are realizing that in many cases substantial benefits can accrue to local people without sacrificing conservation goals. Several questions are relevant here which are discussed in the sections that follow.

Who are the stakeholders?

Apart from the local communities, the other stakeholders in wildlife conservation are the state agencies involved in its management, the non-local rural/urban, national/international communities, various NGOs, and other market and industrial interests. However, the local communities (or sections within them) often fulfill most criteria to be defined as 'primary' stakeholders (as discussed in a section earlier). All other beneficiaries could perhaps be considered secondary stakeholders, since their dependence on this particular area is not as fundamental, i.e., alternate avenues are available to them.

What are the benefits to local communities?

Areas and species which are being conserved can provide benefits of several kinds:

Subsistence benefits: The most critical relationship of local rural communities with their natural surroundings is dependence for subsistence (fuel, fodder, food, fibre, fertilizer, medicines and other products) and for livelihood. Unfortunately, restriction of access to such resources has often been the result of official conservation programmes in most countries (see also, section on legal issues). The ability to securely meet their subsistence requirements can create a lasting stake for conservation; indeed, this may have been one major driving force behind the evolution of sustainable resource use traditions in local communities (see the earlier section on local community knowledge). Recognizing this, government agencies in many countries are providing legitimate access to such resources (Suri 1996; Das and Christopher in this book). In the Annapurna Conservation Area in Nepal, graziers with pasture privilege (living in the forest with their cattle), are responsible for looking after the forest in exchange for the benefits they get (Gurung 1995).

Economic benefits: *(i) Forest and other produce:* Collection and sale of non-timber forest produce (NTFP), and aquatic and grassland produce, are an important source of economic survival. It is estimated that NTFP income accounts for 55 per cent of the total employment in the forestry sector in India (Saigal et al. 1996). In the Harda forests of central India where JFM is relatively successful, the revenue generated by the village-level forest protection committees (FPCs) from the sale of grass was calculated to be INR 1,00,000 (US $3,000) in one year (Rathore 1996). It is no wonder then that attempts to restrict access to NTFP and other produce in PAs have met with considerable hostility. In India, a recent amendment to the Wild Life (Protection) Act has deemed all activities which are not in the interests of wildlife as being prohibited; many states are interpreting this to mean that virtually no activity can be carried out, including any form of NTFP collection and fishing.

Government agencies and NGOs in many places are exploring possibilities of integrating livelihood options with conservation programmes, including within PAs (with appropriate zonation to demarcate areas for use and those which are off-limits) (Suri 1996). In Rajaji National Park, northern India, an area of severe conflict between the Forest Department (FD) and local communities, a recent agreement is allowing the harvesting of *bhabbar* grass (for ropemaking) from within the Park, by villagers. In turn, recipients will be forming FPCs (Vania 1997). Bhatt, as also Lele et al. (in this book) describe several experiments where value-enhancement through enterprise is being used as a means of ensuring a better livelihood as also creating a stake in conservation.

(ii) Employment: Conservation and habitat/species regeneration efforts can generate substantial employment, though this is as yet rare. In Khunjerab National Park in Pakistan, 80 per cent of new employment opportunities are reserved for local people (Slavin 1993); in India, at Keoladeo and Corbett National Parks, local youths are employed as wildlife guides after training.

(iii) Tourism revenues: Even conservationists who are against exploitation of resources from within PAs (Singh 1994), are open to the idea of tourism being one potential way of returning benefits to local people. This does not at present happen, as Pimbert and Pretty (in this book) point out. However, if handled sensitively, tourism can provide substantial benefits to the local people, as illustrated in the case of Nepal's Annapurna Conservation Area (Gurung 1995; Wells 1994). Having said this, we would hasten to add a strong note of caution. There is very little monitoring of the ecological and social impacts of tourism around the globe, but available studies indicate that the effect is usually negative. Strict procedures for environmental and social impact assessment are needed for every major tourism project. Rather than imposing the project on local communities, their participation in planning and implementing it should be mandatory from the start.

(iv) Returns from outside commercial activities: In some limited cases, local communities could actually be linked to industries and research agencies which use biological resources that can be sustainably harvested from conservation areas. Linked to this is the increasing demand that the larger society, including industry and the formal research sector, pays for or returns appropriate benefits for the commercial use of LCK. This demand has received a worldwide boost due to provisions of the United Nations Convention on Biological Diversity. A number of tricky issues related to this are to be sorted out: what kind of returns are appropriate, who should decide them, who should the returns go to and how, how will communities ensure that they are getting a fair return and that their knowledge/resources will not be misused?

In India, the Tropical Botanical Garden Research Institute (TBGRI) in Kerala has helped an ayurvedic institute develop a drug based on the knowledge of the Kani tribe of southern India, in return for which the institute will pay royalties to TBGRI and the tribals (Pushpangadan 1996). Unfortunately, there is a lack of clarity regarding the participation of a section of the Kani tribals in the process, and there is no mechanism to secure tribal rights to forest produce (including to the plant being used for the drug) (Anuradha 1998). Nevertheless, it represents a step forward in the recognition that the contribution of local communities deserves returns from the wider society.

But as Darshan Shankar (in this book) discusses, there are a number of drawbacks in these arrangements. In many cases the contractors and middlemen with the intention of reducing the buying cost encourage poor people to extract forest produce illegally, which by virtue of being illegal leads to faulty and unsustainable extraction. Organizing such collectors and eliminating the contractors and middlemen can prove beneficial to both the community and the ecosystem. In some areas, government agencies like the Large-scale Adivasi Multipurpose Society (LAMPS) and the Tribal Development Cooperative Corporation (TDCC) have been created to eliminate these middlemen but are often known to act more like the middlemen themselves (Fernandes et al. 1988).

(v) Compensation for wildlife damage and opportunity costs: Damage to crops, livestock and human health/life by wild animals is a common complaint in areas of conservation, particularly PAs (for India, see Kothari et al. 1989; for Sri Lanka, see DeCosse and Jayawickrama, and Jayatilake in this book). Wildlife legislation in most countries makes it difficult or impossible for local people to take the retaliatory measures they would traditionally have taken, therefore causing considerable resentment against conservation policies. Compensation is usually tedious to claim and tardy in coming. In addition, compensation is not given for crop damage in most situations (except for damage by elephants in India and Sri Lanka).

There is a clear case for enhancing the compensation to local people for such damage. In India, the concept of crop and livestock insurance has been mooted; in addition, villagers have demanded that to avoid bureaucratic delays, recognized village-level committees (e.g., the FPCs set up in many areas) should be able to verify claims for compensation and forward them to the government. There could also be a permanent fund set aside for this purpose. This should also be accompanied by the responsibility of regular vaccination of livestock, to prevent the spread of diseases among wild animals.

Finally, since there may be certain sacrifices being asked of the local communities (e.g., the lost opportunity of being able to exploit certain resources, or converting forests into non-forest uses and wetlands into fields), society as a whole (which benefits from these sacrifices) should be willing to pay the cost being borne by them. Direct payment for protecting habitats and species is one option; others are appropriate development inputs (see later). This could be financed through international, national or sub-national funds, fed by taxes on the biodiversity-based industry (pharmaceuticals, seeds, tourism, etc.), consumer taxes, donations and other sources.

(vi) Development inputs: In recognition of the fact that communities within conservation areas have as much of a right to appropriate developmental inputs as those outside, and given that some such inputs may offset unsustainable demands on natural habitats, several countries have gone in for 'integrated development' or 'eco-development' programmes. These programmes are supposed to provide inputs which, in ecologically appropriate ways, could meet the survival and livelihood needs of local people.

Unfortunately, most such programmes continue not to tackle the basic issues of inequities, devolution of decision-making powers to local communities, and involving LCK systems. In addition, many developmental inputs remain ecologically or socially inappropriate: medical facilities displace indigenous health systems rather than building on and supplementing them, educational facilities impart knowledge with little local relevance and only help further discredit local knowledge systems; and considerable dependence on outside (including foreign) agencies continues.

Nevertheless, several ongoing efforts demonstrate the potential of an approach which promotes conservation-based development; Khattak, Shrestha and Myagmarsuren (in this book) mention some in Pakistan, Nepal and Mongolia, respectively. In Sagarmatha National Park of Nepal, funds from the fines against the offences in the Park are utilized for community works, such as repair of the village chapel and the maintenance of trails (Sherpa 1993).

Many conservationists have faced hostility or apathy when approaching communities with direct pleas to help in conservation, but have been able to establish some credibility after helping to tackle some seemingly unrelated problem, such as water shortage (Singh in this book), or lack of development opportunities (Khattak and Datta in this book).

Unfortunately, the inter-sectoral approach that this calls for is missing from most governmental functioning. There is little attempt to coordinate the programmes and policies of development departments with those of the conservation agency. In the Melghat Tiger Reserve of western India, an attempt was made by an enterprising administrator to change this; he put together the considerable funds that were available with development departments under him into a common programme oriented towards providing local tribal communities with alternative biomass and livelihood options (Pardeshi 1996).

Social, cultural and political benefits: Protection to the ecosystems often effectively protects many ethnic cultures from sudden or disruptive market or developmental influences, providing time to local communities to assess the desirability of such influences and make appropriate responses. It can also lead to the revival of self-dignity and confidence where it has been lost because of prevalent attitudes outside. In many countries, biosphere reserves are being promoted by UNESCO and other agencies as areas where the ecological and cultural landscape can be conserved as an integrated whole; Ramakrishnan and Krishnan (in this book) have advocated further application of this model.

Equally important, if CBC can help communities to revive or achieve greater self-reliance in decision-making with regard to natural resources, this could lead to a more democratic society overall. Lele et al. (in this book) point out that while economic incentives are important, the legitimacy and political empowerment that may result from communities' involvement in CBC can be an added incentive and often a more effective tool for mobilization. On the other hand, Baviskar (in this book) points out that conservation may form only a part of the struggle for social justice and political determination, especially in tribal communities.

How will beneficiaries ensure conservation?

It is critical that access/rights to benefits go hand-in-hand with responsibilities to ensure that the rights are not misused, and that conservation is achieved. This requires *capacity-building* within local communities to conserve and manage resources and achieve sustainable development in the face of myriad pressures; it also needs capacity-building of the conservation officials to respond to community needs and dynamics, and to help in developmental works. Without such orientation, and without addressing other critical issues like equity (see section on equity issues), devolving the right to use resource to local communities carries the risk of even worse destruction (Western and Wright 1994a).

A further challenge is to define what 'use' is good for conservation. What happens when the livelihood needs of communities clash with conservation goals? The fundamental premise of CBC is that alternatives can and should be found when there is potential conflict between the elements of protection and use. Some contradiction may be seen here between those who are arguing from the perspective of people's rights/entitlement and those advocating the conservation agenda (as brought out by Baviskar in this book), which can only be resolved through dialogue and mutual understanding.

Appropriate planning could facilitate an integration of both these goals. There may be certain areas which because of their vulnerability may have to be kept aside as strict reserves. The Sri Lanka Forestry Sector Management Plan, 1995, designates different conservation categories with different kinds of resource use and community participation (Jayatilake et al. in this book). In the context of the Wild Life (Protection) Act, 1972, of India, there have been suggestions by NGOs for expansion of the existing categories of PAs (Bhatt and Kothari 1997); and for amendments to include clear specifications and elaboration of roles and responsibilities for the state officials and the local communities, as also recording the actual use of the area by the people and then assessing their impacts on the area with the involvement of both community and outside experts (Kothari 1996b).

Further built-in safeguards are required to check that the process of benefit-sharing does not simply become another facade for over-exploitation of resources meant for the outside market economy. This is especially the case with biodiversity-based enterprise, where enhancement of the economic value of a resource could well lead to over-exploitation rather than sustainable extraction. The example of the BRT Sanctuary (see Lele et al. in this book) shows that this can be avoided if required steps are taken and safeguards practised.

What is most important is a process of continuous *monitoring* of the impacts of CBC, which is unfortunately missing in most examples of CBC. Bhatt and Lele et al. (in this book), provide examples of where it is taking place.

Equity

Many communities exhibit internal inequities and differences, based on ethnic origin, caste, class, economic state, religion, occupation, social status, gender and age. These inequities can create profound differences in interest, capacity and willingness to invest for the management of natural resources. Case studies from all over the world, and some of the papers presented in this book (notably Sarin's), strongly imply that if these inequities are not identified and addressed, the process of CBC could become another power game with centralized political and social power changing hands from the state to a few influential sections within the communities. As Pimbert and Pretty (in this book) state, equity should entail the sharing of benefits in a way that is commensurate with the varying sacrifices and contributions made, or damages incurred in the community.

Intra-community inequities

(i) Gender inequities: Inequalities between the sexes exist in all sections of society and all communities. Though unacknowledged, in most resource-dependent communities, women do the bulk of collection for commercial and domestic use, especially of resources like fuel, fodder and NTFP. Yet in most communities, women have very little say in decision-making with regard to the management of these resources.

Field observations have indicated the difference in the way men and women often view natural resources. In Gujarat, western India, men wanted to sell off the entire bamboo harvest from their protected forests while women wanted the village requirements to be fulfilled first (AKRSP 1995). In some parts of the Indian Himalaya, where the Chipko Movement has been active, self-empowered women's committees have successfully opposed the selling of the civil forests to the contractors, by their own menfolk. Indirectly, gender inequities can lead to dissatisfaction and even offences (risking social dignity) by the women, when hard-pressed for resources. At Kailadevi Sanctuary, Rajasthan, western India, the decision not to allow axes in the village forest (in order to stop the cutting of live trees), was taken by the all-male village FPCs; for the women, this represents further hardship (Das and Christopher, in this book).

Some state governments in India have made participation of women mandatory in their JFM programmes. While in many places this representation has remained a mere formality (with women only attending meetings as a token, remaining silent or not being taken seriously), in others it has made a difference (see Sarin et al. in this book). The impact of this involvement on conservation is not clear. Clearly, how gender inequities affect conservation, and vice versa, needs to be examined further in all countries.

(ii) Caste/class/ethnicity/other inequities: The concentration of power amongst the upper castes and classes, and the relatively disprivileged status of the lower castes and classes, and of tribals in mixed communities, has serious impacts on the viability of conservation programmes. Experiences from JFM in India and elsewhere, for instance, show that in many self-initiated village FPCs, either the entire population of lower castes, or tribes (if any), are excluded or their representation is nominal (Campbell 1996; Majumdar 1994; Sarin et al. in this book). Several decisions which were imposed on them by upper caste-dominated FPCs such as restrictions on entry into forests, led to further alienation and disregard for the forest/ecosystem health.

In addition, the official forest management policies and practices have usually favoured higher revenue generation over the basic needs of the economically disprivileged class. JFM in many states of India, for instance, is oriented towards the production of timber, and the interests of the poorest of the society who are more dependent on NTFP for their subsistence and livelihood requirements, are given less importance (Fernandes et al. 1988).

Conservation strategies also need to be sensitive to the unequal land distribution within communities, to avoid being unjust, and at times short-lived. In many JFM areas of West Bengal in India, prohibition of tree-felling by landowners caused a serious economic loss to landless people who depended on sale of timber for survival. The programme was doomed but for the timely land reforms by the state government, which softened this loss (Poffenberger 1994). Community management experiences in the five villages in the lower Shiwalik range of the Haryana and Punjab states, North India, revealed that the programme was successful only in the villages where private assets were more equitably distributed (Murty 1996).

A single powerful individual can also undermine the process of conservation, especially if he/she enjoys connections with powerful outside forces. In many tribal societies, chieftains who have contacts with traders or politicians from outside have sold off the forests that their communities used to depend on.

Religious differences within communities, such as among Hindus, Muslims, Christians, etc., also play a very important role in defining the social structure of a community, and the interaction of this community with the surrounding ecosystem. Available studies do not, however, seem to deal with this aspect.

Inter-community inequities

Differential access to a common resource is a problem between communities too, e.g., if one community has greater connections with the market, or with politically powerful people from outside. Such inequities are also common between tribal and non-tribal communities, and of course between rural and urban populations.

Attempts at conservation can fail unless these inter-community inequities are tackled. For example, the village Mordungri from Ranthambhor Sanctuary in Rajasthan, western India, tried to protect the adjacent forests many times without success, as outsiders (including friends and relatives) always sent their cattle to graze in their forests in monsoons (Desai et al. 1996). It was only when they got an 'informal' sanction from the FD to keep the outsiders away, were they able to protect their forests successfully.

As a number of observers have suggested, the most important step is to identify the actual users of the resources sought to be conserved, and to ensure their complete participation in the entire process of planning and management of these resources, regardless of age, sex, class, caste, power, etc. This links up to the identification of primary stakeholders, discussed earlier. Appropriate legal and policy structures would aid considerably in this, but are on their own not adequate; what is also needed is empowerment of the disprivileged sections through the provision of information and skills that would enable them to participate fully. NGOs and other committed outside agencies and individuals (private or official) can play a catalytic role in this process, though the ability of communities to resolve their own inequities should also not be underestimated.

Institutional structures

The need for change

It is abundantly clear from the experience of all countries that government agencies on their own cannot efficiently conserve wildlife. The most significant indication of this is perhaps the recent shift in policies that India, Sri Lanka, Pakistan, Nepal and other countries in the region have witnessed (as discussed in the country papers in this book, and summarized in the section on legal issues in this paper). This is for several reasons, one or more of which may be in operation in an area at any given time: (*a*) Government agencies tend to be rigid in the application of rules, lacking the flexibility to respond to diverse and dynamic situations; (*b*) These agencies have usually ignored, coopted or sometimes even undermined, indigenous structures and institutions, rather than building on or making constructive use of them; (*c*) These agencies almost always lack the human, financial and technical resources/capacity for conservation; (*d*) Such agencies do not manage to gather the support of local communities or other stakeholders, who are alienated by inappropriate laws, policies, institutions and attitudes; (*e*) Coordination between various divisions of an agency, between hierarchical rungs of this agency, and between this and other agencies, is weak; (*f*) Wildlife/forest agencies are always weaker than development agencies, and have to succumb to their pressures/plans to exploit natural habitats and species; and (*g*) Corruption amongst employees often undermines conservation efforts.

Simultaneously, it is also true that traditional local community institutions (such as village-based common property management structures in all countries) have been eroded or find themselves weak in the face of myriad outside forces. Also, as Pimbert and Pretty point out in this book, the undermining, cooption and suppression of local institutions (e.g., of the *panchayat* in India or local village councils in Nepal) is the most debilitating impact of official bureaucracies. Whereas there have been attempts to remedy the situation (for example, by re-introducing *panchayati raj* in India, and by devolving control over forests to the actual user groups in Nepal; see later), there is in general an absence of linkages between formal governance structures and traditional local community institutions in the context of wildlife management.

Participatory institutional structures

Building truly participatory institutions requires several measures. Most critical is the involvement and empowerment of the actual users (primary stakeholders). Second, to effectively manage a resource, the institution must be transparent, accountable, participatory and fair. It should also have the power to exclude undesirable or unregulated external influences on the resource, as also regulate the exploitation or resources by its own members and constituency (see Raju's paper in this book). The more its members are dependent on the resource being managed, and the more they are tied to each other in reciprocal relationships of various kinds (such as kinship, or barter), the more the institution is likely to be successful (Lele 1996).

Some examples will help to clarify the above points. Most countries of the region have a long history of traditional community institutions for the management of natural resources. While many of these have broken down, some continue in scattered parts. In Nepal, the traditional system of appointing a forest guard (*shingi nawa*), who had the power to enforce communal rules regarding forests, has continued or been revived in some areas (Sherpa 1993).

There is also a more recent and widespread grassroots forest protection movement in India, initiated by village groups. These groups have started protecting forest patches adjoining their villages in response to the forest degradation and hardships they were having to face as a result (Agarwal and Saigal 1996). In certain cases (e.g., Kailadevi Sanctuary in Rajasthan, western India), the community-initiated FPCs (locally called *kulhadi-bandh panchayats* or 'axe-banned' *panchayats*) are constituted entirely of the people (Das 1997; Das and Christopher in this book). In some areas (e.g., Orissa, eastern India) the institutional structures have evolved further in the form of federations of People's Institutions (PIs), which Raju discusses in this book. Some self-initiated FPCs have even forced the government to recognize them as formal institutions under JFM regulations.

The recent Indian Panchayats (Extension to Scheduled Areas) Act, 1996, provides greater scope for people's involvement in resource management. The Act mandates each state to enact a law for scheduled

Box 1.1

Joint Forest Management (JFM), India

The JFM programme has been operationalized mainly through the initiative and commitment of foresters and communities to work together to protect the forests under state control. This began in the 1970s and 1980s, when FPCs (whose members are drawn from the local communities dependent on forest resources), were set up in the states of West Bengal, Gujarat and Haryana. The National Forest Policy, 1988, and circular regarding community involvement in forestry issued by the Government of India in 1990, emphasized the increasing importance of jointly managed forest systems, centred around the needs of forest communities. Through the adoption of JFM in many areas, it has been demonstrated and proved that local communities are capable of protecting the forest patches adjoining their villages, given an appropriate environment. As an institutional structure, however, these FPCs are not without serious shortcomings, including lack of participation by women and disprivileged sections, frequent domination by forest officials, lack of clarity of powers and functions, lack of tenurial security, and others. A precursor to the JFM FPCs were the Van Panchayats (Forest Councils) of the Himalayan forests of northern India, created by an Act as far back as 1931 (Ballabh and Singh 1988; Agarwal and Saigal 1996; Maikhuri et al. in this book).

areas that empowers the *gram sabha* (village assembly) to safeguard the traditions and customs of the people, their cultural identity and communal resources. They are endowed with ownership over minor forest produce. They are also empowered to prevent alienation of land in the scheduled areas (see also Krishnan in this book). Though these provisions do seem to be too broad and vague in their scope, and do not give *gram sabhas* much decision-making power, there is potential to develop these into more concrete provisions for exercising control over access to the resources and knowledge of communities. An example of a successful *gram sabha* experiment is village Seed near Udaipur, Rajasthan. This village in western India, is registered under a unique law called the Rajasthan Gramdan Act, 1971, which gives executive and legal powers to the *gram sabha*. The entire adult population are members of the *sabha*, which has devised rules for protection of the village common lands, such as banned areas for grazing and fuel collection (Agarwal and Narain 1989).

Community forestry in Nepal is another example of participatory conservation (Joshi 1996; Shrestha in this book). The Panchayati Forest and Panchayat Protected Forests Rules were issued in 1978, under the Nepal Forest Act of 1961. They mandate the handing over of accessible forests to Forest User's Groups (FUGs) of local villagers. These FUGs get all benefits from handed over forests. In turn, there has been a transition from the policing role of forest staff to that of being extension officers stationed to assist community forest management activities.

Apart from the above, there are also local institutions sponsored either by the government or by NGOs, for performing more general developmental tasks, to whom the responsibility of forest protection and/or management has also been assigned. Examples from India include: *gram panchayats* in some states, *mahila mandals* (women's committees) in Himachal Pradesh, Tree Growers Cooperatives in Gujarat (all of which are government sponsored); and *gram vikas mandals* (village development committees) in Gujarat (initiated by an NGO, the AKRSP) (Sarin et al. 1996). Another innovative structure exists in Mendha village of Maharashtra, central India; here the villagers have formed their own Village Development Organization with the help of an NGO (Hiralal and Tare 1989). The village also has a study circle open to all local people and individuals from outside, which forms an important preliminary discussion forum for issues which are to be resolved or discussed at the level of the *gram sabha*. In Sri Lanka, Ekaratne et al. (in this book) report an innovative institutional structure for management of coastal areas, the Rekawa Lagoon Fishermen's Cooperative Society. Recently formed, this Society regulates fishing activity, reviews the status of the lagoon prawn resource, and manages the marketing and trade aspects. Elsewhere in Sri Lanka, innovative and complex institutional arrangements have been established to handle difficult coastal resource management issues, as discussed by DeCosse and Jayawickrama, and by Jayatilake et al. (in this book).

Apart from the above, alternative institutional structures have been proposed by several people: e.g., Village Nature, Health and Education Committees (Gadgil and Rao 1994) and Joint Protected Area Management Boards (Kothari et al. 1996).

Issues ahead

There are many lessons to be learnt from the above and other experiences, and several challenges ahead while furthering the creation of appropriate institutional structures for CBC:

(i) Numerous problems would have to be confronted at the attitudinal level: the feeling of 'we know what's best for them' which occurs in many officials, the converse feeling of awe and subservience amongst local communities, the mutual distrust and ill-will built on a history of conflicts, etc. Changing these will require institutional arrangements which demystify symbols of power and authority, build the capacity of officials to work with communities, create mutual confidence and trust, and allow for mutual learning in an atmosphere of openness.

(ii) The functions, duties, responsibilities and powers of respective members of an institution (especially joint management bodies) need to be crystal clear, to leave as little room for misunderstanding as possible.

(iii) Capacity-building programmes with all stakeholders are necessary towards increasing knowledge of partners in CBC, improving understanding of the ecological constraints of the area, clarifying the respective roles, responsibilities, rights and duties of each partner, and developing appropriate leadership (Desai et al. 1996).

(iv) The institutional structure should also ensure that power is devolved to the actual user groups, and is not cornered by the elite in the village.

(v) The institutional structure should be able to respond to the overall developmental and social needs of the community, and not remain restricted to conservation issues.

(vi) Multiplicity of institutions at the local level can be problematic. The possibility of existing formal or informal institutions undertaking the responsibility of management of natural resources should be explored, or else there should be a clear demarcation of responsibilities between the various institutions.

(vii) The role of outsiders to the process needs clarification and regulation. Worldwide, NGOs and other outsider agencies/individuals have played a major role in facilitating CBC (Borrini–Feyerabend 1996; Suri 1996). NGOs should be able to contribute without necessarily imposing their own values and biases on the community, and without forcing themselves into a decision-making capacity.

(viii) Conflict management needs to be built into the institutional (and legal) structure. Where local community/user groups have already been protecting their resources over a period of time, traditional institutions may have evolved territorial demarcations, access controls, and means to tackle intra- and inter-community conflicts. These should be encouraged.

(ix) The conflicts and contradictions between the priorities and programmes of various departments need urgent resolution in all countries. This is particularly so, for instance, between the revenue and forestry or wildlife departments. Examples of inter-sectoral integration and coordination (e.g., some given in the section on benefit-sharing earlier) need to be analyzed for the elements that make them successful.

(x) Financial sustainability of these institutions and their independence from outside agencies is a critical concern in most countries, especially where donors are heavily influential. Programmes for self-financing the conservation efforts need to be devised.

Legal and policy issues

Current law and policy vis-à-vis CBC

Law and policies dealing with wildlife conservation in most countries have sanctioned a 'guns and guards' approach, separating local communities from natural resources and ignoring the traditional institutions, practices and beliefs which encouraged sustainability (Kothari 1996a). Forest laws in most countries of the region were initiated in the colonial period, and were largely exploitative and disregarded local community rights. They have continued virtually unchanged in modern times, till recently. Some historical precedence for this tendency can be claimed from the acts of pre-colonial rulers, many of whom set up exclusive hunting reserves, though it seems that in many of these reserves, local populations were still allowed to continue subsistence activities.

Over the last century or so, the state has progressively strengthened its hold over forests, wetlands and other ecosystems, and simultaneously eroded the access of local communities to the resources contained in these ecosystems. Starting with the colonial administration's declaration of large parts of India's forests as reserved or protected,[2] and going on to the demarcation of several hundred national parks, sanctuaries and other types of wildlife conservation areas,[3] the attitude has been one of saving the environment *from*, rather than *with* or *by* local communities. In Sri Lanka, laws such as the Crown Lands Encroachment Ordinance confiscated large tracts of land customarily owned by communities, and consequently severed the communities' links with them (DeCosse and Jayawickrama in this book). Most of these were converted to coffee plantations. Nationalization of forests also occurred in Pakistan, Nepal and Bangladesh.

The rights of forest-dwellers to NTFP, or of fisherfolk to fish and other aquatic resources, are generally based upon customary and long-standing usage. However, this usage has often not been considered adequate grounds for establishing legally defensible rights. Thus, for instance, the Indian forest legislation and administration put aside the notion of long-respected rights exercised by forest dwellers, and replaced it with the notion of concessions granted by the state. Usufruct (user right) was portrayed by the FD as an act of charity whereas forest dwellers saw it as their legitimate claim (Baviskar 1996).

There are instances where the traditional practices are defined as customs and are recognized as such under modern law. The Constitution of India, for instance, does recognize the status of customary law as law. The Indian Easement Act of 1882 includes a category called 'customary easements' which permit the owner or occupier of a land to continue activities he/she has been carrying out from the past. However, as Krishnan explains in his paper in this book, the laws relating to conservation did not adequately take into consideration customary practices, and prescribed a different scheme of things altogether; first by converting community property (the forests, for instance) into state property, and then steadily restricting or ousting traditional usage altogether.

The result of the dilution of rights has been severe hardships for forest dwellers, their alienation from their surroundings, and hostility towards the government. Communities have even expressed their displeasure by deliberate tree-felling; a dramatic instance of this is related to the well-known Jharkhand movement in eastern India, in which tribals launched a *jungle kato* ('cut the forest') campaign to protest the replacement of native *sal* (*Shorea robusta*) trees by teak (*Tectona grandis*) by the Bihar Forest Development Corporation. Teak, being a purely commercial timber, has no value for the tribals or their cattle (CSE 1982). The movement later expanded into one of India's most powerful voices for a separate state for the tribals.

Conservation laws and policies have also been non-participatory in nature, in that the powers and functions for planning and implementing conservation programmes are almost exclusively held by centralized bureaucracies. Local communities have had virtually no legally enforceable means of involvement, and even where they are involved, it is at the discretion of government agencies.

Finally, conservation laws and policies have not encouraged, recognized, or given protection to the independent efforts of communities in conserving habitats and wildlife. Considerable biological diversity exists outside officially designated protected and conservation areas, including areas which have been under conservation by village communities, as sacred groves, tanks, pastures, or for other land uses, yet the formal legal structure has not recognized them in any way, nor has policy noted the need for such recognition. As noted by Das and Christopher in this book, many self-initiated FPCs find themselves unable to tackle outside offenders (or sometimes even powerful inside ones) in the absence of some formal powers.

The issue of rights and tenure

The most important issue for discussion here is that of tenurial rights to land and water and to resources found therein. What are the legal implications of the right to tenure? In India, the Land Acquisition Act, 1894, defines the term 'land' as including benefits arising out of land and things attached to the earth or permanently fastened to anything attached to the earth.

The International Labour Office (ILO) Convention 169 recognizes the rights of ownership and possession of peoples over the lands that they traditionally occupy, and of the right of use and access to the lands they have traditionally had access to for subsistence (Article 14).

It should be pointed out that there is no uniformity with regard to the characteristics of community-based tenure. Certain broad characteristics can still be drawn from some examples of such tenure. In Sri Lanka, for instance, a form of traditional service tenure was prevalent, which had two basic aspects: first, community-based tenurial rights were legally recognized, and second, strong systems for joint management of natural resources existed (DeCosse and Jayawickrama in this book). Community-based tenure systems are a set of institutional systems that include processes for establishing and allocating property rights to groups and individuals, including tenurial rights to specific agricultural lands, trees or other resources within the community's territory (Lynch and Alcorn 1994). Traditional tenure systems often embody conflict-resolution mechanisms and strategies for defending the local resource base against incursions by outsiders and resolving intra-community disputes. In north-east India, communities have rules of tradition governing the system of cultivation, landholding, village administration, foraging and extraction of resources, many of which continue to exist (Choudhury in this book).

Erosion of community-based tenure has been linked to loss of biodiversity. For instance, some Bangladeshi communities use community property regimes to regulate access to highly biodiverse fisheries in flood plain and wetland waterways, which are also the habitat for many migratory birds (Lynch and Alcorn 1994). This is however being threatened by loss of tenurial security as waterways based upon community-based tenure and management are being appropriated by private enterprises. Chandrashekara (in this book) explains a similar situation in Chinnar Wildlife Sanctuary in South India. In the context of joint community forestry programmes in South Asia, local villagers in the absence of clear tenure live in the fear of government agencies revoking the agreement under any pretext, just before harvesting the forest produce which they have been protecting for years (Poffenberger and Singh 1996).

While the resolution of tenure issues appears to be a critical factor in CBC, Khattak (in this book) justifiably warns against waiting for such resolution before starting CBC processes, especially in very complicated situations such as Pakistan.

Changes in policy/law

In some countries, such changes at the policy and legal levels have already begun:

- In Sri Lanka, the National Forest Policy (NFP) and the Forestry Sector Master Plan (FSMP), both adopted in 1995, recognize the role of local resource users and NGOs in forestry management. A differentiation is made between various categories of areas (PAs, multiple-use natural forests, home gardens/agro-forestry, forest plantations and industrial production areas), and the role of local communities and NGOs in each is specified. These roles range from participation in conservation, by forming partnerships between state agencies and local communities, to authorized utilization of resources. The need to ensure appropriate tenure arrangements in natural forests and forest plantations is also recognized. However, it is not yet clear how these policies and plans are to be implemented (Jayatilake et al., and DeCosse and Jayawickrama in this book). Sri Lanka's Coastal Zone Management Plan (CZMP), 1997, also advocates the concept of collaborative management of coastal resources.
- In Nepal, the Master Plan for the Forestry Sector, 1988, empowers local users to manage forests, provides for the total benefit from such management to go to the community, and stresses the role of forest officials as extension agents rather than a policing force (Joshi 1996). These elements have been codified in the National Forest Act of 1993 (Shrestha in this book).
- In Pakistan (Khattak in this book), the Forest Policy Statement of 1991 stresses the participatory approach and integrated management of natural resources. This has been followed by the National Conservation Strategy of 1992, which supports the creation of local institutional structures for common resource management.
- In India, the National Forest Policy of 1988 linked conservation to meeting the basic needs of the people, and maintaining the intrinsic relationship between forests and tribal and other forest-dependent people by protecting customary rights. A 1990 circular of the Government of India recommended the participation of village communities in the 'regeneration of degraded forest lands', which boosted the move towards JFM. In 1992, the National Conservation Strategy and Policy Statement reiterated the 1988 Forest Policy's thrust. However, no change has

yet been made in the 1927 Forest Act, which remains essentially non-participatory in nature.

The 73rd Amendment to the Constitution of India, and its extension to tribal areas, offers scope for greater involvement of the formal *panchayat* system in ecosystem management. (See the discussion on institutional structures earlier.)

At the international level, a number of instruments concluded at the United Nations Conference for Environment and Development, 1992, such as Agenda 21 (the blueprint and action plan for conservation and sustainable development); Rio Declaration, another product of UNCED, 1992; and the Convention on Biological Diversity, 1992, recognize the special status of local and indigenous communities, and their participation in the management of natural resources.

While these policy-level changes signify a definite change in attitude at the highest levels, the necessary legal changes (and correspondingly the institutional and other conditions on the ground) are far from being conducive to CBC. Lele et al. (in this book) point out from their experience that between any legislation and the field situation are innumerable administrative policies and procedures that serve to effectively marginalize local communities even when the law gives them a central role.

Issues for a law facilitating CBC

Any law/policy dealing with CBC should have at least three basic objectives: (*a*) facilitating the full participation of local, resource-dependent communities in the management and protection of adjoining ecosystems and species; (*b*) ensuring the biomass and other subsistence and livelihood rights of local people; (*c*) regulating human activities to ensure their compatibility with conservation and sustainable livelihood values; in particular, prohibiting destructive commercial–industrial activities in areas of conservation or cultural value.

A number of important issues need to be resolved while incorporating the above in law:

1. Stakeholder identification: Who is the local community? This issue has already been dealt with earlier in this paper; the law would have to contain provisions both for the clear identification of the primary and secondary stakeholders, as also for the demarcation of the rights and roles of these stakeholders.

2. Flexibility: How much should a law prescribe, and how much should be left to site-specific rules? If customary law is to be given validity, formal national or state law should only provide the broad framework which would guide more site-specific rules and regulations.

3. Legal status of benefits/stakes: What would be the legal status of the various possible benefits outlined above (see section on benefit-sharing)? How can secure tenurial rights, coupled with responsibilities, be guaranteed? Legal clarity on these questions is vital, as discussed earlier.

4. Equity considerations: The role of the law in addressing the equity aspect (see the section on equity) is crucial in order to ensure that CBC does not become a means of legitimizing inequities within and between the communities. Specific measures can include reservation of seats for marginalized sections (including women) in local decision-making bodies, and tenure rights to be vested specifically in marginalized sections.

5. Ensuring conservation: How can law help to ensure conservation while providing for livelihood security through tenure and rights? Can the concept of responsibilities and duties be tied up to rights? Who defines conservation; conservation for whom, conservation of what? And then, the question of responsibilities; what happens if right-holders fail to fulfill their responsibilities: are their rights to be revoked?

CBC acknowledges the age-old truism that power has the potential to be misused. It therefore mandates that there has to be a reciprocal system of checks and balances among the parties (the state and the community) involved in operationalizing CBC. The law has an important role in determining how this may be achieved.

Finally, the law has to have strong provisions regulating the activities of the urban-industrial sector, strictly prohibiting those which are clearly at odds with the conservation values sought to be protected. In most countries this has been the most difficult aspect to control, especially in the context of the rapid industrial liberalization that many areas have gone in for. It is imperative that decision-making regarding commercial–industrial projects in areas of conservation interest includes a process of public hearings with full access to information, before the decision is taken.

Conclusion

CBC is not a panacea, nor is it applicable in all sorts of situations, but in many cases it appears to be necessary from the point of view of both conservation and social justice. And yet there are many problematic issues which need resolution, many challenges it faces while becoming an established part of any country's policy and strategy. We have described in this paper how these issues are being brought up, and tackled, in numerous examples of conservation in the South and Central Asian region and elsewhere. Various papers in this book go into further detail than we have been able to in this overview. Of necessity, we have had to take a somewhat cursory look at the whole range of issues, but even such a look brings out both the diversity of approaches as also the common elements between them.

NOTES

1. This paper, as also most other papers in this book, does not deal with agricultural biodiversity, but restricts itself to what can loosely be called wild biodiversity, including natural habitats and species found in the wild. We acknowledge that there is a thin dividing line between wild and agricultural biodiversity, but have for the purposes of convenience and in-depth analysis limited ourselves to the former.
2. The Indian Forest Act was first enacted in 1865, and updated in 1878. This was amended and consolidated by the 1927 Forest Act, which is the law governing forests today. The ownership of a reserved or protected forest declared under the Act is vested with the government, and people who depend on the forest resources for survival are restricted from using its resources. The Act does provide for the formation of Village Forests to be assigned to village communities by the government, for the protection and regulated use of forests. However, this provision has hardly ever been used.
3. Most sanctuaries and national parks were carved out of reserved forests after the enactment of the Indian Wild Life (Protection) Act, 1972. In both cases, a procedure was established for inquiry by the state into the rights of the people dependent on the forest. In sanctuaries, the choice before the collector after the inquiry is to exclude the land from the limits of the PA, or to acquire the same under the Land Acquisition Act, or to allow the rights to continue within the PA. A national park, however, connotes a virtual cessation of human activities. See Krishnan (1996) and Krishnan's paper in this book.

REFERENCES

Agarwal, A. and **S. Narain.** 1989. *Towards Green Villages.* Centre for Science and Environment, New Delhi.

Agarwal, C. and **S. Saigal.** 1996. Joint Forest Management in India: A Brief Review. Draft Paper for Discussion. Society for Promotion of Wasteland Development, New Delhi.

AKRSP. 1995. Soliya Harvesting—Gender Perspective. Mimeo. Ahmedabad. As cited in M. Sarin with L. Ray, M.S. Raju, M. Chatterjee, N. Banerjee and S. Hiremath. 1996. *Who is Gaining? Who is Losing? Gender and Equity Concerns in Joint Forest Management.* Gender and Equity Sub-Group, National Support Group for JFM. Society for Promotion of Wasteland Development, New Delhi.

Anuradha, R.V. 1998. Sharing with the Kanis: A Case Study from Kerala, India. Kalpavriksh, New Delhi. Unpublished report.

Arhem, K. 1985. *Pastoral Man in the Garden of Eden: The Maasai of the Ngorongoro Conservation Area, Tanzania.* University of Uppsala, Uppsala.

Baviskar, A. 1996. Carrying Capacity and Usufruct Rights, in W. Fernandes (ed.). *Drafting a Peoples' Forest Bill.* Indian Social Institute, New Delhi.

Ballabh, V. and **K. Singh.** 1988. Van (Forest) Panchayats in UP Hills: A Critical Analysis. Research Paper. Indian Institute of Rural Management, Gujarat.

Bhatt, S. and **A. Kothari.** 1997. Protected Areas in India: Proposal for an Expanded Set of Categories, in A. Kothari, F. Vania, P. Das, K. Christopher and S. Jha (eds). *Building Bridges for Conservation: Towards Joint Management of Protected Areas in India.* Indian Institute of Public Administration, New Delhi.

Borrini–Feyerabend, G. 1996. *Collaborative Management of Protected Areas: Tailoring the Approach to the Context.* IUCN, Gland.

Campbell, J. 1996. Personal Communication. As cited in M. Sarin with L. Ray, M.S. Raju, M. Chatterjee, N. Banerjee and S. Hiremath. *Who is Gaining? Who is Losing? Gender and Equity Concerns in Joint Forest Management.* Gender and Equity Sub-Group, National Support Group for JFM. Society for Promotion of Wasteland Development, New Delhi.

Centre for Science and Environment (CSE). 1982. *The State of India's Environment.* Centre for Science and Environment, New Delhi.

Das, P. 1997. Kailadevi Wildlife Sanctuary: Prospects for Joint Management, in A. Kothari, F. Vania, P. Das, K. Christopher and S. Jha (eds). *Building Bridges for Conservation: Towards Joint Management of Protected Areas in India.* Indian Institute of Public Administration, New Delhi.

Desai, K., S. Sachdeva, J. Lall, A. Jindal, A. Sherasiya and **M. Joshi.** 1996. Human Resource Development for Joint Protected Area Management, in A. Kothari, N. Singh and S. Suri (eds). *People and Protected Areas: Towards Participatory Conservation in India.* Sage Publications, New Delhi.

Fernandes, W., G. Menon and **P. Viegas.** 1988. *Forest, Environment and Tribal Economy: Deforestation, Impoverishment and Marginalisation in Orissa.* Indian Social Institute, New Delhi.

Gadgil, M. and **P.R.S. Rao.** 1994. A System of Positive Incentives to Conserve Biodiversity. *Economic and Political Weekly.* 6 August, XXIX (32): 2103–7.

Gupta, A. 1996. Rewarding Creativity for Conserving Diversity in Third World: Can IPR Regime Serve the Needs of Contemporary and Traditional Knowledge Experts and Communities in Third World. Working Paper 1339. Indian Institute of Management, Ahmedabad.

Gurung, G. 1995. Indigenous Resource Management System of Manang Village in North Central Nepal: A Case Study and a Project Proposal. Paper Prepared for the Participatory Seminar-cum-Workshop to Develop Projects on Resource Management and Sustainable Livelihoods for Traditional Societies. Faislabad, Pakistan.

Hiralal, M.H. and **S. Tare.** 1989. Forests and People: A Participatory Study on Food, Fuel, Fodder, Fertiliser, Water, and Employment in 22 Villages of Dhanora Tehsil, Gadchiroli. Vrikshamitra, Gadchiroli, Maharashtra.

Joshi, A.L. 1996. Community Forestry in Nepal. Paper presented at the Fifth Asia Forestry Network. Surajkund, India. 3–6 December.

Kemf, E. (ed.). 1993. *Law of the Mother: Protecting Indigenous Peoples in Protected Areas.* Sierra Club Books, San Francisco.

Kothari, A. 1996a. Biodiversity and the Proposed Forest Act, in W. Fernandes (ed.). *Drafting a Peoples' Forest Bill.* Indian Social Institute, New Delhi.

Kothari, A. 1996b. Submissions to Wildlife Act Review Committee. New Delhi.

Kothari, A., N. Singh and **S. Suri** (eds). 1996. *People and Protected Areas: Towards Participatory Conservation in India.* Sage Publications, New Delhi.

Kothari, A., P. Pande, S. Singh and **D. Variava.** 1989. *Management of National Parks and Sanctuaries in India: A Status Report.* Indian Institute of Public Administration, New Delhi.

Kothari, A., F. Vania, P. Das, K. Christopher and **S. Jha** (eds). 1997. *Building Bridges for Conservation: Towards Joint Management of Protected Areas in India.* Indian Institute of Public Administration, New Delhi.

Krishnan, B.J. 1996. Legal Implications of Joint Management of Protected Areas, in A. Kothari, N. Singh and S. Suri (eds). *People and Protected Areas: Towards Participatory Conservation in India.* Sage Publications, New Delhi.

Lele, S. 1996. Environmental Governance. *Seminar* 438. February 1996: 17–23.

Lynch, O.J. and **J.B. Alcorn.** 1994. Tenurial Rights and Community-based Conservation, in D. Western and R.M. Wright (eds). *Natural Connections: Perspectives in Community-based Conservation.* Island Press, Washington, D.C.

Majumdar, S. 1994. Forest Protection Committees: A Process of Empowering the People. Mimeo. Vidyasagar University, West Bengal. As cited in M. Sarin with L. Ray, M.S. Raju, M. Chatterjee, N. Banerjee and S. Hiremath. *Who is Gaining? Who is Losing? Gender and Equity Concerns in Joint Forest Management.* Gender and Equity Sub-Group, National Support Group for JFM. Society for Promotion of Wasteland Development, New Delhi.

Michon, G. and **H. de Foresta.** 1990. *Complex Agroforestry Systems and the Conservation of Biological Diversity.* Proceedings of the International Conference on Tropical Biodiversity 'In Harmony with Nature', Kuala Lumpur (Malaysia), 12–16 June.

Murty, M.N. 1996. Contractual Arrangements for Sharing Benefits from Preservation: Joint Management of Wildlife, in A. Kothari, N. Singh and S. Suri (eds). *People and Protected Areas: Towards Participatory Conservation in India.* Sage Publications, New Delhi.

Nijar, G.S. 1995. *Developing a Rights Regime in Defence of Biodiversity and Indigenous Knowledge.* Third World Network, Malaysia.

Pardeshi, P. 1996. Conserving Maharashtra's Biodiversity through Ecodevelopment, in A. Kothari, N. Singh and S. Suri (eds). *People and Protected Areas: Towards Participatory Conservation in India.* Sage Publications, New Delhi.

Poffenberger, M. 1994. The Resurgence of Community Forest Management in Eastern India, in D. Western and R.M. Wright (eds). *Natural Connections: Perspectives in Community-based Conservation.* Island Press, Washington, D.C.

Poffenberger, M. and **C. Singh.** 1996. Communities and the State: Re-establishing the Balance in Indian Forest Policy, in M. Poffenberger and B. McGean (eds). *Village Voices, Forest Choices.* Oxford University Press, New Delhi.

Poole, P. 1995. *Indigenous Peoples, Mapping and Biodiversity Conservation.* Biodiversity Support Programme, Washington, D.C.

Posey, D. 1985. Indigenous Management of Tropical Forest Ecosystems: The Case of the Kayapo Indians in the Brazilian Amazon. *Agroforestry Systems* 3: 139–58.

Posey, D. and **G. Dutfield.** 1996. *Beyond Intellectual Property Rights.* International Development Research Centre, Ottawa.

Pushpangadan, P. 1996. *Tropical Botanic Garden and Research Institute: People Oriented Sustainable Development Programme.* TBGRI, Thiruvananthapuram.

Rathore, B.M.S. 1996. Joint Management Options for Protected Areas: Challenges and Opportunities, in A. Kothari, N. Singh and S. Suri (eds). *People and Protected Areas: Towards Participatory Conservation in India.* Sage Publications, New Delhi.

Saberwal, V. 1996. Pastoral Politics: Gaddi Grazing, Degradation, and Biodiversity Conservation in Himachal Pradesh, India. *Conservation Biology* 10 (3): 741–49.

Saigal, S., C. Agarwal and **J.Y. Campbell.** 1996. Sustaining Joint Forest Management: The Role of Non-Timber Forest Products. Manuscript. Society for Promotion of Wastelands Development, New Delhi. As cited in Seema Bhatt, paper for this book.

Sarin, M. with **L. Ray, M.S. Raju, M. Chatterjee, N. Banerjee** and **S. Hiremath.** 1996. *Who is Gaining? Who is Losing? Gender and Equity Concerns in Joint Forest Management.* Gender and Equity Sub-Group, National Support Group for JFM. Society for Promotion of Wasteland Development, New Delhi.

Sherpa, M.N. 1993. Grass Roots in a Himalayan Kingdom, in E. Kemf (ed.). *Law of the Mother: Protecting Indigenous Peoples in Protected Areas.* Sierra Club Books, San Francisco.

Singh, S. 1994. Joint Protected Area Management: Some Policy Issues. Paper presented at Workshop on 'Exploring the Possibilities of Joint Management of Protected Areas', Indian Institute of Public Administration, New Delhi. As cited in A. Kothari, S. Suri and N. Singh. Conservation in India: A New Direction. *Economic and Political Weekly*: 2755–66.

Slavin, T. 1993. Survival in a Vertical Desert, in E. Kemf (ed.). *Law of the Mother: Protecting Indigenous Peoples in Protected Areas.* Sierra Club Books, San Francisco.

Suri, S. 1996. Peoples Involvement in Protected Areas: Experiences from Abroad and Lessons from India, in A. Kothari, N. Singh and S. Suri (eds). *People and Protected Areas: Towards Participatory Conservation in India.* Sage Publications, New Delhi.

Vania, F. 1997. Rajaji National Park, Uttar Pradesh: Prospects for Joint Management, in A. Kothari, F. Vania, P. Das, K. Christopher and S. Jha (eds). *Building Bridges for Conservation: Towards Joint Management of Protected Areas in India.* Indian Institute of Public Administration, New Delhi.

Wells, M.P. 1994. A Profile and Interim Assessment of the Annapurna Conservation Area Project, in D. Western and R.M. Wright (eds). *Natural Connection: Perspectives in Community-based Conservation.* Island Press, Washington, D.C.

West, P.C. and **S.R. Brechin** (eds). 1991. *Resident Peoples and National Parks: Social Dilemmas and Strategies in International Conservation.* University of Arizona Press, Tucson (Arizona).

Western, D. 1989. Conservation without Parks: Wildlife in the Rural Landscape, in D. Western and M.C. Pearl (eds). *Conservation for the Twenty-first Century.* Oxford University Press, New York.

Western, D. and **R.M. Wright.** 1994a. The Background to Community-based Conservation, in D. Western and R.M. Wright (eds). *Natural Connections: Perspectives in Community-based Conservation.* Island Press, Washington, D.C.

———. (eds). 1994b. *Natural Connections: Perspectives in Community-based Conservation.* Island Press, Washington, D.C.

2

Diversity and sustainability in community-based conservation

Michel Pimbert* and Jules Pretty

Introduction

Top-down, imposed conservation all too often entails huge social and ecological costs in areas where rural people are directly dependent on natural resources for their livelihoods. A growing body of empirical evidence now indicates that the transfer of 'Western' conservation approaches to developing countries has indeed had adverse effects on the food security and livelihoods of people living in and around protected areas (PAs) and wildlife management schemes (Ghimire and Pimbert 1997; Kothari et al. 1989, 1996; IIED 1995; Wells et al. 1992; West and Brechin 1991). On several occasions, local communities have been expelled from their settlements without adequate provision for alternative means of work and income. In other cases, local people have faced restrictions in their use of common property resources for food gathering, harvest of medicinal plants, grazing, fishing, hunting, collection of wood and other wild products from forests, wetlands and pastoral lands, often turning them practically overnight from hunters and cultivators to 'poachers' and 'squatters' (Colchester 1994). Denying resource use to local people severely reduces their incentive to conserve it. Moreover, the current styles of PA and wildlife management usually result in high management costs for governments, with the majority of benefits accruing to national and international external interests. All these trends may ultimately threaten the long-term viability of conservation schemes

* The views expressed here are those of the author alone and are not necessarily endorsed by his organization.

as local populations enter into direct conflict with park authorities and game wardens.

This deep conservation crisis has led to the search for alternative approaches. 'Community-based conservation' (CBC) and 'peoples' participation' are getting more attention from international and national conservation organizations. There are now several examples of projects which involve local communities and seek to use economic incentives for the conservation and sustainable use of wildlife and PAs (Kiss 1990; McNeely 1988; Sayer 1991; Stone 1991; Wells et al. 1992). However, the practice of CBC remains problematic because of its high dependence on centralized bureaucratic organizations for planning and implementation. Some of these initiatives are nothing more than 'official accommodation responses' to the growing opposition to PAs and conservation programmes. Nonetheless, a few of them are clearly challenging the dominant conservation approaches and seem to be based on more equitable power and benefit-sharing arrangements (for recent reviews see IIED 1995; Borrini–Feyerabend 1996).

Community-based conservation: From blueprints to process

The way conservation bureaucracies and external institutions are organized and the way they work currently inhibit devolution of power to local communities. The methods and means deployed to preserve areas of pristine wilderness largely originated in the affluent West where money and trained personnel ensure that technologies work and that laws are enforced to secure conservation objectives. During and after the colonial period, these conservation technologies, and the values associated with them, were extended from the North to the South, often in a classical top-down manner. Positivist conservation science and the wilderness preservation ethic hang together with this top-down, transfer-of-technology model of conservation (Pimbert and Pretty 1995) (Table 2.1). Conservation usually reflects the priorities of regional, national and above all international interests over local subsistence needs. Local people often express their sense of deep frustration with this by saying that 'people should be considered before animals' (Hackel 1993), and they often view 'wildlife conservation as alien, hypocritical, and as favouring foreigners' (Munthali 1993).

Existing conservation institutions and professionals need to shift from being project implementors to new roles which facilitate local people's analysis, planning and action. The whole process should lead to local institution-building or strengthening, thereby enhancing the capacity of people to take action on their own. This implies the adoption of a learning-process approach in conservation (Table 2.1) and a new professionalism with new concepts, values, participatory methodologies and behaviour.

Table 2.1: Natural resource conservation and management paradigms: The contrast between blueprint and learning-process approaches

	Blueprint	*Process*
Point of departure	nature's diversity and its potential commercial values	the diversity of both people and nature's values
Keyword	strategic planning	participation
Locus of decision-making	centralized, ideas originate in capital city	decentralized, ideas originate in village
First steps	data collection and plan	awareness and action
Design	static, by experts	evolving, people involved
Main resources	central funds and technicians	local people and their assets
Methods, rules	standardized, universal, fixed package	diverse, local, varied basket of choices
Analytical assumptions	reductionist (natural science bias)	systems, holistic
Management focus	spending budgets, completing projects on time	sustained improvement and performance
Communication	vertical: orders down, reports up	lateral: mutual learning and sharing experience
Evaluation	external, intermittent	internal, continuous
Error	buried	embraced
Relationship with people	controlling, policing, inducing, motivating, dependency creating. People seen as beneficiaries	enabling, supporting, empowering. People seen as actors
Associated with	normal professionalism	new professionalism

Table 2.1 continued

Table 2.1 continued

Outputs	1. diversity in conservation, and uniformity in production (agriculture, forestry, …)	1. diversity as a principle of production and conservation
	2. the empowerment of professionals	2. the empowerment of rural people

Source: Pimbert and Pretty 1995.

Reversals for community-based conservation

To spread and sustain CBC, considerable attention will have to be given to the following needs, social processes and policies.

Debunk the 'wilderness' myth and reaffirm the value of historical analysis

Most parts of the world have been modified, managed, and, in some instances, improved by people for centuries. Much of what has been considered as 'natural' in the Amazon is, in fact, modified by Amerindian populations (Posey 1993). Indigenous use and management of tropical forests is best viewed as a continuum between plants that are domesticated and those that are semi-domesticated, manipulated or 'wild', with no clear-cut demarcation between natural and managed forest. Certain large animal species would not occur in forests unmodified by humans, and important 'game' species of mammals such as deer, tapir, monkeys, collared peccary and jaguars reach much higher densities in modified areas. Home gardens planted by indigenous and local communities are particularly attractive to wildlife and several species may have actually increased their populations as a result of crops and fruit trees planted by people.

Biodiversity-rich areas—denser forests, relatively undisturbed grasslands, reefs and waterways—are generally found associated with territories claimed or used by indigenous peoples (Alcorn 1994). The 12 countries with the most biological diversity are also homes to diverse indigenous societies within whose territories much of that biological diversity is conserved. Indeed, many of the landscapes which are often viewed as pristine or 'wild' by outsiders, are in fact human-created and human-modified (Gomez–Pompa and

Kaus 1992; Pimbert and Toledo 1994; Posey 1993). UNESCO introduced the term 'cultural landscapes' to describe this phenomenon (UNESCO 1994).

Designating landscapes and the species they contain as cultural has a number of important implications for CBC and the concept of rights over biological resources. For example, indigenous peoples' organizations point out that where wild species and landscapes are products of nature, local communities can assert no special claim to them, and the national law considers them to be in the public domain, under the sovereign rights of the state. However, if species and landscapes have been moulded or modified by human presence, they are not automatically considered to be in the public domain. Local communities may therefore claim special rights of access, decision, control and property over them. This historical reality should be the starting point of CBC wherever local people have shaped local ecologies over generations. To transcend the 'wilderness myth', CBC must begin with the notion that biodiversity-rich areas are *social spaces*, where culture and nature are renewed with, by and for local people (Ghimire and Pimbert 1997).

Strengthen local rights, security and territory

Colonial powers, international conservation organizations and national governments have a long history of denying the rights of indigenous peoples and rural communities over their ancestral lands and the resources contained therein. This has been one of the most enduring sources of conflicts and violence, both in the developing world and in advanced industrialized nations such as Canada where aboriginal people seek greater self-determination by regaining control over territories now enclosed in the country's PA network (Morrison 1997).

Two immediate priorities to achieve CBC would be to:

1. Reform protected area categories and land use schemes to embody the concepts of local rights and territory in everyday management practice. To better integrate the concept of conservation with sustainable local livelihoods, countries need first to reform their legal and political instruments for PAs. Key reforms relate to communal ownership of lands within PAs, control and management responsibilities and benefit-sharing. The parties to the Convention on Biological Diversity for instance, should facilitate this fundamental rethinking in conservation by preparing a series of recommendations to party countries. They could also request IUCN's World Commission on Protected Areas (WCPA) to develop, in consultation with indigenous and peoples' organizations, a proposal for a new category of PAs more compatible with local priorities, needs, institutions and land use. Currently, the IUCN system has six categories of PAs, however, 'none of the categories are defined to guarantee the recognition of indigenous people's rights to self determination and self development' (Indigenous Peoples, Environment and Development Conference 1995).

In this context, it is particularly noteworthy that the IUCN's newly introduced Category VI allows for the sustainable use of natural ecosystems but, in practice, at least two-thirds of the area must remain in its 'natural state'. Although this category was apparently designed to integrate social development concerns in PA management, human settlements and resource use by local people are only tolerated as exceptions. Moreover, the different PA categories which do allow for some human use are very unevenly represented in the developed and developing countries. For instance, Category V, Protected Landscapes/Seascapes, is 'an area of land, with coast and sea as appropriate, where the *interaction of people and nature over time* has produced an area of distinct character with significant aesthetic, ecological and/or cultural value, and often with high biological diversity. Safeguarding the integrity of this traditional interaction is vital to the protection, maintenance and evolution of such an area' (IUCN 1994a, our emphasis). Out of a world total of 2,273 Protected Landscapes/Seascapes recognized by IUCN (IUCN 1994b), over half the Category V sites are located in Europe, with 1,307 sites covering 6.6 per cent of the land surface. This reflects the view that conservation, in Europe at least, depends on the involvement of people, and therefore places where people co-exist with nature are worthy of special attention. In sharp contrast however, Category V sites are underrepresented in the PA networks of the developing world: four sites for the whole of Central America (0.01 per cent of the land area), 56 in South Asia (0.09 per cent), 20 in Sub-Saharan Africa (0.1 per cent), seven in the Pacific (0.03 per cent) and 175 in South America (1.1 per cent) (WCMC 1994).

Similarly, the concept of 'cultural landscapes' under the World Heritage Convention explicitly recognizes the role of human agents in the continuing, organic evolution of whole landscapes (Phillips 1995). In practice however, the recognition of cultural landscapes 'out there', and the creation of the legal basis for their management, has been an exclusively Eurocentric phenomenon. British conservationists, for example, accepted the vision of nature as part of a process of 'continuity and gradual change, with man at the centre and integral to the rural landscape' (Blacksell cited in Harmon 1991: 34). National parks in Britain and elsewhere in Europe thus recognized existing rights and sought to maintain the established pattern of farming and land use by rural communities. The extension of similar policies and legislation to the developing world is clearly an important prerequisite for sustainable CBC.

2. Strengthen local control over the access and end uses of biological resources, knowledge and informal innovations. There are a variety of legal arrangements that can be introduced by governments to assure local control over resources. The range of choices is not limited to private property of land; communal property of land and/or resources are often more culturally appropriate options in much of the developing world (Bromley and Cernea 1989). Where local communities have been granted secure usufruct rights over neighbouring forests, governments have witnessed clear reversals in forest degradation and its associated biodiversity (Fortmann and Bruce 1988).

Recognition of anthropogenic landscapes and 'wild' species moulded by human agency has important implications for ownership, and consequently rights over access and use of biological resources. Western concepts of private property do not recognize the intellectual contributions and informal innovations of indigenous and rural peoples who have modified, conserved and managed so-called 'wild' plant and animal species (Crucible Group 1994).

Indigenous peoples (some 300 million people), manage or control about 19 per cent of the earth's surface and are currently grouped into 4,000 to 5,000 different cultures. They want governments to recognize their sovereign rights to determine: (*a*) how biological resources should be conserved and managed on their ancestral territories; (*b*) the rules of access to genetic resources; and (*c*) how benefits should be shared for the uses of those resources and the associated indigenous knowledge. Integrating these local views into policies for CBC is a central challenge.

Peoples' participation and professional reorientation in conservation bureaucracies

Despite repeated calls for peoples' participation in conservation over the last 20 years (e.g., Forster 1973; McNeely 1993), it is rare for professionals (foresters, PA managers, wildlife biologists) to relinquish control over key decisions on the design, management and evaluation of CBC.

Seven different types of participation are shown in Table 2.2. The implication of this typology is that the *meaning* of participation should be clearly spelt out in all CBC programmes. If the objective

Table 2.2: A typology of participation

Typology	*Components of Each Type*
Passive participation	People participate by being told what is going to happen or has already happened; the interaction is unilateral. The information being shared belongs only to external professionals.
Participation in information-giving	People provide information, but do not have the opportunity to cross-check what researchers are recording or writing, much less influence proceedings.
Participation by consultation	People participate by being consulted by external agents, who define both problems and solutions. Decision-making is not shared, and professionals are under no obligation to take on board people's views.
Participation for material incentives	People participate by providing resources, for example, labour, in return for food, cash or other material incentives. Much *in situ* research and bioprospecting falls in this category. People have no stake in prolonging activities when the incentives end.
Functional participation	People participate by forming groups to meet predetermined objectives related to the project. Such involvement tends to be only after major decisions have been made.
Interactive participation	People participate in joint analysis, which leads to action plans and the formation of new local groups or the strengthening of existing ones. These groups take control over local decisions, and so people have a stake in maintaining structures or practices.
Self-mobilization	People participate by taking initiatives independent of external institutions.

Source: Modified from Pretty 1994.

of conservation is to achieve sustainable and effective management of biological resources, then nothing less than functional participation will suffice. This implies the use of participatory methodologies, such as Participatory Rural Appraisal (PRA), by the staff of conservation NGOs and government agencies. This calls for a greater emphasis on training in communication rather than technical skills. It may imply a significant shift in technique for conventional trainers, since training for participation must itself be participatory and action-based (Chambers 1992). One practical implication is that conservation agencies set aside time for field experiential learning for their professional staff, so that they, as people, can see, hear, understand that other reality, of local people, and then work to make it count.

However, the adoption of a participatory culture and changes in professional attitudes and behaviour are unlikely to automatically follow when new methods are adopted. Training of agency personnel in participatory principles, concepts and methods must be viewed as part of a larger process of reorienting institutional policies, procedures, financial management practices, reporting systems, supervisory methods, reward systems and norms (Thompson 1995; Absalom et al. 1995). Institutionalizing and operationalizing participatory approaches in conservation bureaucracies will be an arduous task based on trial and error, self-critical reflection and further experimentation and innovation. In that context, inspiration and lessons might be drawn from the few examples of institutional transformation in large-scale programmes dealing with rural development, agricultural research and extension, soil and water management and education (Bawden 1994; Hinchcliffe et al. 1995; Thompson 1995; Scoones and Thompson 1994; Uphoff 1992).

Build on local priorities and definitions

From the outset, the definition of *what* is to be conserved, *how* it should be managed and *for whom*, should be based on interactive dialogue to understand how local livelihoods are constructed and people's own definitions of well-being. This is because:

1. Most professionals have tended to project their own categories and priorities onto local people. In particular, their views of the realities of the poor, and what should be done, have generally been constructed from a distance and mainly for professional convenience.

Many livelihoods, like collecting wild foods and medicine, home gardening, common property resources, share-rearing livestock and stinting, are largely unseen by outside professionals.

2. Many CBC schemes initiated by outsiders have overlooked the importance of locally specific ways of meeting needs for food, health, shelter, energy and other fundamental human needs. Whilst fundamental human needs are universal, their satisfiers vary according to culture, region and historical conditions (Max–Neef et al. 1989).

3. Measures to combat poverty and hardship induced by a PA scheme in a developing country usually fixate on the creation of full- or part-time jobs in, for example, the tourism and crafts sector. The problem is that for most rural people, and particularly for the weak and vulnerable, employment can only be a subset or a component of livelihood. Local definitions of well-being and culturally specific ways of relating to the world and organizing economic life are thus displaced in favour of the more uniform industrial–urban development model of the North.

4. In attempts to reach consensus and get started, outsiders have often overlooked the variability within communities. And yet, local communities are far from homogeneous. Elites are present in all societies. Sometimes they provide much-needed leadership, but frequently they exploit the common folk and further personal interests. (A more detailed discussion on the equity aspect appears in Sarin et al., and Kothari et al. in this book.)

5. The few economic analyses of biological diversity conducted so far have essentially focused on global values and foreign exchange elements and very little on the household use values of, for example, 'wild' foods and medicines (Scoones et al. 1992; Gujit et al. 1995). Simple economic valuations based on direct use values (for consumption or sale) (see Pearce et al. 1989) have often been misleading and too reductionist to provide a sound decision-making basis for policy-makers and land use planners. More participatory and comprehensive local level valuation methodologies, recently developed (Gujit et al. 1995), can help better understand the range of ways biodiversity matters to local people, and how values fluctuate according to season or to the many viewpoints of highly differentiated local communities.

Whilst the above examples of professional biases are also rampant in the wider community of development planners, economists and agricultural scientists (Chambers 1993), the problem is compounded in public and private conservation organizations because they have few, if any, sociologists or anthropologists working in the field or at headquarters.

Build on local systems of knowledge and management

Local management systems are generally tuned to the needs of local people and often enhance their capacity to adapt to dynamic social and ecological circumstances. Vernacular conservation is based on site-specific traditions and economies; it refers to ways of life and resource utilization that have evolved in that place and, like vernacular architecture, is a direct expression of the relationship between communities and their habitats (Poole 1993).

Local systems of knowledge and management are sometimes rooted in religion and the sacred. Sacred groves are common throughout southern and south-eastern Asia, Africa, the Pacific islands and Latin America (Shengji 1991; Ntiamo–Baidu et al. 1992). These pockets of biological diversity could form the basis of more 'culturally appropriate' PAs. Some indigenous peoples and rural communities (e.g., the Awa in Ecuador) have themselves established PAs that resemble the parks and reserves codified in WCPA system and in national PA policies (Poole 1993). Sacred places such as the Loita Maasai's Forest of the Lost Child in Kenya (Loita Naimana Enkiyio Conservation Trust 1994) are also widespread forms of vernacular conservation. Indigenous ways of knowing, valuing and organizing the world must not be brushed aside by so-called 'modern' technical knowledge which claims superior cognitive powers.

Build on local institutions and social organization

Indigenous peoples' resource management institutions include rules about use of biological resources and acceptable distribution of benefits, definitions of rights and responsibilities, means by which tenure is determined, conflict resolution mechanisms and methods of enforcing rules, cultural sanctions and beliefs (Alcorn 1994). Similarly, the literature on common property resources highlights the importance and resilience of local management systems for biodiversity conservation and local livelihoods (Arnold and Stewart 1991; BOSTID 1986; Bromley and Cernea 1989; Ostrom 1990; Jodha 1990; Niamir 1990).

The undermining and suppression of local institutions is no doubt the most debilitating and enduring impact of national and international bureaucracies. International conservation organizations spend a large proportion of their funds on expatriate salaries, policing and travel and meetings. A very small part of the funds managed by these organizations is invested locally in capacity-building and local institution-building.

Increased attention will need to be given to action through local institutions and user groups, such as natural resource management groups, women's associations and credit management groups. Available evidence from multilateral projects evaluated five to 10 years after completion shows that where institutional development has been important the flow of benefits has risen or remained constant (Cernea 1993); when ignored, economic rates of return decline markedly and conservation objectives may not be met. Outside interventions must be designed in such a way that at the end of the project cycle there are local institutions and skills in place to ensure the continuation of natural resource management.

Build on locally available resources and technologies

CBC that seeks to provide benefits for local and national economies should give preference to sustainable and cheaper solutions like informal innovation systems, reliance on local resources, local satisfiers of human needs and local technologies. Agricultural, health, housing, sanitation, and revenue-generating activities (e.g., tourism) based on the use of local resources and innovations are likely to be more sustainable and effective than those imposed by outside professionals. The advantages and skills of professionals (at the micro and macro levels) can be effectively combined with the strengths of indigenous knowledge and experimentation.

Economic incentives and policies for the equitable sharing of benefits

CBC has little chance of success where benefits are not distributed equitably among various members of the community. 'Equity' should entail the sharing of benefits in a way that is commensurate with the varying sacrifices and contributions made or damages incurred in the community (e.g., through lost access to resources, or damage to crops by wild animals).

Eco-tourism: As in the case of classical tourism, eco-tourism schemes are not integrated with other sectors of the national or regional economy; and little earnings generated actually reach or remain in the rural areas (Ghimire and Pimbert 1997; McIvor 1997; Koch 1997). At the same time, traditional livelihood sources and cultures become negatively affected in nearly all cases. Koch, assessing the potential of eco-tourism in the reconstruction of rural South Africa, concludes that generating economic benefits and empowering rural people is only feasible when many wide-ranging reforms such as restoration of land rights to local communities, support for new forms of land tenure, strengthening of community institutions, investment in technical and managerial skills of people, and mandatory impact assessments of all eco-tourism schemes are carried out (Koch 1997).

Biodiversity prospecting and commercial leases: Bioprospecting, the exploration, extraction and screening of biological diversity and indigenous knowledge for commercial use, has become an integral part of the R&D of large industrial corporations which market new natural products such as oils, drugs, perfumes, waxes, dyes, biopesticides (Reid et al. 1993; UNDP 1994; Baumann et al. 1996). It is conservatively estimated that medicinal plants and micro-organisms from the biodiversity-rich developing countries contribute at least US$ 30 billion a year to the developed world's pharmaceutical industry (UNDP 1994). Posey (1990) estimates that less than 0.001 per cent of the market value of plant-based medicines have been returned to indigenous peoples from whom much of the original knowledge came. And whilst various codes of conduct have been developed to ensure greater equity and benefit-sharing (e.g., FAO 1993; Cunningham 1993; Shelton 1995), none are internationally legally binding instruments or a protocol to the Convention on Biological Diversity (CBD).

Whilst the CBD recognizes 'the knowledge, innovations and practices of indigenous and local communities' and specifically 'encourage[s] the equitable sharing of benefits arising from the utilisation of such knowledge, innovations and practices', the Convention and national legislations do not require that bioprospecting agreements be subject to the prior informed consent of local people. Negotiations at the international level are carried out by national elites on behalf of their people, and bilateral agreements signed by the 'contracting parties' make little reference to local farmers, pastoralists, forest dwellers, herbalists and other rural people.

Whilst there is considerable pressure to extend Northern style IPRs through international negotiations in the World Trade Organization (WTO), indigenous peoples' groups and NGOs are making use of the GATT–TRIPS agreement, which calls for the development of *sui generis* legislation for IPRs, to propose more equitable systems of protection and benefit-sharing. Some of these, such as the concept of Traditional Resource Rights (Posey and Dutfield 1996) seek to protect not only knowledge relating to biological resources but also indigenous people's right to self-determination. The original FAO concept of Farmers' Rights is also being reinterpreted to stress farmers' collective right to *directly* control access to and receive benefits from commercial uses of traditional plant and animal resources (GRAIN 1995).

Codes of conduct for outside conservation agencies and professionals

There exists no legal or political framework for local populations to seek redressal against conservation organizations and environmentalists for causing social conflicts and misery. Some communities such as the Kuna of Panama and the Inuit Tapirisat of Canada, have established guidelines to ensure that research carried out on their territories is controlled by the local communities and based on their prior informed consent. Such Community-Controlled Research (CCR) may allow indigenous peoples to better control access and use of, for example, ethnobotanical knowledge which is increasingly targeted by bioprospectors working for pharmaceutical companies (Posey 1995).

The adoption of a policy of reciprocal accountability amongst governments, donors and local communities could potentially open spaces to do things differently in the future. For example, the concept of downward accountability implies shifting more direct control over decision-making and funds to local communities. Local recipients of the funds could then decide what this money should be spent on and by whom. The donors' legitimate demands for accountability could still be met if accountability was framed in terms of long-term process objectives that seek to reconcile conservation with sustainable local livelihoods.

Negotiated agreements and enabling policies for local action

The success of people-oriented conservation will hinge on promoting socially differentiated goals in which the differing perspectives and priorities of community members, and local communities and conservationists, must be negotiated. Signed agreements between external institutions and local community organizations could promote responsible and accountable interaction. Examples include Joint Forest Management (JFM), Joint Protected Area Management (Kothari et al. 1996) and wildlife co-management schemes based on more equitable power and benefit-sharing. Long-term success may depend on culturally sensitive and equitable action in the following areas:

1. In the case of indigenous peoples, national PA and conservation policies need to be brought in line with internationally recognized human rights; they should allow indigenous peoples to represent their own interests through their own organizations and not through consultative processes controlled by conservation organizations. International law and other agreements already provide clear principles for this, such as ILO 169, Chapter 26 of Agenda 21 of the UNCED agreements, and parts of the Biodiversity Convention (Colchester 1994).

2. Attitudinal change and respect for cultural diversity are critical. Joint management schemes for forest use have had notable success in India and elsewhere. But on the whole, the attitudes and behaviour of many forest officers remain paternalistic and profoundly disempowering. For example, informal comments by foresters working at different levels in the Forest Department (FD) hierarchy of the state of West Bengal (India) often describe tribal people and their FPCs as 'ignorant', 'primitive', 'underdeveloped in all respects' and 'economically irrational' (Pimbert 1994). Such negative attitudes clearly undermine the mutual trust needed for successful JFM and other co-management schemes. Today, many local communities dependent on the forests for their livelihoods have insecure rights and are aware that FDs may take back the forests once they are regenerated and productive again.

3. Seeing stakeholders' claims in their historical context is important. Examples from two internationally important wetland sites in Pakistan and India illustrate this. In and around Keoladeo National Park (India) for example, the needs and rights of the tourist industry are not comparable with those of the resident communities. Whereas private firms need only worry about increases in their profits, local peoples' stakes hinge around basic subsistence and adequate nutrition. Moreover, both tourism in Keoladeo National Park and private-owned fishing in the freshwater lake of the Ucchali complex

(Pakistan) take place precisely in those areas from which previous residents have been expelled and denied their prior rights of access and use. Huge differences in the scale of opposing stakes and claims were revealed in PRAs done with these residents, as village voices reconstructed the social and ecological histories of the wetlands. Joint management agreements thus need to acknowledge that some stakeholders' claims to resources are illegitimate, having ignored previously existing rights of long-time local residents. Enabling policies for joint PA management will need to address larger questions of land alienation and land scarcity (Ucchali) and grazing rights (Keoladeo). For the villagers these are *the* crucial policy issues (Pimbert et al. 1996).

Conclusion

Sustainable and effective conservation calls for an emphasis on community-based natural resource management and enabling policy frameworks. These are not easy options. Contemporary patterns of economic growth, modernization, and nation building, all have strong anti-participatory traits. The integration of rural communities and local institutions into larger, more complex, urban-centred and global systems often stifles the capacity for decision-making the local community might have had.

It should be emphasized here that the devolution of conservation to local communities does not mean that state agencies and other external institutions have no role. A central challenge will be to find ways of allocating limited government resources so as to obtain widespread replication of community initiatives. Understanding the dynamic complexity of local ecologies; honouring local IPRs; promoting wider access to biological information and funds; designing technologies, markets and other systems on the basis of local knowledge, needs and aspirations; call for new partnerships between the state, rural people and the organizations representing them.

This requires new legislation, policies, institutional linkages and processes. National regulatory frameworks should be flexible enough to accommodate local peculiarities, should focus on the granting of rights, access and security of tenure to farmers, fisherfolk, pastoralists and forest dwellers, and regulate the resource use activities of the rich and powerful, e.g., timber, bioprospecting and mining companies. Economic policies should include the removal of distorting subsidies that encourage the waste of resources; targeting of subsidies to the poor instead of the wealthy, and encourage resource enhancing rather than degrading activities through appropriate pricing policies.

REFERENCES

Absalom, E., R. Chambers, S. Francis, B. Gueye, I. Guijt, S. Joseph, D. Johnson, C. Kabutha, M. Rahman Khan, R. Leurs, J. Mascarenhas, P. Norrish, M.P. Pimbert, J.N. Pretty, M. Samaranayake, I. Scoones, M. Kaul Shah, P. Shah, D. Tamang, J. Thompson, G. Tym and **A. Welbourn.** 1995. Sharing Our Concerns, Looking Into the Future. *PLA Notes*, 22: 5–10.

Alcorn, J.B. 1994. Noble Savage or Noble State? Northern Myths and Southern Realities in Biodiversity Conservation. *Ethnoecologica*, 2(3): 7–20.

Arnold, J.E.M. and **W.C. Stewart.** 1991. Common Property Resource Management in India. Tropical Forest Papers. No. 24. Oxford Forestry Institute, Oxford.

Baumann, M., J. Bell, F. Koechlin and **M.P. Pimbert.** 1996. *The Life Industry: Biodiversity, People and Profits*. Intermediate Technology Publications, London.

Bawden, R.J. 1994. Creating Learning Systems: a Metaphor for Institutional Reform for Development, in I. Scoones and J. Thompson (eds). *Beyond Farmer First: Rural People's Knowledge, Agricultural Research and Extension Practice*. Intermediate Technology Publications, London.

Borrini–Feyerabend, G. 1996. *Collaborative Management of Protected Areas: Tailoring the Approach to the Context.* IUCN-The World Conservation Union, Gland.

BOSTID. 1986. *Proceedings of the Conference on Common Property Resource Management, April 21–26, 1985, Annapolis, Maryland.* National Academy Press, Washington, D.C.

Bromley, D.W. and **M.M. Cernea.** 1989. The Management of Common Property Natural Resources. Discussion paper No. 57. The World Bank, Washington, D.C.

Cernea, M.M. 1993. Culture and Organisation: The Social Sustainability of Induced Development. *Sustainable Development*, 1 (2): 18–29.

Chambers, R. 1992. Rural Appraisal: Rapid, Relaxed and Participatory. Discussion paper No. 311. Institute of Development Studies, Brighton, UK.

———. 1993. *Challenging the Professions. Frontiers for Rural Development.* Intermediate Technology Publications, London.

Colchester, M. 1994. Salvaging Nature. Indigenous Peoples, Protected Areas and Biodiversity Conservation. UNRISD Discussion paper No. 55, Geneva.

Crucible Group. 1994. *People, Plants and Patents: the Impact of Intellectual Property Rights on Trade, Plant Biodiversity, and Rural Society*. International Development Research Centre, Ottawa.

Cunningham, A.B. 1993. *Ethics, Ethnobiological Research and Biodiversity: Guidelines for Equitable Partnerships in New Natural Products Development.* WWF-International, Gland.

FAO. 1993. *International Code of Conduct for Plant Germplasm Collecting and Transfer.* United Nations Organisation for Food and Agriculture, Rome, Italy.

Forster, R.R. 1973. *Planning for Man and Nature in National Parks*. IUCN Publications, New Series No. 26. Morges, Switzerland.

Fortmann, L. and **J.W. Bruce** (eds). 1988. *Whose Trees? Proprietary Dimensions of Forestry*. Westview Press, Boulder.

Ghimire, K.B. and **M.P. Pimbert** (eds). 1997. *Social Change and Conservation: Environmental Politics and Impacts of National Parks and Protected Areas*. UNRISD and Earthscan, London.

Gomez–Pompa, A. and **A. Kaus.** 1992. Taming the Wilderness Myth. *Bioscience*, 42 (4): 271–79.

GRAIN. 1995. Towards a Biodiversity Community Rights Regime. *Seedling*, 12 (3): 2–14.

Gujit, I., F. Hinchcliffe, M. Melnyk, J. Bishop, D. Eaton, M.P. Pimbert, J.N. Pretty and **I. Scoones.** 1995. *The Hidden Harvest. The Value of Wild Resources in Agricultural Systems*. International Institute for Environment and Development, London.

Hackel, J.D. 1993. Rural Change and Nature Conservation in Africa: A Case Study from Swaziland. *Human Ecology*, 21: 295–312.

Harmon, D. 1991. National Park Residency in Developed Countries: the Example of Great Britain, in P.C. West and S.R. Brechin (eds). *Resident Peoples and National Parks: Social Dilemmas and Strategies of International Conservation*, University of Arizona Press, Tucson.

Hinchcliffe, F., I. Gujit, J.N. Pretty and **P. Shah.** 1995. *New Horizons: The Economic, Social and Environmental Impacts of Participatory Watershed Development*. Gatekeepers Series, No. 50. Sustainable Agriculture Programme, International Institute for Environment and Development, London.

IIED. 1995. *Whose Eden? An Overview of Community Approaches to Wildlife Management*. International Institute for Environment and Development, London.

Indigenous Peoples, Environment and Development. 1995. International Conference, May 1995, Zurich, Switzerland.

IUCN. 1994a. *Guidelines for Protected Area Management Categories*. IUCN-The World Conservation Union, Gland.

———. 1994b. *1993 United Nations List of National Parks and Protected Areas*. IUCN-The World Conservation Union, Gland.

Jodha, N.S. 1990. Rural Common Property Resources: Contributions and Crisis. ICIMOD, Kathmandu.

Kiss, A. (ed.). 1990. Living with Wildlife: Wildlife Resource Management with Local Participation in Africa. Technical Paper No. 130, World Bank, Washington, D.C.

Koch, E. 1997. Ecotourism and Rural Reconstruction in South Africa: Reality or Rhetoric?, in K.B. Ghimire and M.P. Pimbert (eds). *Social Change and Conservation: Environmental Politics and Impacts of National Parks and Protected Areas*. UNRISD and Earthscan, London.

Kothari, A., N. Singh and **S. Suri** (eds). 1996. *People and Protected Areas: Towards Participatory Conservation in India*. Sage Publications, New Delhi.

Kothari, A., P. Pande, S. Singh and **V. Dilnavaz.** 1989. *Management of National Parks and Sanctuaries in India: A Status Report*. Indian Institute of Public Administration, New Delhi.

Loita Naimana Enkiyio Conservation Trust. 1994. *Forest of the Lost Child. A Maasai Conservation Success Threatened by Greed*. Narok, Kenya.

Max–Neef, M., M. Elizalde, F. Hopenhayn, H. Herrera, J. Jataba, H. Zemelman and **L. Weinstein.** 1989. Human Scale Development: An Option for the Future. *Development Dialogue*, 1: 5–80.

McIvor, C. 1997. Management of Wildlife, Tourism and Local Communities in Zimbabwe, in K.B. Ghimire and M.P. Pimbert (eds). *Social Change and Conservation: Environmental Politics and Impacts of National Parks and Protected Areas.* UNRISD and Earthscan, London.

McNeely, J.A. 1988. *Economics and Biological Diversity: Developing and Using Economic Incentives to Conserve Biological Resources.* IUCN-The World Conservation Union, Gland.

———. 1993. *Parks for Life: Report of the IVth World Congress on National Parks and Protected Areas.* IUCN-The World Conservation Union, Gland.

Morrison, J. 1997. Protected Areas, Conservationists and Aboriginal Interests in Canada, in K.B. Ghimire and M.P. Pimbert (eds). *Social Change and Conservation: Environmental Politics and Impacts of National Parks and Protected Areas.* UNRISD and Earthscan, London.

Munthali, S.M. 1993. Traditional and Modern Wildlife Conservation in Malawi—The Need for an Integrated Approach. *Oryx* 27: 185–87.

Niamir, M. 1990. *Community Forestry: Herders' Decision Making in Natural Resource Management in Arid and Semi-arid Africa.* Community Forestry Note 4, FAO, Rome.

Ntiamo–Baidu, Y., L.J. Gyiamfi–Fenteng and **W. Abbiw.** 1992. *Management Strategies for Sacred Groves in Ghana.* A report prepared for the World Bank and EPC Ghana.

Ostrom, E. 1990. *Governing the Commons: the Evolution of Institutions for Collective Action.* Cambridge University Press, New York.

Pearce, D., A. Markandya and **E. Barbier.** 1989. *Blueprint for a Green Economy.* Earthscan, London.

Phillips, A. 1995. Cultural Landscapes: an IUCN Perspective, in B. Von Droste, H. Plachter and M. Rossler with A. Semple (eds). *Cultural Landscapes of Universal Value: Components of a Global Strategy.* Fisher-Verlag, Berlin.

Pimbert, M.P. 1994. Field Observations on Joint Forest Management in West Bengal and Report on an International Workshop on JFM, Co-organised by the Ford Foundation, the WWF-UNESCO-Kew Gardens People and Plants Initiative, the Government of West Bengal, the Indian Institute of Biosocial Research and Development and the Society for the Promotion of Wastelands Development, 7–18 November 1994, WWF-International, Mimeo.

Pimbert, M.P. and **J.N. Pretty.** 1995. Parks, People and Professionals. Putting 'Participation' into Protected Area Management. UNRISD Discussion Paper No. 57. Geneva.

Pimbert, M.P. and **V. Toledo.** 1994. Indigenous People and Biodiversity Conservation: Myth or Reality? Special issue of *Ethnoecologica*, 2(3). 96pp.

Pimbert, M.P., B. Gujja and **M.K. Shah.** 1996. Village Voices Challenging Wetland Management Policies: PRA Experiences from India and Pakistan. *PLA Notes* 27: 37–41. IIED, London.

Poole, P.J. 1993. Indigenous Peoples and Biodiversity Protection, in S.H. Davis. *The Social Challenge of Biodiversity Conservation.* Working paper No. 1, Global Environment Facility, Washington, D.C.

Posey, D.A. 1990. Intellectual Property Rights and Just Compensation for Indigenous Knowledge. *Anthropology Today*, 6 (4): 13–16.

Posey, D.A. 1993. The Importance of Semi-domesticated Species in Post Contact Amazonia: Effects of Kayapo Indians on the Dispersal of Flora and Fauna, in C.M. Hladik, A. Hladik, O.F. Linares, H. Pagezey, A. Semple and M. Hadley (eds). *Tropical Forests, People and Food: Biocultural Interactions and Applications to Development.* Man and Biosphere, Vol. 13, UNESCO, Paris.

———. 1995. Indigenous Peoples and Traditional Resource Rights: a Basis for Equitable Relationships? Proceedings of a workshop held at the Green College Centre for Environmental Policy and Understanding, 28th June 1995, Oxford, UK.

Posey, D.A. and **G. Dutfield.** 1996. *Beyond Intellectual Property Rights: Towards Traditional Resource Rights for Indigenous and Local Communities.* International Development Research Centre, Ottawa, and WWF-International, Gland.

Pretty, J.N. (1994) Alternative Systems of Inquiry for Sustainable Agriculture. *IDS Bulletin* 25 (2): 37–48, IDS, University of Sussex.

Reid, W.V., S.A. Laird, C.A. Meyer, R. Gamez, A. Sittenfeld, D.H. Janzen, M.A. Gollin and **C. Juma.** 1993. *Biodiversity Prospecting: Using Genetic Resources for Sustainable Development.* World Resources Institute, Washington.

Sayer, J. 1991. *Rain Forest Buffer Zones: Guidelines for Protected Area Managers.* IUCN-The World Conservation Union, Gland, Switzerland.

Scoones, I. and **J. Thompson** (eds). 1994. *Beyond Farmer First: Rural People's Knowledge, Agricultural Research and Extension Practice.* Intermediate Technology Publications, London.

Scoones, L., M. Melnyck and **J.N. Pretty.** 1992. *The Hidden Harvest: Wild Foods and Agricultural Systems. A Literature Review and Annotated Bibliography.* International Institute of Environment and Development, London, and WWF-International, Gland.

Shengji, P. 1991. Conservation of Biological Diversity in Temple-Yards and Holy Hills by the Dai Ethnic Minorities of China. *Ethnobotany*, 3: 27–35.

Shelton, D. 1995. *Fair Play, Fair Pay: Laws to Preserve Traditional Knowledge and Biological Resources.* WWF-International, Gland.

Stone, R.D. 1991. *Wildlands and Human Needs: Reports from the Field.* World Wildlife Fund, Washington, D.C.

Thompson, J. 1995. Participatory Approaches in Government Bureaucracies: Facilitating the Process of Institutional Change. *World Development* 23: 1521–54.

UNESCO. 1994. Operational Guidelines for the Implementation of the World Heritage Convention. Paris, UNESCO.

UNDP. 1994. *Conserving Indigenous Knowledge: Integrating Two Systems of Innovation.* United Nations Development Programme, New York.

Uphoff, N. 1992. *Learning from Gal Oya: Possibilities for Participatory Development and Post-Newtonian Science.* Cornell University Press, Ithaca.

WCMC. 1994. Data sheet compiled by the World Conservation Monitoring Centre, Cambridge.

Wells, M. and **K. Brandon** with **L. Hannah.** 1992. *People and Parks: Linking Protected Area Management with Local Communities.* World Bank, WWF-US and US Agency for International Development, Washington, D.C.

West, P.C. and **S.R. Brechin** (eds). 1991. *Resident People and National Parks: Social Dilemmas and Strategies of International Conservation.* University of Arizona Press, Tucson.

PART 2

country overviews

3

Community-based conservation in India

Arvind Khare

India: Diversity, inequity and politics

The 329 million hectares of land area of India, a unique mosaic of biological, anthropic, cultural, physiographic and climatic diversity, is home to more than 900 million human beings, who are as diverse as its geography. Approximately 300 million of them live in extreme poverty; in the terminology of economists, below the poverty line. Amongst them one can count one-fourth of the world's tribal people, the indigenous inhabitants of India. At the same time there are people who are as rich and as consumerist as any in the Western world. There is little wonder then that the political organization of a country of this size, diversity and inequity is not without its problems.

India is a federal democratic republic consisting of 25 states and seven union territories. There is a clear delineation of powers between the central government and the state governments. The recent 73rd Amendment to the Indian Constitution has given a chance to the local village to assert itself as a third level of governance. Current trends also show that state governments are becoming increasingly more powerful and have a greater say and influence on the government at the centre.

This has led to an ambivalent response from Indian conservationists. While they applaud the devolution of powers to the local councils (*panchayats*) following their belief in the decentralization of power, they are very apprehensive of a similar kind of devolution of powers to the state governments. In fact, many of them are quite happy with centralization of powers by the central government through policy measures (dealt with later).

There is, however, little difference in the conservation orientation of the central or the state governments. The ambivalence of Indian conservationists has more to do with their own predilections of carefully nurtured contacts in Delhi coming unstuck. Almost similar is the case with the dispensation of justice on environmental matters. While the cases of millions of tribals and other rural poor relating to usurpation and settlement of their rights, and dispossession of their lands languish in lower courts for years without a hearing, there is little in the 'judicial activism' of the higher judiciary to rejoice them. Except for the places where grassroots environmental activists are organizing them, they are increasingly turning to the protection of the local underworld or seeking solace in the violent methods of extremist groups. The recent step-up in the activities of the Peoples' War Group in Andhra Pradesh, where forest rights are at the heart of the struggle, is an example. These instances of the dichotomy in Indian society continue to influence the environmental debate.

Biological and anthropic diversity

Due to its physiographic and climatic conditions as also its location at the confluence of three major biogeographic realms, the Indo-Malayan, the Eurasian and the Afro-Tropical, India is endowed with bewildering biodiversity. It has 10 biogeographic zones: Trans-Himalayan, Himalayan, Indian Desert, Semi-arid, Western Ghats, Deccan Peninsula, Gangetic Plains, North-East India, Islands and Coasts. These biogeographic zones represent a broad range of ecosystems, containing 6 per cent of the world's flowering plant species and 14 per cent of the world's birds. There are over 45,000 identified plant species, and one-third of its 15,000 flowering plants are found only in India. It has 81,000 identified species of animals. Some 14 per cent of its 1,228 bird species, 32 per cent of its 446 reptile species and 62 per cent of its 204 amphibians are unique to India (MOEF 1994). There are no firm statistics about the biodiversity loss in India. It is however estimated that at least '10% of flowering plants, over 20% of mammals, and about 5% of birds are at various stages of threat' (Kothari 1997).

More amazing is the diversity of India's people/communities. The People of India project of the Anthropological Survey of India has identified 91 eco-cultural zones in India inhabited by 4,635 communities, speaking 325 languages/dialects. A community of India is best understood in terms of its relationship with resource endowments of such micro-regions and its relationship with various other communities in the control and exploitation of such resources (Singh 1992).

Forest, poverty and tribal nexus*

India's existing forests are primarily concentrated in three regions: the Himalayan band stretching from the north to the north-east; the central forest belt with its nexus in the Chhotanagpur Plateau of Orissa, Bihar and Madhya Pradesh and the north–south belt of the Western Ghats. Significantly, the location of India's predominant tribal populations is closely superimposed on the nation's forest tracts. With the greatest economic dependence on forest resources, it is not surprising that perhaps tribals possess the most extensive knowledge of India's forests, as well as the strongest motivation to ensure the continuity of these ecosystems. Barring a few isolated patches, the tribal communities co-exist with other local communities, whose production systems exhibit a close linkage with forest biodiversity. These combined local communities (estimated population 200 million) therefore constitute the critical segment of the Indian population whose survival depends on the sustainability of forest biodiversity. There are also strong correlations between the locations of tribal people, forests and India's concentrated poverty areas (Poffenberger et al. 1996).

Forests are an important source of food specially for the tribals and the rural poor. Reportedly, 60 per cent of non-timber forest produce (NTFP) is consumed as food or as a dietary supplement by forest dwellers. In Bastar district of Madhya Pradesh, about 75 per cent of forest-dependent people supplement their food by tubers, flowers and fruits all the year round. In the Andaman and Nicobar islands, several tribes wholly subsist on the food derived from forests and the sea. In a survey of 216 households (tribal and caste) it was found that, of the 122 uses of plants or their parts listed by the people, the maximum were for food (44), followed by fuel (39) and medicinal purposes (18) (Malhotra et al. 1991).

NTFP is potentially obtainable from around 3,000 species found in the Indian forests. Apart from food and other domestic uses, NTFP is an important proportion of income which varies from state to state, ranging from 5.4 to 55 per cent.

* This article is confined to forest biodiversity and its relationship with people.

Seventy per cent of the rural and 50 per cent of the urban people use fuelwood for cooking purposes. Forests meet nearly 80 per cent of the rural energy requirements (according to India's submission at UNCED). Apart from subsistence fuelwood needs, 'headloading' of fuelwood is also an important source of income for many poor families specially during the lean agricultural season.

Rural communities require *timber*, bamboo and grass for house construction, bullock carts, agricultural implements, fencing, etc. Most of these needs are fulfilled from forests. The consumption of bamboo alone for this purpose is estimated to be around 1.6 million tonnes per annum.

Fodder from forests is yet another critical survival resource. A study shows that 66 per cent of small and marginal farmers in Andhra Pradesh would not be able to cultivate at all in the absence of forest resources as they would not be able to maintain a pair of bullocks. It is therefore not surprising that the Forest Survey of India recorded widespread grazing in forest areas across the country (FSI 1987). Of the 174 protected areas (PAs) surveyed, 67 per cent of national parks and 83 per cent of the sanctuaries reported grazing incidence (Kothari et al. 1989).

Forest-based enterprises provide 1,623 million person-days of employment of which 1,063 million days are created in the small enterprises consisting of collection, gathering and NTFP processing in cottage industries. Of this, 5,178 million days of employment are that of women.

Threats to biodiversity and community knowledge systems

Of the many historical and current threats to forest biodiversity and community knowledge systems, the following need special mention:

1. Habitat destruction: There are three main processes leading to habitat destruction:

(*a*) Development projects like agricultural expansion, dams, roads and mining have destroyed more forests than any other single cause. Approximately 1.883 million hectares of forests were lost for these purposes in India between 1952 and 1980. In the first two decades after independence, serious concern about food deficit and inability to implement genuine land reforms led to a large-scale diversion of forest lands for agriculture purposes. Even after the passing of the Forest Conservation Act in 1980 (which prohibits transfer of forest land for non-forestry purposes without central government permission), 1,14,809 hectares of forest lands were transferred for other uses.

(*b*) Modern economic processes have subjected the traditional economies to the requirements of production and consumption of distant places and populations, hence breaking the fragile balance between supply and demand. Urban and industrial consumers' demand (national and international) has no relationship with the carrying capacity of forests and no societal control, as it is independent of the ecosystems which are harvested to satisfy it.

(*c*) The invasion of commercial forces with exclusive interest in particular species displaces biodiversity with monocultural industrial plantations. After the enactment of the Forest Conservation Act in 1980 this process has slowed down considerably. Yet destructive activities continue, as Forest Departments (FDs) supply raw material to industries at concessional rates, and tribal/local community rights to certain quantities of timber are exploited for commercial purposes leading to massive deforestation in north-east India and other areas. There is also sustained pressure (so far unsuccessful) on the government to lease forest land to industries. Monocultural commercial plantations are of very little use to local communities whose livelihood strategies are dependent on multiple outputs from biodiverse forests.

2. Tenurial insecurity: With the advent of colonialism there was a continuous onslaught on the rights of forest dwellers. 'The right of conquest' converted people's 'rights' into 'privileges' and 'concessions', and in a number of places to forceful eviction from their forest homes. Alienation of local communities from their resources and passing of control to commercial interests ensured that the forest resources were exploited by those with the least understanding of the very nature of these resources. The experience of a number of developing countries suggests that protection to natural forests can be effectively provided by the local communities whose very sustenance is dependent on them, provided they are ensured of their tenurial rights. Once the security of tenure is assured, forest-dependent communities develop their own institutional mechanisms for social control, self-abnegating rules for conservation, and harvesting practices that are sustainable.

3. Disempowerment of communities: Political processes in India have invariably passed on the control and regulation of natural resources from the hands of people into the hands of those who have the right to use them but no responsibility for guaranteeing their sustainability. Not to suggest that all traditional social structures are desirable or equitous or without problems, it is a fact that instead of building on the essentially self-governing traditional systems, political processes have continuously taken the governing functions away from the community.

4. New international developments: The TRIPs provision in GATT 1994, and the 1991 revision of UPOV are some of the international developments which pose a potential threat to the knowledge systems of forest-dependent communities. The developing countries are under extreme pressure to harmonize their legislation with global IPR standards. These provisions could end up conferring monopolistic ownership rights to products made in laboratories from the knowledge of indigenous peoples and local communities.

A recent example illustrates this phenomenon. This relates to the species *Phyllanthus niruri*, or *jar-amala* or *bhuin amala* as it is locally called in India. A patent application has been filed by the Fox Chase Cancer Center of Philadelphia, US, at the European Patent Office for the manufacture of a medicament for treating viral hepatitis B derived from this plant. It is one of the many plants effectively used by the formal systems of Ayurveda Siddha and Unani for treating jaundice or yellowness of the skin and eyes.

The plant grows wild throughout the hotter parts of India. The chemical constituents of *P. niruri* have also been investigated and reported in numerous scientific journals, including the publications of the Indian Council of Medical Research from as early as 1969. Even though the use of *P. niruri* for treatment of all kinds of hepatitis has been an ancient and well-recorded innovation in the Indian systems of medicine, the patent claim of the Fox Chase Cancer Centre states: 'In so far as is known *P. niruri* had not been proposed for the treatment of viral hepatitis infection prior to the work done by the inventors of the present invention.'

If one were to take a look at who is patenting and what is being patented then the implications for the countries of the South become clear. It is not surprising that the overwhelming majority of patent claims originate in the industrialized world, and the South is virtually unrepresented despite the fact that much of the patented germplasm originates there.

Conservation policy: History and current status

Current conservation policy and practice in India have not been able to break out of their colonial roots. In the colonial era, it was a section of forest officials, hunter-naturalists and both the British rulers and Indian aristocrats, who were at the forefront of conservation efforts. Their major concern was the preservation of their hobby of 'game' hunting. The colonial government also needed to maintain the sustained supply of commercial timber. One of the first steps in this direction was the system of Reserved and Protected Forests established in 1878, which entailed a major takeover of common lands by the state.

> The forest hierarchy was empowered to establish restrictions on local use of the forests wherever necessary to assure regeneration of timber supplies…. By 1900 the Forest Department regulated access to the Reserved Forests not only for wood and fodder but also for game. In the same years, a movement was launched to declare certain Reserved Forests closed to all human exploitation, so as to preserve endangered game species from extinction. This fledgling national parks movement saw little legitimate place for any human or livestock residents in the vicinity of endangered game…
>
> By 1900 another powerful Indian component of the conservation movement emerged: several leading 'Native Princes', *rajahs* who began to realise that game species were being rapidly depleted in the ancestral hunting reserves…. Several major *rajahs* began adopting British India's game laws to their own hunting reserves…. Being autocrats, rooted in medieval times, they tended to levy harsh and instant punishment…. One of the most powerful, the Maharaja of Kashmir, summarily removed the human population from the Dachigam deer reserve in the high mountains in about 1910…
>
> In 1934 the Indian National Parks Act became law, embodying many years' experience of game laws and their implementation… the act defined which human visitors could enter the game reserves, with which licenses and at which seasons. But this focused on sports hunters; regarding resident tribal and peasant populations, it was almost silent. A year later Corbett Park, India's first modern national park, was established.
>
> The United Provinces Wildlife Preservation Society, assisted by the Bombay Natural History Society, organised a major conference in New Delhi in 1935. The familiar theme dominated each province's report. Endangered

> species must be preserved at all costs. More national parks were needed. Budgets for control of poaching must be increased.... The leading Indian conservationists, whether the *rajahs* or the urban-based naturalists in Bombay... together with Nehru [India's first prime minister], constructed a new and far more ambitious legal and administrative structure for wildlife preservation (Tucker 1991; bracketed text added).

It is clear from the above that the colonial conservation policy and practice was: (*a*) governed by an elitist concern for some game species and the colonial need for timber; (*b*) increasingly relied on exclusion of humans and human activities from reserved areas; and (*c*) was (barring a few exceptions) apathetic to the plight of tribal and other communities whose livelihood depended on such reserved areas.

Now a look at the current scene. The basic elements of the current conservation policy and practice, insofar as they affect CBC, can be assessed through policy pronouncements, legal structures, and the implementation of these pronouncements and structures.

Policy: The *National Forest Policy, 1988* states the basic objective very clearly: 'Conserving the natural heritage of the country by preserving the remaining natural forests with the vast variety of flora and fauna, which represent the remarkable biological diversity and genetic resources of the country.' The policy, which is otherwise distinguished by a pro-people approach in the management of forests, reversing a hundred years' anti-people trend, however treads the beaten path of conservation and restricts itself to suggesting: (*a*) The conservation of total biological diversity, the network of national parks, sanctuaries, biosphere reserves and other protected areas should be strengthened and extended adequately (para 3.3); and (*b*) The forest management should take special care of the needs of wildlife conservation, and forest management plans should include prescriptions for this purpose. It is specially essential to provide for 'corridors' linking the PAs in order to maintain genetic continuity between artificially separated sub-sections of wildlife (para 4.5). The forest policy has hardly anything to recommend for the people living in and around the suggested PAs and the proposed corridors.

The 12-point *National Wildlife Action Plan* (Department of Environment Undated) also does not have a focus on the role that communities may play in the conservation of wildlife. Its main focus is on the establishment of a network of PAs. The best that is offered to the local communities is: 'Where appropriate, habitat restoration should be accompanied by welfare and development measures to replace or ameliorate the dependence of local people on natural resources within the protected areas.... Identify the surroundings of protected areas for ecodevelopment and undertake community development programmes through the concerned agencies to elicit the support and involvement of the local people.'

Law: It should be noted that almost all the colonial provisions relating to creation and declaration of Reserved and Protected Forests and other provisions of the Forest Act, 1927, continue to be in force. The major development after independence was the 42nd Constitutional Amendment, bringing forest and wildlife conservation into the concurrent list of subjects over which both the central and state governments have power to make law, with central laws judicially superseding the state laws. Subsequently, the Wildlife (Protection) Act, 1972, afforded varying degrees of protection to wildlife species under different schedules, and enabled the creation of national parks and wildlife sanctuaries. The Act was later amended in 1982, 1986 and 1991, and is at the time of writing under major revision.

A word about the difference between national parks and sanctuaries would be appropriate. National parks are given a higher level of protection, with no grazing and no private landholding or rights permitted within them, while sanctuaries are given a lesser level of protection, and certain activities may be permitted within them for the better protection of wildlife or for any other good and sufficient reason. This difference is notional as far as the communities are concerned. According to this law, the communities living inside the national parks have to be evicted and those having any relationship with forest resources of the park are prohibited from using them. In the case of sanctuaries some of these activities may be permitted, the permitting authority being the Chief Wildlife Warden. Not many cases are known where formal rights have been granted for resource use. The net result is that while many human activities continue both in the parks and sanctuaries, laws become a major instrument in the hands of foresters to harass the people living in and around these PAs.

Implementation: What has happened in the implementation of these policies and laws is even more appalling. We may first take a look at the rate of expansion of PAs, which was the single most important conservation strategy that was followed.

Between 1975 and 1980, the number of national parks and sanctuaries increased from 131 to 224, the area trebling from 24,000 sq. km to 76,000 sq. km; thereafter, up to 1995, the number shot up to 521, the area doubling to 148,000 sq. km. This gives an expansion rate of 620,000 hectares per annum, from which peoples' rights are virtually extinguished or considerably curtailed. This could not have happened, at least not on this scale, without the backing of political power. The decade of the 1970s was in this sense an unusual one and was marked by both an increase in the grassroots environmentalism and the increased activities of elite environmentalists who boasted about and utilized their access to the then political powers. The two developments had hardly any connection: the grassroots people fought for rights of access and conservation of natural resources in order to protect their livelihoods while the elite–bureaucrat–politician combine showed their concern for the environment by reserving more and more areas to exclude these very people. This sounds very similar to the colonial days' combine of rajahs, urban-based naturalists and their British rulers. Consider the following evidence (Rangarajan 1996):

> The Indian Board of Wildlife had its first meeting at Mysore in 1952, with the Maharaja being the host. Among the key personalities in policy formulation in the early years were English tea and coffee planters... foresters and the princes.... It was no great surprise that even attempts to control shooting of tigers in the breeding season in Rajasthan were shot down. As the author of the proposal later recounted, virtually all the members of the advisory board that considered the issue were keen hunters...
>
> What is important is the fact that the small but influential wildlife lobby had won significant gains in a short span of time due to support from the Prime Minister...
>
> New initiatives include the end of all tiger-hunting, the creation of core zones in the tiger reserves where commercial forestry was halted and the expansion of the protected area system.... In the process, however, the wider human dimensions of wildlife conservation, especially the vexed issue of local rights, received little attention.

Conservation and people

The result of these 120 years of exclusion of people or their rights from the PAs can be seen in the deteriorating relationship between the park managers and the people and the immense suffering that the communities have had to go through and experience everyday. It did not even have the desired results as far as conservation benefits are concerned. The current situation in the PAs has been succinctly summed up in the following statistics: PAs having human populations living within them (> 55 per cent of PAs) and around them (< 80 per cent); traditional rights and leases (40 per cent); traditional grazing by livestock (> 40 per cent); fodder extraction (> 15 per cent); timber extraction (> 16 per cent); and NTFP extraction (> 35 per cent); those used by other government agencies (> 55 per cent); public thoroughfares (> 45 per cent); plantations (> 45 per cent); illegal occupation and use (> 8 per cent); encroachment (> 7 per cent); and poaching (> 55 per cent). In terms of PA management, some have plans (> 30 per cent) with zoning (> 20 per cent); research and monitoring usually by external persons (> 23 per cent); exotic species introduction (> 15 per cent); captive breeding programmes (15 per cent). In terms of land use and environmental effects, some PAs are affected by forest fires (> 20 per cent); floods (> 35 per cent); droughts (50 per cent); and/or water pollution (> 40 per cent). Some have trained personnel (> 35 per cent) and honorary wardens (> 35 per cent); and very few have any association with NGOs (> 15 per cent) (Kothari et al. 1989).

There are many reports detailing the suffering of people due to the current conservation policy and management practices:

(i) Several thousand people have been displaced (Kothari et al. 1996) and many more, specially those living inside the boundaries of national parks, live as stateless citizens with the constant threat of displacement.

(ii) From a sample of 39 national parks and 167 sanctuaries, 14 (36 per cent) and 49 (29 per cent) respectively, reported incidents of injury or death of humans due to attacks by wild animals (Kothari et al. 1989). Of a total of 379 cases of attack reported from national parks and 250 reported from sanctuaries, 87 and 62 per cent respectively were fatal. Between 1956 and 1983, an yearly toll of about 20 persons is reported in Sundarbans. In the case of deaths due to elephant attacks, in some states, 'the family of the person killed has the solace of a princely sum of Rs 5,000 obtained after much difficulty and red tape… if the elephant so killed has tusks, it can fetch a very handsome income, often equivalent to several years of earnings for a landless tribal or small farmer' (Sukumar 1989 quoted in Gadgil and Guha 1995). Compensation is sometimes paid and sometimes not, and even when paid, it is low and only obtained after going through a maze of bureaucratic red tape.

(iii) A ban on grazing continues to be one of the major sources of conflict between the PA authorities and settled and nomadic communities. The famous case of Van Gujjars in the Rajaji National Park remains unresolved. Questions regarding the scientific validity of a total ban are best illustrated by the example of Keoladeo National Park in Bharatpur, Rajasthan. This partially artificially created wetland harbours numerous resident and migratory birds, including the endangered Siberian crane *Grus leucogeranus*. In the early 1980s, a total ban on grazing was imposed. In the absence of any alternative the local people protested. In the ensuing violent conflict between the police and the local communities, seven people were killed and several injured. But the ban on grazing was enforced. 'The results have been disastrous for Keoladeo as a bird habitat. In the absence of buffalo grazing, Paspalum grass has overgrown, choking out the shallow bodies of water, rendering this a far worse habitat, specially for wintering geese, ducks and teals, than it ever had been before' (Vijayan 1987).

(iv) In the earlier part of this paper, some details were provided about peoples' dependency on forest resources. Creation of a PA severely affects this dependence, threatening the very survival of these communities. Exclusion or restrictions on resource use, at the rate of 620,000 hectares per annum, are bound to create conflicts. 'These conflicts have been reported from all over the country, from the Rajaji National Park in the north to the Nilgiri Biosphere Reserve in the south, and from the Betla tiger reserve in the east to the Nal Sarovar Sanctuary in the west. Sometimes they can have tragically destructive effects, as when a hostile local population is moved to burn large areas of a National Park which they perceive to be against their interests—acts of arson which they have resorted to in two important national parks, Kanha in Madhya Pradesh (in 1989) and Nagarhole in Karnataka (in 1992)' (Gadgil and Guha 1995).

It will be however unfair to say that there has been no attempt by the government to respond to this situation. The two most noteworthy efforts are: (*a*) ecodevelopment, which attempts to respond to the hardships faced by the people in and around the PAs; and (*b*) joint forest management (JFM), which is concerned with the degraded forest areas but has conservation benefits.

Ecodevelopment: Ecodevelopment as a concept finds mention in the Wildlife Action Plan and was supported by a centrally sponsored scheme. It gained greater prominence when the Government of India negotiated a large project with a loan component from IDA of US$ 28 million, and a GEF grant of US$ 20 million. This project aims to conserve biodiversity by addressing both the impact of local people on PAs and the impact of PAs on local people, improving the capacity of PA management, involving local people in PA planning and protection, developing incentives for conservation, and supporting sustainable alternatives to the harmful use of resources. It supports collaboration between the state FDs and local communities in and around ecologically valuable areas (World Bank 1996).

Moving beyond these broad statements into the project content reveals the actual nature of ecodevelopment. Apart from the usual project activities dealing with improved PA management, education, impact monitoring and project management, the essence of the project is contained in what is called village ecodevelopment which deals with reducing negative interactions of local people on biodiversity and increases collaboration of local people in conservation by '(i) conducting participatory microplanning and providing implementation support, (ii) implementing reciprocal commitments that foster alternative livelihoods and resource uses to be financed by a village ecodevelopment program and that specify measurable actions by local people to improve conservation, and (iii) special programs for additional joint forest management, voluntary relocation, and supplemental investments for special needs' (World Bank 1996).

The implementation is in very early stages in the seven project areas and it is difficult to make any assessment on the basis of results. There are, however, a number of difficulties with the concept which are enumerated here:

(*i*) The ecodevelopment concept accepts the existing legal and policy framework relating to PAs. It has already been stated that this framework is based on the exclusion of people from these areas and has resulted in unprecedented conflicts. It is therefore difficult to see what it is that the people would be able to contribute by being involved in 'PA planning and protection'.

(*ii*) The main element of ecodevelopment is to substitute people's dependency on PAs by sustainable alternatives outside the PAs. This element is based on the assumption that either sufficient land resources are available in the immediate vicinity of the PAs or opportunities exist for alternative non-land-based occupations and the only input that is required to make the switch is a set of incentives. This assumption flies in the face of facts. Given the fact that the most scarce resource in India is land and that hardly any forest land (surrounding the PAs) exists that already does not have recorded rights of communities, it is difficult to see how any of the land-related dependency (fuel, fodder, food, medicine, etc.) can be shifted outside the PAs. As far as alternative non-land-based occupations are concerned, the efforts of hundreds of voluntary agencies across the country, with the best participatory skills and incentives, prove that at best it succeeds in some places and mostly it is not sustainable.

(*iii*) The third element, which has evoked considerable resentment from human rights groups, relates to the business of 'voluntary relocation'. If it is a voluntary relocation it will take place with or without the project. Therefore it is in fact an attempt to determine the quantum of incentive that would be sufficient to move the people out of the project area without resorting to more violent methods. Thus ecodevelopment also supports the whole notion that biodiversity and people cannot subsist together. 'Yet in most of the world there is very little true wilderness where wildlife does not coexist with human and domestic populations. Given this elementary fact, it is puzzling that a strategy that encompasses domestic as well as wild animal populations has taken so long to evolve' (Tucker 1991).

Joint Forest Management: A significant step has been taken by the Government of India in 1990, through its thrust on JFM of degraded forest areas. This policy shift was a culmination of ceaseless struggle of communities to assert control over their productive resources, experimentation by visionary forest officers, and the work of NGOs with rural communities. The policy aims at recognition of rights of organized communities over a clearly defined degraded patch of forests. The benefits accrue to the community on fulfillment of certain responsibilities for protection and conservation of the forest patch. The rights and responsibilities regime is implemented through the mechanism of a local institution. This policy initiative was taken up by the state governments, 19 of whom have so far come out with their state orders on JFM, laying down the rules for people's participation in forest management. The community response has been tremendous. Currently, approximately 20,000 communities are protecting over 2.5 million hectares of forests.

This has not happened only due to the assurance of economic benefits offered to the communities. 'In most states today, the rights now being offered under joint forest management (JFM) government resolutions actually offer little more, and sometimes even less, than what these villages enjoyed under earlier settlement acts and *nistar* agreements. JFM agreements, however, are helping to forge new relationships between rural communities and the Forest Department. By formalizing and further legitimizing prior or existing rights, they provide the framework for an essential psychological security heretofore unknown, enabling communities to invest their labour and time in patrolling, protecting and managing the forests' (Poffenberger and McGean 1996). It may be mentioned that the whole programme of JFM does not enjoy statutory protection, and yet the formal legitimization of community control over degraded patches of forests has enthused not only the communities but a whole range of NGOs, researchers and foresters.

The evidence collected by researchers points out that (*a*) community-protected forests are more biodiverse, (*b*) community extractions from their forests have been less than ecologically permissible limits, and (*c*) the annual outputs from forests are more important for communities than their terminal timber value.

JFM is not without its problems. Increasingly, those associated with the movement feel its restricted framework of dealing with only degraded forests, inability to provide a more solid foundation for community rights, inequity between the two partners—the FD and the communities, internal inequities in the community organizations and absence of a greater role for the communities in management functions rather than only protection functions (see also Sarin et al., and Raju in this book).

The experience of JFM, however, provides many valuable lessons for CBC: (*a*) tenurial security is a more effective incentive for the communities than monetary incentives; (*b*) if the communities are organized into their own institutions they exhibit an exemplary resource prudence and are able to exert considerable social pressure on the erring members of the group; (*c*) biodiversity provides a greater assurance of livelihood security than the timber value of forests; and (*d*) assurance of rights elicits responsible behaviour.

Status of community-based conservation

In addition to the above mentioned efforts involving communities in the conservation efforts, mostly initiated by the government, a number of other efforts, initiated by the communities themselves or by NGOs, also exist.

Folk traditions of conservation: Many communities practising traditional resource use systems have also developed a systematic body of knowledge regarding the natural environment, the functioning of the ecosystem and different habitats and how to manipulate these for human use without damaging the natural processes and cycles. This is not to say that all such communities have done so; those which have not succeeded have either been destroyed or have moved to new localities on the exhaustion of the resources. However, the very survival of thousands of communities which are directly dependent on the utilization of natural resources indicates that they had accumulated knowledge of natural processes.

Traditional knowledge, like modern 'science-based' knowledge, is also a system of knowledge, in that it is based on the accumulation over thousands of years of human observation and practice and the working out of inter-relationships and cause-and-effect relations of different processes. Such knowledge apparently seems to be more observational than modern knowledge, but spans multidimensional aspects of natural processes and is more holistic because it is accumulated through the process of human intervention *in situ*.

Furthermore, these knowledge systems are handed down through oral tradition as well as through various religious rituals, cultural practices and beliefs in which they are embodied. There is increasing evidence suggesting the adaptation of these systems to the changing ecological, social and economic conditions, although it is not really known how new experiences and effects of changing conditions become assimilated into the 'accepted' body of knowledge. The cultural practices in the Indian subcontinent indicate a number of traditions of restraints on the exploitation of wild plant and animal resources that reflect a detailed knowledge of the functioning of the ecosystem and the need to preserve biological diversity. (For more details, see Gadgil, and Ramakrishnan in this book.)

Unfortunately, most such conservation practices are under serious threat due to processes of alienation and modernization that outside society has imposed on local communities. Some examples of these practices are:

(i) The Changpas of Ladakh are the herders of Pashmina goats which produce the famous Pashmina wool. Allocation of pastures and pasture routes, cyclical change in routes, a mechanism to settle disputes through the institution of 'Goba', societal controls like polyandry and the culling of excess animals, helped in turn by their customary rights and equity in resource allocation, have led to excellent maintenance of pastures through centuries.

(ii) Sacred groves are one of the finest instances of traditional conservation practices. To a great extent, sacred groves in the hilly and mountainous regions of India are a legacy of shifting cultivators (see Ramakrishnan in this book).

(iii) Restraint on periods of harvest of wild plants and animals was very common. In Jakhol–Panchgai area of the Uttarkashi district in the Himalayas, the tubers of a plant, locally known as *nakhdul* may be harvested only at the time of a religious festival; this is also true of the flowers of *brahmakamal*, a herb of alpine meadows near the Nandadevi peak in the Chamoli district in the Himalayas.

(iv) The Cholanaickens have well-defined principles that allow the members to gather and extract minor forest produce (MFP) within their respective territories. While there is no restriction on gathering edible items for self-consumption from the entire forest region, MFP can be collected only from the territory allocated to each family, with sons inheriting this right from the father but sons-in-law being excluded.

(v) Only one member of each household gathers fuelwood once a week from the village forest of Gopeshwar in Chamoli district of the Uttar Pradesh Himalayas—hence the village forest is still well-preserved, although most of the neighbouring land has been completely deforested.

Self-initiated conservation efforts: Throughout history, local communities have responded in their own way to the threat posed by colonizing and centralizing forces. One of the best examples of this process is the existence of a number of self-initiated forest protection groups which are reported to be protecting more than 200,000 hectares of forests on both state and community lands near their villages in the eastern states of Orissa and Bihar and, on a smaller scale, in parts of Rajasthan, Gujarat, Karnataka and Punjab (Sarin 1996). Their efforts appear to be very effective, particularly where they have successfully negotiated area boundaries and access control rules with other settlements. Local institutions are also marked by various levels of organizational complexity. The *shamilat* (common property) forests in Punjab not only involve collective management by several villages but also reciprocal agreements with nomadic pastoral groups (Sarin 1996).

In many areas, villagers are themselves protecting habitats with an explicit rejection of any government involvement. Inhabitants of some villages in the Alwar district of Rajasthan have declared 1,200 hectares of forests as the Bhairodeo Dakav 'Sonchiri', promulgated their own set of rules and regulations which allow no hunting, and are zealously protecting the area against any outside encroachments. Several patches of forest in the Himalayas have obtained strict protection from the communities associated with the Chipko movement (Kothari 1997).

NGO-supported community-based conservation: NGOs are supporting community struggle for conservation of their resources in two ways: one is to fight the attempts of entrenched classes to do any more damage and the second is to devise positive solutions for conservation. For example, Sakti is a federation of 23 community groups who have been protecting their forests for more than 10 years, much before the advent of JFM, and, because of their strength, are in a position to consolidate their conservation efforts by utilizing JFM provisions. NGOs have also been in the forefront of the attempts to stop mining in Radhanagri Sanctuary, Maharashtra; Sariska Tiger Reserve, Rajasthan; and the misuse of Bhitarkanika Sanctuary, Orissa, for commercial purposes.

Proposals for the future

In the current situation most CBC efforts fighting against the historically entrenched vested interests are losing control over their resources, facing breakdown of their institutions and are far too small to be effective except in their localized area of influence.

In a populous and diverse country like India, with a long tradition of co-existence of wilderness with human beings and the heavy dependence of local communities on forest resources for their livelihood, the whole paradigm for conservation has to change. It has to shed the arrogance of elite environmentalists who assume that they alone can think of the conservation of flora and fauna and the local communities, left to themselves, would not take a day in destroying the entire biodiversity. On the contrary, since the survival of local communities depends on the biodiversity of forest resources they have the greatest vested interest in preserving it. The communities, therefore, have to be the focal point of the new strategy.

A beginning must be made with the existing PAs. For each of these areas, the communities must determine a wilderness area that they would ensure remains inviolate. The remaining areas must be segmented by the communities allowing progressively greater utilization. The largest segment will consist of the area, determined by the community, which will be used only for the collection of NTFP and fallen twigs. Some areas will be silviculturally manipulated to assure greater benefits for communities' subsistence and some portion can be earmarked for controlled grazing. The PA, in this fashion, can be divided in as many segments as required so long as it fulfills the twin objectives of conservation and community subsistence needs. The communities on the other hand must assume the responsibility for maintaining the ecological integrity of all such segments and engaging in active collaboration with the Forest Department in elimination of poaching and smuggling and prevention of fire.

The same principle can be extended to biodiversity outside the PAs, for which currently there seems to be no plan. If extended to the other areas, the total inviolate wilderness area will probably be much larger than can ever be expected under the present dispensation. This change, however, calls for substantial change in policies and the law and more importantly in the attitudes towards the resident communities who must be viewed as partners rather than enemies of the PAs. It will also pose several problems in terms of defining communities, determining their rights, evolution of their institutions and their relationship with FDs. These very problems confronted JFM in its early stages of implementation, but have so far been successfully handled by constantly analyzing every step, through hands-on experience, field research, and involvement of many outside groups. There is no reason why this should not be the case with conservation of biodiversity.

The required change in the law and policy will take time. However, one need not wait. The current provision in the Wildlife Protection Act, Section 33 (c), provides ample scope for serious experimentation. It says that the Chief Wild Life Warden 'may take such measures, in the interests of wildlife, as he may consider necessary for the improvement of any habitat'. This can be a beginning to the radical changes required in India's conservation policies and practices.

Box 3.1

CBC in Non-forest Areas

Editorial note: Apart from forest areas, on which the author has concentrated, considerable activity towards CBC is taking place in freshwater wetlands, grasslands, marine areas and agricultural and pastoral lands. In the Himalayan pastures of Uttar Pradesh, communities are reviving protection of *bhugiyals*, grasslands which were once considered sacred or used with strict restrictions (Bhatt, personal communication). In the Arvari catchment area of Alwar District, Rajasthan (see also Singh in this book), reservoirs created by community-built checkdams are being protected as habitats for fish and other wildlife. Across India's coasts, traditional fisherfolk are fighting against the introduction of large-scale destructive technologies like trawlers, and in places trying to protect or revive customary rules regulating fishing activities. Low-intensity, small-scale aquaculture, building on traditions, is emerging as a viable alternative to the polluting commercial-scale aquaculture which is being promoted by the government and donor agencies (Shiva and Karir 1997). Hundreds, perhaps thousands, of farmers are continuing to use, or are reviving, agricultural biodiversity in their fields, in a move to make farming more sustainable, self-sufficient and fulfilling than what the Green Revolution model has offered so far (Alvares 1996; Kothari 1997).

Box 3.2

Building Bridges

Editorial note: The conflicts between conservation agencies (governmental and non-governmental) on the one hand and local communities and social activists on the other, has considerably reduced the chances of either conservation or sustainable livelihoods being achieved. Over the 1990s, attempts have been made by activists and academics to establish a dialogue between these contending parties, to explore their points of agreement, and to put together a joint front against the industrial-commercial economy which is the main threat to both wildlife and local communities. A 45-day journey through 16 protected areas in western and central India, in 1996, in which a core team of NGOs and villagers took part, initiated a dialogue at several sites. A series of national and state-wise workshops have developed this dialogue further and wider (Kothari et al. 1997). Government officials in many places have also, on their own initiative or through the urging of NGOs, established more harmonious relations with local people through offering surer access to livelihood resources and involving them in some aspects of management (Rao and Sharma 1998).

REFERENCES

Alvares, C. (ed.). 1996. The Organic Farming Sourcebook. The Other India Press, Mapusa, Goa.

Bhatt. Personal communication with Om Prakash Bhatt, Dashouli Gram Swarajya Mandal, Chamoli, Uttar Pradesh.

Department of Environment. Undated. *National Wildlife Action Plan*. Government of India, New Delhi.

FSI. 1987. *The State of Forests 1987*. Forest Survey of India, Dehra Dun.

Gadgil, M. and **R. Guha**. 1995. *Ecology and Equity*. Penguin Books, New Delhi.

Kothari, A. 1997. *Understanding Biodiversity: Life, Equity, and Sustainability*. Orient Longman, New Delhi.

Kothari, A., N. Singh and **S. Suri** (eds). 1996. *People and Protected Areas: Towards Participatory Conservation in India*. Sage Publications, New Delhi.

Kothari, A., P. Pande, S. Singh and **D. Variava.** 1989. *Management of National Parks and Sanctuaries in India: A Status Report*. Indian Institute of Public Administration, New Delhi.

Kothari, A., F. Vania, P. Das, K. Christopher and **S. Jha** (eds). 1997. *Building Bridges for Conservation: Towards Joint Management of Protected Areas in India*. Indian Institute of Public Administration, New Delhi.

Malhotra, K.C., D. Deb, M. Dutta, T.S. Vasulu, G. Yadav and **M. Adhikari.** 1991. Role of Non-Timber Forest Produce in Village Economy. A Household Survey in Jamboni Range, Midnapur District, West Bengal. Mimeo. Institute for Biosocial Research and Development, Calcutta.

MOEF. 1994. *Conservation of Biological Diversity in India: An Approach*. Ministry of Environment and Forests, Government of India, New Delhi.

Poffenberger, M. and **B. McGean.** 1996. *Village Voices, Forest Choices*. Oxford University Press, New Delhi.

Poffenberger, M., B. McGean and **A. Khare.** 1996. Communities Sustaining India's Forests in the Twenty-First Century, in M. Poffenberger and B. McGean (eds). *Village Voices, Forest Choices*. Oxford University Press, New Delhi.

Rangarajan, M. 1996. The Politics of Ecology: The Debate on Wildlife and People in India: 1970–1995. *Economic and Political Weekly* 31 (35, 36, 37), September 1996.

Rao, K. and **S.C. Sharma.** 1998. Collaborative Management of Protected Areas in India: A Country Perspective. Paper presented at the workshop on Collaborative Management of Protected Areas in the Asian Region, Chitawan National Park, Nepal, 25–28 May.

Sarin, M. 1996. *Joint Forest Management: The Haryana Experience*. Environment and Development Series, Centre for Environment Education, Ahmedabad.

Shiva, V. and **G. Karir.** 1997. *Towards Sustainable Aquaculture: Chenmmeenkettu*. Research Foundation for Science, Technology and Ecology, New Delhi.

Singh, K.S. 1992. *People of India: An Introduction*. Anthropological Survey of India, and Laurens and Co., Calcutta.

Sukumar, R. 1989. *The Asian Elephant: Ecology and Management*. Cambridge University Press, Cambridge.

Tucker, R. 1991. Resident Peoples and Wildlife Reserves in India: The Prehistory of a Strategy, in P.C. West and S.R. Brechin (eds). *Resident Peoples and National Parks: Social Dilemmas and Strategies in International Conservation*. University of Arizona Press, Tucson.

Vijayan, V.S. 1987. *Keoladeo National Park Ecology Study*. Bombay Natural History Society, Bombay.

World Bank. 1996. *India Ecodevelopment Project*. Global Environment Facility and World Bank, Washington, D.C.

4

Conservation by local communities in the Maldives

Muhammed Zuhair

Profile of the country

The Republic of Maldives consists of a chain of coral atoll formations 80–120 km wide, stretching approximately 860 km in length, from latitude 7°6′ N to 0°41′ S and longitude 72°32′ to 73°45′ E, in the centre of the Indian Ocean close to the south-western tip of the Indian subcontinent. There are 26 geographic atoll formations, divided into 20 separate administrative units. Malé, the capital city island of the country forms the 21st administrative component. Although the maritime area of the country's exclusive economic zone (EEZ) amounts to more than 1 million sq.km, the dry land area estimated on the basis of inadequate admiralty surveys however does not exceed 0.001 per cent of the total area of the country. This makes the country one of the most watery nations on the face of the planet. The dry land is divided among 1,190 single coral islands of which only 200 are currently inhabited, 74 have been developed as tourist resort islands and the rest are uninhabited.

The islands also vary in size (from 0.5 sq. km to around 6 sq. km) and shape, ranging from small sand banks with sparse vegetation to elongated stripe islands; many have storm ridges at the seaward edges with swampy depressions at the centre. The maximum height above sea level recorded is around 3m, although over 80 per cent of the land area is approximately 1m above mean high tide.

The Maldives experiences a tropical climate with a mean annual temperature of 28°C. There is little seasonal variation in the temperature. Annual average rainfall is around 1.9m. The weather is dominated by two monsoonal periods, namely the SW monsoon from

April to November, and the NE monsoon from December to March, when winds blow predominantly from either of these two directions.

Like most archipelagic low-lying island nations, the economy and the lifestyle of the Maldives is essentially maritime and marine-based. Until recently, the lifestyle of the Maldivians had little direct impact on the environment. The main environmental impacts now include pressure from population growth, over-exploitation of limited resources such as mining of coral and coral sand in the absence of other building materials, deforestation induced by the need for fuelwood, and reef resource exploitation mainly for food and livelihood. These activities were earlier generally sustainable since human populations were relatively small and stable.

Over a couple of decades, the population of the islands has grown fast: from 180,088 in 1985 to 244,644 in 1995 (Population and Housing Census 1995). The current population growth rate of 2.75 per cent per annum would result in a doubling of the population by the first decade of the next century. This rapid growth, especially in Malé, has resulted in critical environmental problems. At present, 25.7 per cent of the total population lives in Malé in a land area of just 1.7 sq. km, giving it one of the highest population densities in Asia.

During the last two decades, the country has experienced rapid socio-economic development, particularly in the tourism industry, which is the main foreign revenue earner. There are now over 74 tourist resorts in operation and others either proposed or under construction. The fast growing population and rapid socio-economic development could lead to unsustainable development if integrated environmental management and planning options are not taken into consideration. The fisheries industry also plays a vital role, as fish are the main staple diet and in economic terms form the major export base (particularly the canned tuna). Agricultural crops are limited due to restrictions of land area and fruits and vegetables are grown only for local consumption.

Biological diversity

The extent of biological diversity present in the islands of the Maldives is not adequately documented or thoroughly researched. As the Maldives is an island nation, biological diversity is confined to the tropical island environments. Many species show resemblance to their ancestors from the Indian mainland, but have evolved through time to adapt to the conditions of the Maldives.

Webb (1988) has discussed that most of the fauna has arrived on the archipelago using the Laccadives and the direct drift from the NE monsoon winds; the majority of the wildlife originates from India and Sri Lanka. He also suggests that possibilities of colonization from the direction of Africa should not be discounted as the fauna of the southern area demonstrates. The country has a diverse vegetation composition, particularly in islands of the south such as Fuamulaku and Hithadoo.

Terrestrial diversity

The terrestrial floral diversity is relatively depauperate. Forests of *Psonia grandis* probably covered much of the land before clearing for settlements and agriculture occurred. Based on published plant species lists and vegetation descriptions, 583 species of plants have been recorded and, of these 55 per cent are cultivated species (MPHRE 1994). Also, over 30 species are known to have medicinal values which are being utilized for traditional medicinal practices.

In comparison to the rich terrestrial faunal diversity of the region, the Maldives demonstrates a rather small proportion of the representatives. Webb (1988) states that islands of the Maldives are not noted for their abundant wildlife, but do display diversity. This takes into account the absence of huge land masses, forests and associated ecosystems unlike the rest of the region.

A total of 180–200 bird species have been recorded from the Maldives, most of which are seasonal visitors, migrants, vagrants, introductions and imported as pastime pets. There are very few residents, mostly seabirds. A complete study on the ornithology of the Maldives has not been undertaken though.

Seabirds are widely seen throughout the country and are extremely important to the local communities as they are directly related to fishing. Tuna schools chase small fish and other marine life such as shrimps up to the surface where they are preyed upon by several species of seabirds. Flocks of seabirds are thus important indicators of the presence of tuna schools. Perhaps as many as 90 per cent of the tuna schools are located this way (Anderson 1996). At least 40–50 species of seabirds are seen from the Maldivian waters, of which only 13–15 are known to nest and breed in the country. Terrestrial birds are very minimal compared to other tropical islands and most are probably introductions. Reptilian fauna include two geckoes, two agamid lizards, two species of snakes, one snake skink, one frog and one toad species (Webb 1988).

Over 130 species of insects, including 67 butterfly species have been identified (Holmes 1993; Webb 1988). The only native mammals endemic to the country are the two species of fruit bats, *Pteropus gigantus ariel* and *Pteropus hypomelanus maris*. The latter one is very rare and has only been recorded once from the Maldives from Addu Atoll (Holmes 1993). The other mammals, all probably introductions, are the house mouse, black rat, Indian house shrew and cat (Webb 1988).

Marine diversity

In contrast to the terrestrial biological diversity found in the country, marine biological diversity shows an outstanding richness. There are over 250 species of hermatypic corals, and over 1,200 reef fish species (Pernetta 1993). Webb (1988) states that the Maldivian reefs are bursting with a variety of lesser creatures whose sheer numbers defy description. For instance, there are as many as 5,000 species of shells, around 100–200 species of sponges, over 1,000 species of marine crustaceans, and over 100 species of echinoderms. Marine algae including some 21 species of *Cyanophyceae* (blue-green), 163 *Rhodophyceae* (red), 83 *Chlorophyceae* (green) and 18 *Phaephyceae* (brown) have been recorded (Pernetta 1993). There are five species of sea turtles, all of which are endangered. Also, a variety of sharks, eels, rays, dolphins and whales are commonly seen throughout the country.

Importance, status and special characteristics of biodiversity

Due to the lack of other natural resources and wealth, biodiversity, particularly marine biodiversity, is the most vital resource base for the country. The main livelihood has traditionally been marine-based, which acts as the major generator of food, earnings, employment, protection and shelter. Fishing provides about 60 per cent of the export earnings, employs about 25 per cent of the workforce and provides the primary source of dietary protein (Pernetta 1993). Tourism also plays a vital role in the economy as the main earner of foreign revenue, but this entirely depends on the health of the marine habitats. With the boost to the tourism and fishery industries during the last two decades, every constituent of marine diversity has become surprisingly important to the economy as well as to the local inhabitants.

Pernetta (1993) also describes the importance of live bait fish for the tuna industry and of some gastropod molluscs such as *Strombus gibberulus, Atactodea striata, Turbo argyrostomus* as food items. Larger gastropods such as the bivalve *Athrina vexillum* are occasionally collected. Other items such as giant clams, sea cucumbers, dried shark fins, are important items for the export industry. Edwards (1986) described that around 100 species of small reef fishes are collected for the aquarium trade. Some marine invertebrates such as the larger anemones, *Radianthus* and *Stoichactis* together with their symbiotic clown fish, and smaller starfish such as *Linckia* and *Fromia* are also exported. In recent years, certain corals, sponges and other reef constituents have been identified as having medicinal value.

Certain subspecies of fauna like the Maldivian little heron (*Butorides striatus didii phillipsi*) and central Maldivian little heron (*Butorides striatus albidulus*) are unique to the islands. The only known native plant species are the five species of *Pandanus*; more are expected to be discovered as further studies go on.

Not much is known about the status of most of the species found in the country due to lack of technical and scientific assessments or comprehensive studies. However, there are certain species which have been studied extensively mainly due to their economic, commercial and ecological values of importance. For instance, turtles have been over-exploited in the past for making ornamental and souvenir items. The government has now banned turtle-catching and the sale of turtle products. The only native terrestrial mammal that has been identified as either endangered or rare is the fruit bat, *Pteropus hypomelanus maris* (Holmes 1993).

The species that are locally at risk due to exploitation for various uses in the past include: The Whale shark (*Rincidon typus*), Napoleon wrasse (*Cheilinus undulatus*) and Giant triton (*Charomia tritonis*). Currently, these species are banned from trade, export and utilization as their populations have declined in numbers due to over-exploitation (MOFA 1995). Some bird species, particularly the seabirds, are also found to be locally threatened. The sole bird that has been given official protection status is the White tern (*Gygis alba monte*), found only in Addu Atoll. Other species are likely to get protection soon.

Major threats

Pernetta (1993) states that the greatest threats are population growth and migration. Internal migration occurs mainly for educational, health and job opportunity purposes. Some natural vegetation and other habitats have been cleared for settlement purposes. Other islands are reclaimed by pumping sand from the atoll lagoon floor onto reef flats, destroying the reef flat communities.

Increased population has also led to the increased mining of corals, the basic construction material in the country. Coral mining also has adverse impacts on the reefs itself, on islands as well as on biodiversity. Coral reefs act as a strong coastal protection layer against ocean currents, waves and tides; mining of corals has resulted in the destruction of this layer in some islands causing considerable amount of beach sand to wash away from the island into the sea. As the protection layer is destroyed, waves and tides directly enter into the island causing damage to the vegetation and intruding into the freshwater aquifer. The other associated impacts on the reefs include: loss or migration of residential reef fish communities and other living organisms, loss of baitfish that are important for the local tuna fishery, and reduced coral percentage cover. These reefs may take several years to recover.

The possible impacts of climate change and subsequent sea-level rise, ozone depletion, domestic pollution, introduction of non-indigenous species, over-fishing of certain species, clearance of natural vegetation for housing and agricultural activities, are identified as other significant causes of the loss of biodiversity in the Maldives. The use of uninhabited islands to supply fuelwood and construction timber and for agricultural activities, including coconut plantations, has significantly altered natural vegetation patterns on many islands (Pernetta 1993).

Local communities dependent on biodiversity

Diversity of communities

As a nation that is entirely surrounded by sea, the Maldives has always been exposed to different cultural influences. Ellis (1995) described that these came mostly from neighbouring maritime cultures, those countries bordering the Indian Ocean. Therefore, traces of Africa, Malaysia and Indonesia as well as of Arabia and India are to be found in the people and culture. Also, some books written on the Maldives state that the people of these islands are of the Aryan race, who were originally Indo-Europeans with Indo-Iranian subgroups who spread from south Russia and Turkistan in about 2000 B.C. to Mesopotamia and Asia Minor, invading India in about 1500 B.C. (Ellis 1995). People may have started settling in the Maldives some 2,000–2,500 years ago.

Apart from the capital island of Malé, most of the inhabited islands are rural. Depending on the population and size of the islands, some smaller islands may have just a single community while larger islands have at least three to four communities. One feature common to all these rural communities is that they essentially depend on fishing as a source of food, employment and earning. On smaller atolls with comparatively large islands, such as those found in the south, agriculture is also a major community activity. Most crops are cultivated for local community consumption while others are sent to Malé for trade. There are also communities dependent on mining corals and sand for construction purposes, making handicrafts from natural products, extracting local coconut juice and making honey from it, and cultivating coconuts and other products. All this indicates that local communities are substantially dependent on natural and biological resources.

Major sources of livelihood and extent of direct dependence on natural resources

It has been discussed throughout this paper that the source of livelihood in the Maldives essentially originates from marine and associated environments. Almost every industry including fishing, tourism, export, construction has its roots in the marine natural resources. Since soils are poor, agricultural production is not high. Total cultivable land is around 46,766 acres, of which most is used for subsistence production. Unlike many other developing countries, agriculture accounted for only 9 per cent of GDP in 1990, having declined from around 18 per cent in 1981 (MPHRE 1994). In contrast, fishing and tourism have shown increase in their contribution percentages in GDP.

The fisheries sector contributed around 15 per cent of the GDP in 1990 and 1991 and around 80 per cent of export income during the same years (MPHRE 1994). Fishing has been expanded through the mechanization of the traditional fishing fleet, fuel distribution systems and fish collection facilities. The main product is skipjack tuna and 73 per cent of the total annual catch of skipjack was exported in 1991 (MPHRE 1994). The SoE Report (MPHRE 1994) also stated that the total catch in 1992 was 82,000 metric tonnes, of which 46,000 metric tonnes were exported.

At 18.4 per cent of the GDP which provides over 25 per cent of government revenue, tourism is the second largest contributor to the economy and is increasing in importance year by year. This trend can only be sustained by maintaining the environmental quality. Most of the present visits relate to marine-based activities such as scuba diving in the coral reef environments and visitors expect a high level of water quality and pristine reef environments. Therefore, the Ministry of Tourism enforces regulations and standards on resort islands to maintain their environmental and natural quality.

Changes taking place under external and internal forces

Currently, Malé is the only urban centre in the country. The majority of government activities, including government ministries and offices, education and health facilities, communication centres and other services are located in the capital island. Also, the island is the home of economic, commercial and trade activities with most of the imported goods directly coming to Malé and then later to other islands. Viewed from the sea, Malé has altered its appearance over the last 200 years or so with large-scale land reclamation projects having enlarged the area of the island and altered its shape (Webb 1988). People from other islands requiring job and service opportunities often migrate to Malé. This way there is a high rate of internal migration within the country particularly to Malé. Thus, Malé is one of the areas in the whole country where the natural vegetation is depleting at a faster rate and terrestrial biodiversity is almost negligible.

In order to reduce this trend of internal migration and to mitigate the current situation in Malé, the government has selected some larger islands from different areas for urban development. A few of them are at present reaching up to an urban level while others are still at a pioneer level. The current population distribution into small islands across the country enhances environmental problems such as clearing of natural vegetation for settlements, mining of coral and sand for construction, building of jetties and other coastal structures for easy access, etc. in each island. However, it is hoped that with the development of a few urban centres in the country, environmental and biodiversity threats will be minimized accordingly with most of the previously inhabited islands starting to recover after the pressure has been diminished.

The development of islands into tourist resorts also has some implications although it is very localized. Most of the natural vegetation and habitats have been cleared for building rooms, restaurants and other tourist facilities while others are imported as ornamental plants to look more attractive. However, the reef is usually maintained as the main attraction area due to its abundance of marine biodiversity. Fishing, anchoring, netting, etc. are prohibited to protect the area from being exploited.

Official efforts at conservation

History

Conservation efforts in the country were initiated during the 1970s after the approval of the Fisheries Law by the Citizen's Majilis. As the fisheries industry started expanding and became one of the most important economic activities in the country, the Fisheries Law was revised during the mid-1980s, making greater provisions for the conservation of marine resources. Its 10th clause states: 'In the event of a special need for the conservation of any species of the living marine resources, the Ministry of Fisheries shall have the right to prohibit, for a specified period, the fishing, capturing or the taking of such species or the right to establish special sanctuaries from where such species may not be fished, captured or taken.' Under this law, regulations have been developed and enforced to protect marine species such as turtles, whales, dolphins and certain fish.

The environment sector was formally recognized as a government entity during the 1980s. In 1984, the Council for the Protection of the Maldives Environment was established under the Ministry of Home Affairs and Social Services, to act as an advisory body to the government on environmental measures. A small Environmental Affairs Division was created and gradual strengthening of this division occurred within the Home Ministry.

In late 1988, environment was given elevated status, being combined with the then Ministry of Planning and Economic Development to form a new Ministry of Planning and Environment. In 1990, the Environment Research Unit was formed within the Ministry, and in 1993, the Ministry was incorporated with the human resources sector to form the present Ministry of Planning, Human Resources and Environment (MPHRE).

Current conservation policies, laws and programmes

Through its adoption of the National Environment Action Plan in 1989, the government is committed to the concept of utilizing its resources in a sustainable manner by undertaking various measures and policies, and seeking appropriate mechanisms for translating this goal into concrete action.

Recognizing the environmentally damaging effects of coral mining, the government has initiated a number of measures to restrict and control this activity. Initial controls banned the mining of coral on reefs surrounding inhabited islands; subsequently, mining was restricted to designated sites selected by the Environment Division in consultation with the Atoll Chiefs (MPHRE 1994).

In April 1993, the Citizen's Majilis approved the Environmental Protection and Preservation Act which empowers the MPHRE to draft guidelines for the protection of the environment and make it responsible for the identification and designation of PAs and natural reserves (MPHRE 1994).

Currently, MPHRE is undertaking observations and studies to designate protection and conservation sites and species. This relates to the recently protected bird, White tern (*Gygis alba monte*), and 14 marine PAs. A biodiversity project has been approved by the government to draw up appropriate action plans and strategies for its conservation and designate PAs throughout the country. A study has been undertaken by a UNESCO mission to recognize the possibilities of the Maldives joining or being a party in the activities of the Convention on International Trade in Endangered Species. The Maldives has also actively participated in various international and regional fora and has signed and ratified the UN Convention on Biological Diversity at a very early stage.

Extent and range of protected areas

The existing and proposed Marine Protected Areas (MPAs) in the Maldives are protected mainly due to their outstanding diversity of corals, reef fish, sharks, rays, eels and other organisms ranging from sponges to molluscs and bivalves. These areas are banned from all anthropogenic activities such as coral and sand mining, fishing, collecting, netting and anchoring, except baitfish fishing which is important for the tuna fishery. Baitfish can only be collected in these areas by using methods which do not damage or harm any living organism. These MPA systems will be extended across the country in the future in terms of protecting and conserving the rich marine biodiversity found in the country. There are potential areas that need to be carefully observed for the purposes of conservation and protection of biological diversity. The Environment Division of the MPHRE and Marine Research Section of the Ministry of Fisheries and Agriculture (MOFA) is currently undertaking research for the establishment of certain areas and species to be given the status of protection by identifying their distribution, features, status, uniqueness and so on.

Structure of official conservation agencies

The Environment Division of the MPHRE is identified as the focal point for all national and international conservation actions such as national legislation, policies, guides and international conventions, e.g., the UN Convention on Biological Diversity. Also, there is the National Commission for the Protection of Environment (NCPE) which is represented by top officials from all government departments and members from the private sector. NCPE is an independent advisory body for the Environment Minister on matters dealing with protection of environment, sustainable resource utilization and conservation of biodiversity.

MOFA is responsible for the implementation of the Fisheries Law, regulations and development of sectoral policies relating to the development of fisheries in the country. The Marine Research Section (MRS) of the MOFA undertakes mainly research work relating to marine natural resources, sustainable utilization and management of these resources and protection of vulnerable, threatened marine species. The MOFA works in close collaboration with MPHRE.

Independent community/NGO efforts at conservation

History

Independent efforts at conserving biodiversity did not start until very recently. Although there are many local communities and NGOs involved in creating environmental awareness, not many of them are working on conserving biodiversity. Currently, Eco-care Maldives and Bluepeace are the most actively involved local NGOs in the field of conservation. Atoll development communities such as Addu Development Community have also played a key role.

Ongoing and new efforts

Eco-care Maldives specializes in threatened species and habitats. They launched a project on saving sea turtles in early 1996, which addressed the various threats to turtles and publicized this among locals and tourists. Due to their initiative, catching of turtles and selling of their products have been completely stopped. They have created mass public awareness with respect to community participation and involvement on the issue of conserving sea turtles in the Maldives.

Bluepeace has a proposal for undertaking measures for conserving seabirds in the Maldives, recognizing their threatened status and their importance to the livelihoods of fisherfolk. The proposal includes protecting seabirds that are important for the tuna fishery, public awareness among local fishermen, and protecting breeding and nesting sites for these bird populations. Bluepeace also has a campaign for saving sea turtles in the Maldives.

Eco-care Maldives is also planning two new projects for biodiversity conservation: one will deal with the protection of sharks, and the other with protection and conservation of birds that breed and nest in the islands of the Maldives.

Considering that there are limited financial and human resources for individual communities and NGOs to undertake protection and conservation measures on biodiversity in the country, projects undertaken so far could be regarded as appropriate and successful. Although only the first phase of the turtle protection and conservation project has been implemented by Eco-care Maldives, it gave the general public a broader perspective on the issues such as ecological importance, threats and the ways in which turtles can be protected. Tourists also credited the project with being an important measure for the development of eco-tourism in the country. Tourists would pay a lot of money to watch wild turtles while most of the tourists would not like to see tufted turtles hanging from the corners of the shops and buying products made out of turtle shell. Thus, the project achieved its aims and goals.

The initiation for the protection and conservation of the White tern came from the Addu Atoll Development Community. In the country, these terns are only found in Addu Atoll and their numbers have been declining due to human exploitation and destruction of their habitats. After official declaration and registration of White tern as a protected species by the MPHRE, the Addu local community including schoolchildren became involved in undertaking protective measures. This was the first major CBC effort undertaken by an individual community in the country. Although the effort is currently ongoing, it has shown that active involvement of community members in the processes of conservation is vital as they are the major users.

Generally, the government encourages efforts undertaken by individual communities and NGOs in order to protect and conserve entities of biological diversity. However, the government should be notified of any project or an effort relating to conservation. The government also advises and gives technical assistance where appropriate to such efforts.

The major weakness that has been identified in these efforts is the lack of technical know-how and information regarding biological diversity. There is a great demand for technical strengthening of involved personnel which would ease the difficulties during the operation processes. And there is also demand for creating and developing technical knowledge in the field of conservation among local communities and NGO personnel.

Constraints and opportunities for community-based conservation

Existing and potential constraints

As CBC of biological diversity is a new phenomenon to the country, there are several existing and potential constraints. One of the major constraints is the lack of technical personnel both in the government and the private sector in the field of conservation. A prerequisite for effective CBC activities is the development of human resources. Second, financial incapabilities result in limitations to the extent of efforts that can be undertaken. For most activities undertaken so far, external financing was provided. Comprehensive and long-term activities also require additional financing for their implementation and management. Finally, lack of appropriate laws and regulations which facilitate conservation measures make the efforts difficult to enforce. Therefore, strengthening and development of laws and regulations relating to conservation is an imminent necessity.

Existing and potential opportunities/advantages

The concept of CBC is spreading throughout the country. One of the effective methods of managing, sustaining and conserving biodiversity is by involving local communities or those who utilize the resources. The process can easily be integrated into the island communities, due to their having lesser number of individuals in contrast to the communities found in the rest of the region. As less financing would be required to undertake community-based small projects, the communities could arrange self-financing mechanisms by initiating various fund-raising activities within the island. The advantages of CBC lie within the conservation and management side of the biological resource. If not all, most of the islanders have a great tendency for utilizing this life-sustaining resource mainly as a source of food and income in a manner that is not either sustained or conserved. Thus, introduction and practice of CBC would result in modifications of present-day resource-use patterns into a more manageable, sustainable and conservationist way of utilizing resources.

Proposals for the future

In order to develop and strengthen CBC processes in the country, certain measures need imminent action. There should be an authentic dialogue between the government, communities, NGOs and private sector stakeholders for the development of the process. Issues to be discussed include the propriety of CBC efforts in the country, their administration and management, the exchange of information, financing mechanisms, etc. A scheme for developing and strengthening technical and scientific knowledge of the involved personnel of either government or private sector need greater emphasis. This should involve providing adequate education and training in the field of conservation. A translucent plan of action for conservation measures needs to be developed giving CBC priority. The government and private sector including communities should collaborate with each other for the strengthening and change of legal aspects. There should also be a scheme for sharing benefits arising from such efforts.

Although there is no law specific to CBC in the country, the stress on environmental protection and conservation in the Environmental Act and Fisheries Law of the Maldives could be used as an interim measure, till new appropriate laws and regulations are developed. Eventually, an orientation towards CBC needs to be given in the reformulation of legal and policy measures, and the reviewing of the Environment Action Plan.

Some slight differences in viewpoint may be present amongst the government, NGOs and local communities. For instance, the official view emphasizes the need for sustainable resource use practices and management of the limited resource base by addressing present trends and levels of resource exploitation in the country; NGOs may deal mainly with creating mass awareness with regard to conservation of biodiversity components already experiencing major negative forces; while communities might view official efforts as implying that biodiversity conservation means total abandonment of the resource use practices. Actually, conservation efforts are a strategic basis of conservation, sustainability and benefit-sharing. Therefore, to minimize the difficulties that would arise from these different views, effective coordination is essential.

REFERENCES

Anderson, R.C. 1996. *Seabirds and the Maldivian Tuna Fishery*. Marine Research Section, Ministry of Fisheries and Agriculture, Malé, Maldives.

Edwards, A.J. 1986. Preliminary Report on the Aquarium Fish Export of the Republic of Maldives. Centre for Tropical Coastal Management Studies, University of Newcastle upon Tyne.

Ellis. 1995. *Guide to the Maldives*. Novelty Press Ltd., Malé, Maldives.

Holmes, M. 1993. The Maldives Archipelago, Indian Ocean: A Report on an Investigation of Fruitbats and Birds. Malé, Maldives.

MOFA. 1995. Protected Marine Life and Prohibited Marine Product Exports from the Maldives. Posters of the Ministry of Fisheries and Agriculture, Malé, Maldives.

MPHRE. 1994. *State of the Environment Report*. Report Prepared for the Global Conference on the Sustainable Development of Small Island Developing States. Ministry of Planning, Human Resources and Environment, Malé, Maldives.

Pernetta, J.C. (ed.). 1993. *Marine Protected Area Needs in the South Asian Seas Region, Volume 3: Maldives*. IUCN Marine and Coastal Areas Programme, Gland, Switzerland.

Population and Housing Census. 1995. Preliminary Results. Ministry of Planning, Human Resources and Environment, Malé, Maldives,

Webb, P.C. 1988. *People and Environment of Maldives*. Novelty Press Pvt Ltd, Malé, Maldives.

5

Towards community-based conservation in Mongolia*

D. Myagmarsuren

Introduction

Mongolia is located in Central Asia between China and Russia, and covers three time zones. The total land area of the country is approximately 1.565 million sq. km. It is divided into four large physiographic zones, namely: the Altai mountain area, the Khangai and Khentei mountain areas, the East Mongolian steppe areas and the southern Gobi desert areas. The climate is dry and continental with sharply defined seasons, due to its geographical location, topography and elevation above sea level. The winter is cold with temperatures stooping down to –40°C, and thick frost, while summer is short and hot with temperatures going up to 40°C.

Mongolia's vegetation types range from the arid desert in the south to the moist taiga forest in the north, and from rolling steppe grasslands in the east to alpine terrain and glaciated peaks in the west. This varied terrain contains a wide array of ecotypes, many exhibiting unique characteristics found nowhere else on the globe. At the same time, Mongolia's population density is among the world's lowest; the territory of Mongolia is among the least disturbed by human activity of the northern temperate regions of the world. This unique, varied, and substantially undisturbed territory supports a wide diversity of living organisms, many of which are endemic to Mongolia.

Socio-economic profile

The total human population is 2.2 million people. About 50 per cent of the entire population is semi-nomadic, based on livestock breeding,

* The original paper has been substantially added to by the editors, based on Mearns 1995.

following essentially the same lifestyle as Chengis Khan's ancestors. Agriculture (of which livestock production accounts for three-quarters of the value added) directly contributes over 25 per cent of GDP (IBRD 1994). After decollectivization began in 1991 (Mongolia was a communist state till then), 43 per cent of the total population became private herders; by 1994, 90 per cent of the livestock were privately owned (Mearns 1995).

Legal status of the resources

All land in the country is currently under state ownership. The new Constitution which came into effect in February 1992, admits the principle of private ownership but reserves for the state the right of 'eminent domain' over all land (Whytock 1992; Mearns 1993a). In 1994 the Parliament took a decision to hand over private ownership of about 1 per cent of the total land area, which included land under cultivation, urban areas and industries. Pasture land has been excluded from the land to be privatized. Pastures comprise 75 per cent of the total land area of Mongolia which is managed by the state as 'common land' through the local provincial and district authorities. However, these pastures which make the largest common grazing areas in the world, *de facto* remain largely under the control of the local herders as they have been throughout the communist phase in the country (Mearns 1993b).

Traditional practices of conservation

The tradition of protecting nature has a long history in Mongolia. There were closed seasons for hunting gazelle, saiga, deer and rabbits at the time of Marco Polo. Later, the laws of Khalkh Juram in the 1778, set aside 16 mountains which had been protected since the 12th century, as holy mountains.

In addition, research has shown that the sustainable management of pastures or common grazing land (most important ecosystems both socially and ecològically) has not so much been a result of the formal organizations of the state as of the informal, kinship- and residence-based community groups. These groups have their own internal rules and regulations, both to ensure equitable use and avoid unsustainable use. These regulations are mutually respected and agreed upon (Fletcher 1986; Szynkiewicz 1977, 1982; Vreeland 1962).

Traditionally, the *khot ail* or the nomadic herding camp is the basic unit of a herding community. The size of a *khot ail* varies from season to season and ecosystem to ecosystem. Most often they have a core group of families staying together, thus overcoming the problem of both labour scarcity and conflicts over the resources. In times of natural resource scarcity they split into smaller fractions and move to different geographical areas. In times of need they form joint parties in search of animals, clearing the pastures of snow, etc. At the same time they have strict rules about 'opening' a certain area for pastures, and other traditional and cultural arrangements for use and management of natural resources (Fernandez-Gimenez 1993; Humphery 1978; Mearns 1993b; Purev 1990). There are no fixed groups, and the sizes and composition of these groups vary. However, at times there might be some groups who may have lived in close vicinity for generations, and share ascriptive customary use rights in the pastures (Bazargur et al. 1992). These neighbourhood groups often have an acknowledged leader specifically for resolution of conflicts (e.g., over land and water resources) (Mearns 1995).

Current official efforts towards biodiversity conservation

After 70 years of the centrally planned system, the country has adopted a new democratic government as well as a free market system since 1991. This transition period carries many problems which have a direct impact on the economic situation of the country. The new government is responsible for implementing the National Development Concept of Mongolia, which was approved by the previous Parliament in June 1996. The Development Concept specifies that 'an economically and ecologically balanced development will be ensured on the basis of rational utilization and protection of natural resources, and prevention of natural calamities and environmental degradation', and that 'the diversity of animals, plants and microorganisms will be protected from adverse effects resulting from nature or human activities'. It also refers to the government's policy of decentralization, and expansion of powers and strengthening the independence of local governments.

Mongolia ratified the Convention on Biological Diversity in September 1993. It also recently (March 1996) became a signatory to the Convention on International Trade in Endangered Species (CITES). The country has also participated in international discussions on the

Convention to Combat Desertification. Complementing these global agreements, Mongolia has entered into important regional agreements on the conservation of biodiversity, especially with its neighbours, China and Russia. These include the Cooperative Agreements on Environmental Protection with China (May 1990) and Russia (February 1994). The three countries also signed the Dornod International Protected Areas agreement in June 1994. Efforts to enhance implementation of these agreements are underway.

The Mongolian Environmental Action Plan 'Towards Mongolia's Environmentally Sound, Sustainable Development' was approved by the Cabinet in February 1995. Principal environmental issues outlined in the plan include land management, land degradation and desertification, rangelands and overgrazing, conservation of biological diversity and management of PAs. To address these key issues, the plan proposes to build capacity, develop an environmental monitoring and ecological information system, enhance public awareness and participation in environmental protection, and strengthen the role of NGOs and cooperation with international organizations and institutions.

Following the approval of the Environmental Action Plan, and after extensive consultations with scientists, PA managers, international experts, relevant government ministries, NGOs, local governments and people, the National Biodiversity Conservation Action Plan was finalized in April 1996 and was approved by the Cabinet in July of the same year. The plan outlines objectives and actions to (*a*) establish a research programme that improves knowledge of biodiversity and relevant threats; (*b*) establish a programme for national education and training for biodiversity conservation; (*c*) establish public information programmes to improve people's knowledge of biodiversity and the importance of conserving it; (*d*) control hunting; (*e*) prevent pasture deterioration; and (*f*) establish land use planning to protect biodiversity.

At the Earth Summit in Rio in 1992, the Government of Mongolia, after much internal discussion, suggested that Mongolia as a whole be considered as a biosphere reserve.

Mongolia recognizes the importance of its biodiversity heritage as a significant component of the world's biological diversity, and as an economic resource. The goal, then, is to reconcile the twin objectives of economic development and biodiversity conservation in the country's development programme. It is required to take not less than 30 per cent of the country's territory under some form of protection for the purpose of establishing an environmentally sound socio-economic system and sustaining Mongolia's existing range of wild species and ecosystems. In 1994, the Ministry of Nature and Environment (MNE) established criteria for selection of PAs, which include ecosystem specifics, wilderness extent, biodiversity conservation needs, scientific, historical, cultural and economic significance, impact of economic activities, and local people's opinions.

Since 1990, a new period has started in the expansion of the PA system of Mongolia. Before this, there were only strict PAs of one category. The MNE has developed a new law of PAs, a Protection Regime, and regulations for selection of PAs. This has been approved by the Parliament and the government and has come into force. According to these new documents there are four categories of classification: Strictly Protected Area, National Park, Natural Reserve, and Natural and Historical Monument. These were established on the basis of the international categories of PAs, adjusted to the particular conditions in Mongolia. After 1990, 22 separate areas totalling more than 10 million hectares were taken under protection. Now there are altogether 35 PAs covering 15.3 million hectares (about 10 per cent of the whole country) including 11 Strictly Protected Areas (SPAs), six National Parks (NPs), 12 Natural Reserves (NRs), six Natural and Historical Monuments (NHMs). (See below for definitions.) There are 10 specialized Administrations which are responsible for organization and implementation of conservation activities, proper use and research in PAs. The Protected Areas Bureau is responsible for making an integrated national policy on the management of PAs, regulation of training, public awareness and other activities in particular fields. Given below is an introduction to the PAs classification in Mongolia:

Strictly Protected Area (SPA): This refers to an area of particular importance for science and human civilization, which can represent specific features of nature, natural zones and wilderness. It is taken under the protection of the state with the purpose of ensuring its ecological equilibrium. Allowed activities are research, tourism, hunting for purposes of research and population control, soil and flora rehabilitation, clearing dead and fallen trees, gathering forest products for non-commercial purposes, and collecting fodder. Prohibited activities are to alter the pristine wilderness, any polluting activities, sport or subsistence hunting, reintroducing and expatriating wildlife species for any other than research purposes, gathering forest products without permission, building fences and shelters for animals other than protected species.

Each SPA is divided into three zones: Core area (Virgin Zone), Buffer area (Protection Zone), and Transition area (Limited Zone), and appropriate management activities are carried out in each zone. Roughly, the Virgin Zone receives the heaviest protection and Limited Zone has the most allowed uses. In the Virgin Zone, use for research works is allowed; in the Protection Zone, in addition to the above, protection activities, forest sanitation, and biotechnological measures are allowed; in the Transition Zone, in addition to the above, economic activities by indigenous people are allowed.

The majority of the administrative bodies of these areas are newly established and are facing serious financial constraints due to the country's economic difficulties.

National Park (NP): This is an area under state protection with relatively pristine wilderness and of historical, cultural, scientific, cognitive, ethical, ecological and educational importance. Governing Administrations responsible for protection activities and research works are in operation in the Govi Gurvan Saikhan, Gorkhi–Terelj, Khovsgol and Khangai Nuru NPs.

Each NP is also divided into three zones (Special Zone, Tourism Zone, Limited Zone). Allowed activities are: research work, tourism along established routes, traditional uses such as worshipping at sacred mountains, cleaning out of dead and fallen trees, low economic activities in service areas, gathering forest products for private purposes, reintroduction of animals, hunting or fishing for private purposes, organizing public gatherings, building temporary dwellings, tents, campsites, and preparing fodder for livestock. Prohibited activities are: disturbing the earth, using explosives, recovering mineral resources (including sand and gravel), commercial or live timber harvesting, any polluting activities, hunting or otherwise disturbing animals, constructing permanent dwellings, reforming economic activities, possessing weapons for any other than research and species regulation purposes.

Natural Reserve (NR) and Natural and Historical Monument (NHM): There are four types of NRs in Mongolia: the Ecological Reserve, an area conserving rare, endangered species of flora and fauna and providing adequate conditions for their reproduction; Landscape Reserve, an area of aesthetic importance; Natural Resource Reserve, an area of geological importance; and Paleontological Reserves.

These areas may be designated to conserve and rehabilitate natural resources and an ecological balance for a certain period of time.

Admitted activities in the NR include economic activities and research work, which does not adversely affect the values for which the reserve was established, or the area's ecological balance, and which are not expressly prohibited. Activities are only allowed at the discretion of the management authority. Prohibited activities are: disturbing the soil, mineral recovery, timber cutting, any activity causing pollution, commercial fishing, hunting, or collecting of pharmaceutical plants, construction of permanent structures for anything other than conservation purposes.

NHM is an area taken under protection for its natural historical and cultural interest, with the potential for research and sightseeing. It may contain unusual structures of scientific, historical and cultural interest. Any kind of activity not expressly prohibited is allowed in the area. If the activity is admitted, it must not have an adverse impact on the national monument and must be approved by concerned organizations. Prohibited activities are: building constructions, disturbing the earth, building roads which alter the natural appearance of the area and disturbing the natural and cultural relics.

PAs in Mongolia are not supposed to contain human settlements. This has been taken into account while developing management plans for these areas. The major strategy is to approve a livestock limit and a limit on use of natural resources. By this regulation it is expected that people from inside the PAs will move to the surrounding zones.

Protection and research activities in the NR and NHM except the Hustain Nuru (Hustain mountain range) are to be implemented by the local government. At this time, these are supervised by one or two rangers and there is no other administrative unit in operation. The Hustain Nuru NR is an area established for the purpose of reintroduction and gene-pool conservation of Takhi Horses (Przewalski's horses). The protection and research activities in this reserve are carried out by the Mongolian Association for Conservation of Nature and the Environment (MACNE) with the collaboration of the Ministry of Nature and Environment (MNE). The research works include reintroduction and protection of the Takhi horse in a captured condition, protection of natural wilderness and complex studies of nature.

The ultimate objectives of the existing PAs are to protect biodiversity and to replenish its resources. The territory of Mongolia represents a wide range of the world's specific natural and climatic features which can be found in a vast area between the southern edge of the Siberian Taiga and the northern edge of the Central Asian Desert. That is why the PAs of Mongolia have been identified taking into account the six categories of the international classification of PAs laid down by the World Conservation Union (IUCN).

Community involvement in areas around PAs

An attempt has been made by the UNDP/GEF Biodiversity Project in recent times to start small-scale economic development projects in areas adjoining four PAs, and to link these projects to biodiversity conservation by demonstrating economic benefits of PAs. Nine such projects have been financed with grants for initial start-up costs; with the aim that they help them to become self-sufficient. These projects have had a positive response from local people wherever implemented extensively. An important component of the project is awareness-building amongst the public for nature conservation in Mongolia. However, the benefits of these programmes for the PAs have to be still assessed.

In another project on surrounding zone management financed by the German Agency for Technical Development (GTZ), a feasibility study has been carried out to assess the potential for improving the living conditions of the people living near Khan Khenti SPA, Gorkhi–Terelj NP and Govi Gurvan Saikhan NP, and involving them in the management of these PAs. Yet another project around the Hustain Nuru NR, implemented by the (MACNE), is investigating ways of improving living conditions for people in the surrounding area.

In the above mentioned cases, possibilities of co-management have been examined in a portion of the surrounding zones of PAs. The purpose was to bring together local communities and conservation organizations in an effort to combine conservation goals with community development goals, and to encourage inter-institutional collaboration to achieve these goals. The participatory process focused on the authorization of local leaders and the community organization to take more control of their destiny and to explore other opportunities they may not have been aware of previously. The process was facilitated by researchers and guided by the community groups and local administration. The planning process began with community members identifying values they held about their local environment. These values included a clean environment, natural beauty, and availability of water and pasture land. Next, they set acceptable criteria (or boundaries) on potential change that might result from development in the community and affect the values they had identified. The process culminated with a workshop that brought the community into contact with representatives of donor and other assistance agencies in order to encourage collaboration. The agencies represented a wide range of organizations, including national and international NGOs, government agencies, and private-sector agencies.

Public awareness for protection of natural resources and role of NGOs

Protection of the environment in Mongolia gets substantial support from citizens. For example, in the mid-1980s, people resisted the Khovsgol phosphorus mine proposal, considering its damaging impacts on the environment. Mongolian scientists supported this and the government postponed the project.

About half of Mongolia's populace are nomads and have a direct link with nature. Ethnic groups such as the Zaalan, Buryat, Turian and Urainkhai have a different approach to obtaining livelihoods than the Mongols do. These ethnic groups' traditions of nature conservation lie with their individual religious beliefs.

Populations of both cities and rural areas take active measures for protection of their local wildlife and endeavour to reveal any shortcomings of the laws or violations of the laws. There are strong feelings among herdsmen against people who come from cities and towns to hunt local wildlife. Today there are about 10 NGOs in Mongolia which are taking action for the conservation and protection of the natural resources.

The first environmental association, known as MACNE, was established in Mongolia in June 1975. Since 1990 it has become more independent of the government. It is concerned with training, public information and conservation projects such as the Takhi horse reintroduction and Snow leopard research.

Other Mongolian environmental associations include: (*a*) The Mongolian Gazelle Society which is concerned with conservation of the Mongolian gazelle; (*b*) The Environmental Law Society, which was established in 1994 to advise the public on environmental law issues and to lobby for environmental issues on behalf of private citizens; (*c*) The Green Movement, which was formed in 1994 to lobby for and carry out activities in support of environmental protection; (*d*) The Green Party, a political party campaigning on an environmental platform, which has intensive influence today on the environmental policy of the new government; (*e*) The Mongolian Ecotourism Society, which was established in 1995 to build a strategy of sustainable tourism development and to introduce Environmental Impact Assessments for tourism activities. The Society is also taking steps for establishing environmentally responsible courses for tour guides.

According to the Law on Environmental Protection, environmental NGOs can implement environmental legislation, carry out reviews, demand alleviation of shortcomings determined during such reviews, submit their suggestions and recommendations, and suggest methodologies for conservation and restoration of the environment. A new law was adopted in March last year which lays a basis for cooperation between government institutions and NGOs. The functions that could be delegated to NGOs cover a wide range of possibilities. Most NGOs are actively implementing nature conservation projects which are funded by different international organizations. Some projects have guidelines for involving the local community in developing and implementing management plans.

The World Wide Fund for Nature has been carrying out joint surveys with the MNE and has been working in the field of public information, particularly in the eastern steppes, the wetlands of the Great Lakes Depression, and north-eastern Dornod; the Christian Oswald Foundation, the Baumann Foundation and other donors are assisting the National Commission on the Conservation of Endangered Species to establish wild populations of Przewalski's horse in Gobi B, part of Great Gobi SPA and Hustain Nuru NR; the Asia Foundation has been working with the MNE and with local NGOs in the field of environmental law; and the International Snow Leopard Trust is supporting a reward programme for local herdsmen who refrain from killing Snow leopards in their ranges. The New York Zoological Society/The Wildlife Conservation Society (New York) is supporting research into Snow leopards and the Institute for Zoo

Biology and Wildlife Research (Berlin) has been supporting the survey and publicity work of the Mongolian Gazelle Society.

Constraints and opportunities

One major constraint in the way of CBC is a weak policy base for it in the country. In Mongolia, official conservation efforts have recently started, on the other hand the country has a continuing tradition of natural resource management through community groups. There seems to be little or no conflict till now between the authorities and the communities over the issue of natural resource use and management. This situation can be used for much more participatory management of resources (as is being tried for the areas surrounding the PAs). This will also help overcome the problems associated with the limitation of human and monetary resources for conservation in the country. However, certain conditions are necessary for community involvement in natural resource conservation:

- Recognizing the existing knowledge, management systems and organizational capacity of the communities, especially the nomadic pastoralist community and build on it the national conservation policy for these areas, with their involvement. The traditional system which has its roots in the social structure of the community is much more widely and readily acceptable by members of the community.
- Clarifying the new land legislation law which is currently very unclear and is a potential source of conflict in the near future over access to, and control over, the common grazing lands. Considering that common land makes up almost 70 per cent of the country, the law needs to be fair and clear (Mearns 1995).
- Clearly defining the surrounding zones of the PAs which has not been done so far in the Mongolian Law on Special Protected Areas. The Agency for Protection of Nature and Environment (APNE) is responsible for designating the boundaries of the surrounding zones after consultation with the local administration. However, none have yet been established and the scope of citizens' involvement in protection as also sharing the benefits has not been defined. Here the PA managers have to take into account the needs and aspirations of the people living in and around these areas. Surrounding zone development involves inter-community

coordination and participation of the grassroots at both local and central organizational levels. Links between these two levels are often weak and need strengthening. The people also need to understand the importance of biodiversity conservation; however, if people's livelihood is threatened then conservation will hold no meaning for them.

- The probable constraints and trade-offs also need to be clarified with all actors although not necessarily agreed upon. This is time-consuming, therefore, the project phasing needs to be reasonable, with extra time accounted for.
- Conflict is part and parcel of the negotiation process. Projects in surrounding zones have to take this into account and facilitate mechanisms for permanent communication and coordination between all parties. They also need, as part of a project design, to develop strategies for conflict resolution.
- Technology development, identification of incentives and the definition of a legal framework need to evolve in parallel. These are all aspects to be considered by any such project.

REFERENCES

Bazargur, D., B. Chinbat and **C. Shiirev-Adiya.** 1992. Territorial Organisation of Mongolian Pastoral Livestock Husbandary in the Transition to a Market Economy. *PALD Research Report* 1, IDS, Brighton.

Fernandez-Gimenez, M. 1993. The Role of Ecological Perception in Indigenous Resource Management: A Case Study for the Mongolian Forest-Steppe. *Nomadic Peoples* 33: 31–46.

Fletcher, J. 1986. The Mongols: Ecological and Social Perspectives. *Harvard Journal of Asiatic Studies* 46 (1): 11–50.

Humphery, C. 1978. Pastoral Nomadism in Mongolia: The Role of Herdsmen's Cooperatives in the National Economy. *Development and Change* 9 (1): 133–36.

IBRD. 1994. Mongolia Country Economic Memorandum: Priorities in Macroeconomic Management. The World Bank, Washington D.C.

Mearns, R. 1993a. Pastoral Institutions, Land Tenure and Land Policy Reform in Post-socialist Mongolia. *PALD Research Report* 3, Brighton: IDS.

———. 1993b. Territoriality and Land Tenure Among Mongolian Pastoralists: Variation, Continuity, and Change. *Nomadic Peoples* 33: 73–103.

———. 1995. Community, Collective Action and Common Grazing: The Case of Post-Socialist Mongolia. Institute of Development Studies: Discussion Paper DP 350. IDS Publications, Brighton, England.

Purev, B. 1990. Traditional Pastoral Livestock Management in Mongolia, in *Proceedings of the International Workshop on Pastoralism and Socio-economic Development*. Ulaanbaatar, Mongolia.

Szynkiewicz, S. 1977. Kinship Groups in Modern Mongolia. *Ethnologia Polona* 3: 31–45.

Szynkiewicz, S. 1982. Settlement and Community Among the Mongolian Nomads: Remarks on the Applicability of Terms (1). *East Asian Civilization* 1: 10–44.

Vreeland, H.H. 1962. *Mongol Community and Kinship Structure*. New Haven, Connecticut.

Whytock, C.A. 1992. *Mongolia in Transition: The New Legal Framework for Land Rights and Land Protection*. IRIS Country Report 7. IRIS Centre, University of Maryland at College Park.

6

Involving local communities in conservation: The case of Nepal*

Bharat Shrestha

Introduction

With a physical area of 141,000 sq. km, Nepal extends between India and China from 80°15′ to 88°10′ E longitude and 26°20′ to 30°10′ N latitude, mainly along the south slope of the Central Himalaya. While the northern part includes the major ecological zones of High Himal, High Mountains, Middle Hills and Siwaliks, the southern part is the *tarai* ecological zone, an extension of the Gangetic plains. Altitude ranges from less than 100 msl in the southern plain to more than 8,000 in the Northern Himalayas. Due to sharp altitudinal differences and climatic conditions (subtropical to alpine), distribution of forests, farming systems and lifestyles of people vary greatly in the country.

Socio-economic profile

The population of Nepal increased from 1971 to 1981 with an unprecedented compound growth rate of 2.66 per cent per annum, whilst from 1981 to 1991 the growth rate was recorded at 2.1 per cent per annum. Total population based on the 1991 Census is 18.5 million (CBS 1991).

* I am grateful to Dr John V. Kingston, Director and UNESCO Representative to Nepal, for sponsoring this paper for the workshop. Village-level case study was undertaken with the help of Mr Thir Bikram Basnet, Chairman; Mr Nawa Raj Neupane, Secretary; and Ms Tara Devi Gautam of Thulo Ban Community Forestry Users' Group of Pinda VDC in Dhading district. I am thankful to them and the beneficiary households of Thulo Ban forestry group. My special thanks go to Mr Ashish Kothari, Ms R.V. Anuradha and Ms N. Pathak, IIPA, New Delhi for their invaluable comments in improving this paper. I acknowledge the help rendered by Ms Devi Rai in producing this outcome, and Mr M.D. Joshi, Forestry Specialist, for his suggestions in improving this paper.

Agriculture is the mainstay of over 92 per cent of the population and contributes over 65 per cent of the total Gross Domestic Product (GDP). Less than 8 per cent of the total population lives in urban areas. Nepal remains one of the poorest countries in the world despite efforts towards development in the last four decades. The per capita income (US$ 210) reflects the extent of poverty in the country (World Bank 1996). Life expectancy for men is 56 years while for women it is about 53 years, making Nepal one of the few countries in the world where female life expectancy is less than that of males (CBS 1991). Population density is 1,349 persons per sq. km of total land and 447 persons per sq. km of arable land (HMG/N 1995). Nepal's average literacy rate is 24 per cent and is even lower for females (CBS 1991).

The country is divided into five development regions, 14 zones, 75 District Development Committees (DDCs) and, over 3,900 Village Development Committees (VDCs). A VDC is the lowest political and administrative unit. The country is now governed by a multiparty system re-established in 1990.

Biodiversity profile

Nepal possesses a great diversity of flora and fauna in habitats ranging from the dense tropical monsoon forests of the *tarai*, with highly productive paddy fields and warm waters, to deciduous and coniferous forests of the subtropical and temperate regions, and finally to the sub-alpine and alpine pastures and snow-covered Himalayan peaks, with their cold streams, glaciers and lakes. Over 35 different forest types have been identified in Nepal. Approximate estimates indicate that there are over 6,500 species of flowering plants, over 1,500 species of fungi and over 350 species of lichens in these forests. About 370 species of flowering plants are considered endemic to Nepal and around 700 species are known to possess medicinal properties. The fauna of Nepal includes about 175 species of mammals, 850 species of birds, 180 species of fish, 640 species of butterflies, 143 species of moths (though some estimates put this figure at more than 1,000) and 180 species of dragonflies (NPC/IUCN 1993).

Natural resources and livelihood

The predominant livelihood for the rural people is subsistence agriculture, which depends on a continuous supply of forest products for varying uses among which leaf materials for fodder, litter for animal bedding/composting, fuelwood for energy and timber for housing and implements are the prime ones.

Nepal's forestry sector has, therefore, been one of the major components of the GDP, contributing 15 per cent to it; it also acts as a pivotal point of entire agricultural systems in view of forestry–agriculture linkages for over 90 per cent of the farming households (Wyatt–Smith 1982); it plays a central role in the economic and social life of 99 per cent of the rural population by providing full-time employment to over 1.36 million people (HMG/N 1988); and it supplies about 75 per cent of the country's total energy requirement in the form of fuelwood (SAARC 1992).

Threats to wild biodiversity

In recent years there has been increased forest degradation and destruction. At present, the forest area of Nepal is estimated at 5.5 million hectares which amounts to about 37.4 per cent of the total land area. Loss of forests during 1965–79 was 0.4 per cent per year (equivalent to 38,000 hectares), though this rate has been reduced on an average to 27,000 hectares per year since 1979 (WECS 1992). Currently, 26 mammals, nine birds and three reptiles have been legally classified as endangered. The protected wildlife species list under the National Parks and Wildlife Conservation Act (1973) is not comprehensive and several identified endangered species are not included. If Nepal loses its remaining humid tropical forest, it has been estimated that 10 species of highly valuable timber, six species of fibre, six species of edible fruit trees, four species of traditional medicinal herbs and some 50 species of little-known trees and shrubs would be lost forever. In addition, the habitat for 200 species of birds, 40 species of mammals and 20 species of reptiles and amphibians would be severely affected (NPC/IUCN 1993).

In recent years, efforts have been made to conserve this diversity by establishing a network of protected areas (PAs) in ecologically important regions. PAs cover 10 per cent of the total land mass of the country including national parks (NPs), wildlife reserves (WRs), conservation areas (CAs) and hunting reserves (HRs). Such a system can protect and conserve biological diversity *in situ*. However, outside PAs, where most wildlife habitat occurs, wild animals and plants are virtually unprotected. Poaching of wildlife, including birds, and capturing of live animals for trade poses a serious threat. Similarly, aquatic species including fish, receive little direct protection.

Traditional practices of natural resource management

Prior to 1950, most of the hill forests were effectively under the control of local headmen called *talukdars* and *mukhiyas*. Although traditional ways of managing forests as common property were already in existence, their documentation and analysis are limited. Among many practices existing in the hills and mountains, a few examples from various research studies are:

(*a*) Among the earliest accounts of indigenous systems of forest management is the account of C. von Furer–Haimendorf (1964) of the Sherpas of the Khumbu region in eastern mountains. The Sherpas appointed forest guards (*shingo naua*) who were responsible for protecting forests and allocating forest products. The position was held for one or two years and then handed over to someone else. Such traditional systems have been documented in several tribal communities studied by several researchers.

(*b*) The pine forest at Ghasa of Mustang district in the western mountainous region has been managed by a Thakali (an ethnic group) committee. Grazing of sheep and goats is prohibited. The cutting of firewood or timber for construction is forbidden except with the permission of the committee, for public use. Pine needles and litter are collected during a nine–10 day period each year. This has a double function: it reduces fire risk and makes bedding available for animal stalls (Messerschmidt 1987).

There were many other traditional management systems in middle hills where forests were managed and protected either through the users' groups or by appointing the forest watchers named locally as *chitadar, chaurasi*, etc. on a *mana-pathi* system (the system of paying fixed amount of paddy or other foodgrains. Usually four *mana* or a *pathi* (equivalent to 2.25 to 5 kilograms of produce) is given by each of the households per year to cover the remuneration).

Although traditional land and forest-holding systems were officially abolished after the fall of the Rana (traditional ruling class) regime in 1950, they appear to have survived in hidden forms. These types of traditional management of land resources prove that the user group concept is not at all a recent phenomenon. These systems are designed to be user-managed, and management practices have gradually evolved to cope with new situations. These systems, with small landholdings, fall under the user's private/community domain.

Official efforts towards conservation and policy adoption

History and current policies

In a review of forest and agricultural practices in Nepal till the early 1980s, Bajracharya (1983) points out that the policy has been dominated by resource exploitation and greed, not conservation. The Department of Forests (DOF) was established in 1942 primarily for exporting timber from the *tarai* forests. Nepal enacted the Private Forest Nationalization Act, 1957; Forest Act, 1961; Forest Protection (Special Arrangement) Act, 1967; and Forest Product (Sale and Distribution) Rules, 1970, which placed all the forests in the government's ownership with a view to conserve, protect and better manage the forest resources. Unfortunately, due to the government's inability to replace the traditional control with effective forest protection and management, these Acts succeeded in undermining existing, indigenous systems of managing forest resources, and increased the process of forest depletion, especially in the hills and mountains.

The adverse impacts of the Nationalization Act (1957) were later realized and the government introduced the Panchayat Forest (PF) and Panchayat Protected Forest (PPF) rules in 1978 to involve the local communities in forest management practices. The primary objective of PF and PPF rules was to provide certain areas of degraded forest or deforested land to the local *panchayat* under an official management plan for protection and utilization of forest products through peoples' participation.

A narrow approach was taken in PF and PPF with regard to conservation, where the emphasis was on maintaining and planting trees and the *panchayat* was given the right to generate revenue from the forest products. Forest management thus had a very restricted purpose of either preventing access to forestry by local people, or to maintain and increase the stock of trees where access was allowed. There has, however, been a recent realization that this approach to forest management has failed. The Master Plan for the Forestry Sector (MPFS) (HMG/N 1988), endorsed by the government in 1989, presents a comprehensive strategy for forestry in Nepal. The strategy focuses on procedures to enable a handover of forests to user groups and the private sector, based on a partnership between the Ministry of Forest and Soil Conservation (MFSC) and local forest users. Accordingly, it also encourages management of not just trees, but all aspects of forests, including shrubs, grasses and other plants.

Despite the Forest Act of 1961 that made provisions for classifying existing natural forests into community, leasehold and national forests for the purpose of enhancing private sector participation in forestry development, it was only in 1978 that the rules governing community and leasehold forestry development, and in 1983 that the rules governing private forests, came into force. Drawing on this legislation, the MPFS proposed two primary development programmes to implement sound management practices effectively: (*a*) the community and private forestry programme; and (*b*) the national and leasehold forestry programme. However, a review of the MPFS implementations so far, as well as the continued deterioration of the country's forests, suggests that immediate reforms are needed in some policies and actions to support effective forest management.

The recent Forest Act of 1993 tried to alleviate many of the contradictions and restrictions posed by past legislations that have prevented a rapid handover of national forests to community management and other private enterprises under leasehold arrangements. This has been attempted by specifying the elements of the law to be implemented in practice under the new structure of conservation agencies.

The current policy of the government is to promote community forestry in the hills, where forests form an integral part of farming systems and are of high environmental value in terms of stabilizing soils and protecting watersheds. However, in the *tarai* plains, because of commercial potentiality, forests will be managed for production either through leasehold arrangements with the private sector or as national forests managed by MFSC. Although recent legislation mentions leasehold forestry as an alternative to government management, there are concerns that an effective system of control governing the rights and responsibilities of potential leasehold concessionaires has not yet been fully developed. Most large forests in the *tarai* are therefore likely to remain under government management until an appropriate system is finalized.

In consideration of forest conservation and development, the Nepal Environment Policy and Action Plan (NEPAP) (NPC/IUCN 1993) proposed several actions, summarized in Tables 6.1 and 6.2.

Table 6.1: Action plan for forest management

Policies	*Recommended Actions*	*Institutional Responsibility*	*Time Frame*
Improve forest management by implementing the findings of the MPFS	Finalize the by-laws for the implementation of the Forest Act, 1993, ensuring they are consistent with HMG forest policies stated in the MPFS and EFYP	MFSC	Immediate
Encourage community participation in forest management	Continue to promote community forestry schemes in the hills	MFSC	Immediate, continuous
Encourage greater private sector involvement in managing national forests	Develop an appropriate system of incentives and regulations governing private sector management of forests	MFSC	Immediate
	Review the present system of open-ended subsidies for the purchase of wood by the District Forest Products Supply Boards, which prevents the proper evaluation of forests and undermines private sector involvement	MFSC	Short-term
Reorient forestry research	Develop programmes to provide information (including utilization of so far unused or lesser known forest species) for users' groups, forest industries and private individuals	MFSC	Short-term
Raise awareness of the importance of forest conservation	Develop forest extension agents' role based on promotion and persuasion rather than enforcement and coercion	MFSC	Short-term
Improve the basis on which land use is decided	Adopt a national land use policy classifying areas by their suitability for alternative uses	NPC, MFSC, MOA	Long-term

Table 6.1 continued

Table 6.1 continued

Minimize adverse environmental impacts of forest-related projects	Finalize EIA guidelines for the forestry sector	NPC, MFSC	Short-term
Promote research and development of alternative energy sources to reduce dependence on biomass	Finalize the energy sector strategy study to incorporate alternative energy development and promotion as an integral part of this strategy	NPC, WECs	Short-term

Source: NPC/IUCN 1993.

Table 6.2: Action plan for biodiversity conservation

Policies	*Recommended Actions*	*Institutional Responsibility*	*Time Frame*
Strengthen the capacity of DNPWC to act as the main institution responsible for PAs	Reassess the role of the army as park protectors to minimize 'people–park' conflicts; develop an alternative protection force	DNPWC, RNA	Underway
	Commission a study to resolve the problems of overlapping jurisdiction in PAs and to recommend a simplified procedure for handling various activities affecting PA management	DNPWC, MTCA, DOT	Short-term
Ensure adequate representation of Nepal's major ecosystems in the PA system	Review the representativeness of the existing PA system	NPC, MFSC, MOA	Immediate
Involve local people directly in the management of parks	Develop mechanisms for benefit-sharing with people whose livelihoods are adversely affected by parks	DNPWC	Short-term
	Effectively harness the efforts of NGOs to test and develop appropriate models of park management	DNPWC, NGOs	Short-term, continuous

Table 6.2 continued

Table 6.2 continued

Policies	*Recommended Actions*	*Institutional Responsibility*	*Time Frame*
	Set up a Task Force to prepare guidelines for the development of management plans	DNPWC, NGOs	Immediate
	Enact and enforce necessary legal and regulatory measures to implement major international treaties and conventions, as well as to control illegal wildlife trade within the country	MFSC	Short-term
	Promote tourism in PAs, consistent with conservation objectives	MFSC, MTCA	Short-term
Preserve endemic and endangered species and their habitats	Identify and take actions to protect marshes, wetlands and waste bodies significant to biodiversity conservation	MFSC, MWR, NEA	Short-term
	Develop management plans to conserve biodiversity, while providing for people's basic needs	DNPWC, NGOs	Immediate, continuous
	Mount a study to assess the status of biological diversity of endemic plants and animals, both terrestrial and aquatic, occurring outside PAs in farmlands, pastures, rangelands, forests, rivers, lakes and ponds	MFSC, MOA, NARC	Immediate
Promote private and public institutions for biological resource inventory and conservation	Collate and disseminate data on biodiversity from various existing databases, and establish a national biodiversity database	DNPWC, MFSC, NARC, DOB, TU, NGOs	Short-term
	Identify and strengthen institutions responsible for research, education and training in biological resource management	DNPWC, TU, NGOs	Short-term

Source: NPC/IUCN 1993.

Extent and range of protected areas

Nepal has developed an impressive network of PAs as a means of preserving its valued forests, genetic resources and natural habitat. The conservation movement through PAs was effectively launched in 1970 with the establishment of Royal Chitwan National Park in Siwalik and Langtang National Park in the mountains. The National Parks and Wildlife Conservation Act was officially promulgated in 1973.

Today there are eight national parks, four wildlife reserves, two conservation areas and one hunting reserve, occupying approximately 1,094,320 hectares (10 per cent) of the country (see Table 6.3).

Table 6.3: National parks and reserves in Nepal

Ecological Region	*Parks and Reserves*	*Total Area (hectares)*
High Himal	Shey Phoksundo (part) 1980	347,870
	Langtang (part) 1976	109,900
	Sagarmatha 1976	114,140
High Mountains	Langtang (part) 1976	61,120
	Khaptad (part) 1984	19,980
	Rara 1976	10,570
	Dhorpatan 1987	80,820
	Shey Phoksundo (part) 1984	7,640
	Sagarmatha (part) 1976	660
Middle Mountains	Dhorpatan (part) 1987	51,680
	Khaptad (part) 1984	2,530
	Shivapuri 1987	14,480
Siwalik	Royal Bardiya (part) 1976	60,140
	Royal Chitwan (part) 1973	93,200
	Parsa (part) 1984	29,500
Tarai	Royal Suklaphant (part) 1976	15,500
	Royal Bardia (part) 1976	36,670
	Parsa (part) 1984	20,410
	Koshi Tappu 1976	17,500
Total	19	1,094,310

Source: Parks and People Project 1996.

The PAs lying in Tarai Siwalik are facing social and political pressures for grazing, collecting grasses, extraction of firewood, etc. while in the mountains and hills, there is a lot of stress due to trekking and mountaineering teams indulging in excessive extraction of firewood and dumping of garbage. Grazing, collecting forest products, and poaching of wildlife are some persistent problems in both the areas.

There are several issues pertaining to the management of PAs. Many PAs either have no management plans or adhere to outdated plans that rely mostly on exclusion through tight policing rather than encouraging local participation. In some areas, conflicts have developed between the park authorities and local residents regarding the latter's right to extract essential products such as fuelwood and fodder. To enforce PA regulations, authorities rely on the assistance of the Royal Nepal Army. Although the army has helped reduce poaching within PAs, it has created its own difficulties. Administrative authority has become divided, leading to some confusion about who has overall authority for parks. Also, soldiers are not given any special training and consequently lack an adequate appreciation of their new, non-military role. Parks guarded by such armed forces and people driven by hunger, often confront each other. The restrictions on the access to parks and reserves have in many cases resulted in economic and social hardship for local people. Successful protection has allowed animal populations to expand. As a result, some species are having a serious impact on local communities, causing injuries (and occasional fatalities) of humans, livestock losses and the destruction of crops, thus adding to the conflict between people and parks.

In the past, communities have not been involved in management decisions and have not been entitled to a share of locally generated revenues from park activities. Local people have therefore tended to regard parks with suspicion. Parks have not only meant reduced access to areas where once they were free to go, but they have also not received adequate compensation for various losses that they suffer because of conservation. Reduced access has led to a perverse effect on areas in the immediate vicinity of PAs. Because of the restrictions inside the PAs, extractive activities have been intensified in the surrounding areas, causing severe damage to ecosystems.

Involving local communities and institutional arrangements

Linking PA management with local communities is a strategy for successful conservation. Therefore, in the recent years, there has been a growing realization that the ultimate success of PAs depends upon the cooperation and support of local people. In the case of some

PAs, autonomous, non-governmental committees have been created and authorized to collect funds, including tourism revenues, and use them for sustainable community development. Channelling tourist revenues into community development has helped increase people's participation in national parks and reserves.

There is also a wider recognition of the fact that people living within and adjacent to the PAs need to be compensated for their loss. The compensation has been in the form of community development works (including agriculture), supported by funds raised from tourism, government allocations, as well as international sources. Forests within the Conservation Areas have been handed over to the local people under community forestry programmes and they are supported in all subsequent activities.

New approaches are being tried in the Annapurna Conservation Area (ACAP), Makalu–Barun Conservation Project (MBCA) and Royal Bardiya National Park, and some success at involving local people directly in project planning and implementation has been achieved. These approaches are, however, innovative and experimental, and the extent to which they may be replicated for other areas is yet to be assessed (Shrestha 1995). ACAP in particular is a very interesting example, being one of the few large PAs in Asia being managed entirely by an NGO with widespread community involvement.

The National Parks and Wildlife Conservation Act (1973) mainly governs the conservation of ecologically valuable areas and their wildlife. Very recently, it has been amended to introduce the concept of revenue sharing too. According to the amendment (Buffer Zone Regulations), the local communities in areas surrounding the PAs have been entitled to retain 30 to 50 per cent of locally generated revenues for development projects. This revenue is to remain in a common fund and utilized for community welfare and development activities to meet common goals.

The main responsible agency for PA management is the Department of National Parks and Wildlife Conservation (DNPWC). Its major constraint in managing and protecting areas is the lack of financial and human resources. There is a strong need to substantially improve the institutional capacity of DNPWC. It may also be desirable to reconsider the role of the army as park protectors. Closer inter-departmental coordination is only possible if a forum exists for discussion and if responsibilities are modified to reflect DNPWC's role as the lead agency for PAs. Other institutions directly involved in conservation include the DOF, which is responsible for taking care of all flora, fauna and timber species outside the PAs. Both the DNPWC and the DOF are under the Ministry of Forests and Soil Conservation.

Community forestry in Nepal

Among the most successful programmes towards people's involvement in Nepal has been the Community Forestry Programme. Nepal is one of the first countries to embrace full community participation in forestry, as a main strategy of its national forest policy. This thrust formally commenced in 1978 with the enactment of progressive legislation—the Panchayat Forest Rules, 1978 and Panchayat Protected Forest Rules, 1978. The objective of these legislations was to return the ownership of forest resources to the people. Since then, many bilateral and multilateral donors have provided financial and technical assistance towards developing community forestry programmes across the country.

The essence of community forestry philosophy in Nepal is the development of a partnership between the local communities and the Department of Forests (DOF) for the management of local forests. Initially, these partnerships were established with the local government administrative unit, the *panchayat* (called the Village Development Committee after 1990), but as these did not represent the real user of a particular forest effectively, the social unit was later changed to a Forest Users' Group (FUG). The HMG/N (1988) has clarified the following key policies relating to community forestry:

- Promotion of community forestry, entrusting forest protection and management to actual users.
- Priority to community forestry in the allocation of research and resource development.
- Phased handing-over of all the accessible hill forests to the communities to the extent that they are able and willing to manage them.
- The users to receive all the income from the sale of forest products.
- Orientation of the entire staff of the Ministry of Forests for their new role as advisors and extension workers.
- Formulation of a simple management agreement.
- Planning and rapid implementation of community forestry according to decentralization principles.

- Ensuring local people's benefit if they protect natural forests or plantations.

Studies show the potential of the extension of such community forests to be 3.5 million hectares, 61 per cent of the total forest area. The procedure for handing over a forest to a community consists of (*a*) formation of a user group, following an identification process; (*b*) demarcation of a forest as a community forest; (*c*) preparation and approval of an operational plan; and (*d*) handing over the forest to the user group and implementation of the operational plan. A formal handover of each forest is required before the users can officially begin their management. An operational plan, acceptable both to the users and to the FD, has to be prepared before the handover can take place. In 1990, the government produced operational guidelines for these processes.

The focus of the community forestry programmes is now on natural forests, because the villagers prefer to take over these rather than establish new plantations, due to the quick benefits associated with the former. User groups are provided opportunities to discuss ways and means of managing community forests through networking in districts and at the national level. The district-level forestry staff are encouraged to plan community forestry work through range-level planning, by a method of Participatory Rural Appraisal.

The user groups receive a cash subsidy as an incentive for plantation development and protection. This subsidy is being reduced and gradually withdrawn to make the programme sustainable. The user groups have been trained to manage nurseries, thus gradually handing over not only forests but also the technology of management to user groups. It must, however, be noted that actual ownership of forest land is not transferred to user groups but remains with the government. By 1996, a total of 4,698 FUGs had been formed in the country, 4,627 in the hills and 71 in the *tarai* (CFDP 1997). Though the establishment of FUGs and negotiation of operational plans alone are not good indicators of the success of community forestry, the fact that a process like this has been initiated and is continuously being developed based on actual ground-level experiences, needs to be acknowledged.

Constraints of community forestry

Lack of trust: Frequent changes in government policy, beginning with the Forest Nationalization Policy of 1957, have contributed to the development of a feeling of distrust among people towards these policies. Top-down enforcement of forest protection rules and regulations by forest officers is almost certainly another reason for this distrust. Above all, the inappropriately accomplished process of identifying real users has created problems as in Dolakha, Ramechhap, Dhading, Gorkha and many other districts, and has resulted in a lack of satisfaction and participation within the communities.

Weak participation: Experience suggests that there are two basic reasons for the lack of popular participation. Many villages are simply not aware that management responsibility and use rights are available to them. This is generally because forest office staff have not been able to take on extension work on a large scale. It may also be due to a lack of confidence on the part of the government foresters in the capacity of the local communities to carry out these activities. Second, the long time gap between investment and return in forestry enterprises also has a lot to do with the low level of participation. In the case of existing natural forests, people begin to benefit from some sort of return in the form of twigs and fuelwood immediately, but in the case of plantations, they have a long wait before harvesting.

Weak community institutions: There exist national operational guidelines for the handing-over process, but they are flexible enough to be site-specific. Depending upon the situation, the District Forest Officers (DFOs), empowered to hand over forests, can function by either forming a local-level committee first and then helping its members to prepare and implement the plan, or help the users to prepare a plan and then form a committee to implement it.

However, as local institutions in most places are very weak, community organization takes a long time and lots of effort. Therefore, most DFOs choose to form a committee first. This has a potential danger of committee members influencing the plan to reflect vested interests. The end result may consequently contradict the basic philosophy of the government's forest policy, which clearly states that priority will be given to poorer communities and people in community forestry (Anonymous 1990).

The process leading to the preparation of an operational plan requires the field staff to undertake a series of participatory investigations and negotiations with the community. The lack of implementing this process has ultimately resulted in weak community institutions and has raised questions about the sustainability of these institutions.

Inadequate finances: Except in some cases where the users' groups raise funds generally from the sale of products, membership, donations, etc., a majority of the forests are only under regeneration and user groups have not been able to obtain forest products. Therefore, the groups constantly face financial difficulties. In some districts, both national and international NGOs working with local users' groups have moderately strengthened their fund-raising but this is not a long-term solution.

Local disputes: Disputes raised within the users' group are about the identification of actual users, level of participation, inequitable benefit-sharing and leadership or disputes with regard to forest boundary and forest product use. Often, these have been resolved by the groups and extension functionaries (i.e., the staff of the MFSC). To avoid disputes being created by communities other than the actual beneficiaries, a clear definition of primary, secondary and tertiary users and their access to the resources is required (see Kothari et al. in this book). Disputes between the users' group and the DFOs pertaining to the deviation between the operational plan and policies and actions during implementation also need to be seriously addressed.

Existing and potential opportunities

Joint forestry management scheme: Community forestry is aiming at alleviating rural poverty, mobilizing rural people to make themselves self-reliant for basic needs and simultaneously conserving forest resources. Implementation of such management schemes in totality has resulted in improvement in biodiversity conservation, and has helped many communities to achieve self-reliance.

Group dynamism among community people: The primary objective of community forestry is also to bring cohesiveness among people at the village level. Working in groups with discussion and mutual understanding has led to collective bargaining and capacity building. Newly formed groups may work together in the future for several other community development programmes and income-generating activities. In addition, NGOs have emerged as an important link between the government agencies and the local communities.

Proposals for future improvement

There have been subsequent amendments to improve the programme since it was first initiated (e.g., in 1993). A further need is to communicate the process of handing over forests to the communities, most of whom are not aware of it. This could be done through the involvement of the VDCs.

Nepal is a multi-ethnic society, and equity issues are very important in any development process including community forestry. Issues of equitable distribution of benefits thus need to be clearly spelt out in a plan and practically applied by all involved. Moreover, gender issues are of particular importance in forest management, as fuelwood and fodder collection is mainly done by women and children (see Sarin et al. in this book). Assistance from some advocacy groups could help in constituting equitable user groups. The role of NGOs and VDCs in forming user groups and preparing operational plans is crucial.

The process of cadastral survey needs to be expedited in order to resolve conflicts over forest and resource ownership. It will also help reduce the conflicts arising from the encroachment of forest areas by politically backed individuals. Not only is it important to clearly identify a user group, it is also important to clearly demarcate the area under its management.

The state of the forest staff involved in these activities (especially lower staff) also needs to be critically assessed. Their general problems are low salaries, low incentives, low support, no proper evaluation, and no clear-cut posting and transfer policy. These factors result in low commitment and dedication to community forestry, and to the profession and its institutions.

Extension materials produced so far are mostly descriptive and non-informative, and hence of not much use for rural people. Field staff are sent to villages for extension without enough extension materials or written information. Similarly, few field staff have been reoriented to their new role as community foresters. Because of lack of experience and knowledge of community management, the quality of operational plans is very poor. As a result, many users need to come to the DFO even for minor instructions.

There is a need for participatory research on identification of indicators for social and ecological sustainability, which will help assess the true success of this effort.

Often, it is seen that the national forests surrounding the community forests are highly degraded as the members of the FUG extract resources from these forests while protecting their own resources.

This needs to be urgently dealt with and needs to become a part of the operational plan for the area.

This may be equally true for wildlife and PAs too. It should be recognized that while the community forestry programmes are directly related to meeting local subsistence needs, the primary objective of the national forestry programme should be to meet national needs of conservation. Selected blocks of national forest should be nationally managed, putting them above district-level politics, and strictly controlled against any free use by the local people.

REFERENCES

Anonymous. 1990. *Operational Guidelines for the Community Forestry Programme.* Ministry of Forests and Soil Conservation, Kathmandu.

Bajracharya, D. 1983. Fuel, Food or Forests? Dilemmas in a Nepali Village. *World Development* 11: 1057–74.

CBS. 1991. *Statistical Handbook.* Central Bureau of Statistics, National Planning Commission. Kathmandu, Nepal.

CFDP. 1997. *Community Forestry Development Project.* Department of Forests, Nepal.

Furer–Haimendorf, C. von. 1964. *The Sherpas of Nepal.* University of California Press, Berkeley.

HMG/N. 1988. *Master Plan for the Forestry Sector.* Ministry of Forests and Soil Conservation, His Majesty's Government of Nepal, Kathmandu.

———. 1995. *Statistical Handbook.* Central Bureau of Statistics, His Majesty's Government of Nepal, Kathmandu.

Messerschmidt, D.A. 1987. Conservation and Society in Nepal: Traditional Forest Management and Innovative Development, in P.D. Little and M.M. Horowitz (eds). *Land at Risk in the Third World: Local Level Perspectives.* Westview, Boulder.

NPC/IUCN. 1993. *Nepal Environmental Policy and Action Plan.* National Planning Commission and The World Conservation Union. Kathmandu, Nepal.

Parks and People Project. 1996. Buffer Zone Development Bulletin. Department of National Parks and Wildlife Conservation, Kathmandu.

SAARC. 1992. Regional Study on the Causes and Consequences of Natural Disasters and the Protection and Preservation of the Environment. South Asian Association for Regional Cooperation, Kathmandu.

Shrestha, B. 1995. Rural Communities in Conserving Forests of Makalu Barun Area: An Eco-Institutional Appraisal of User's Group Participation. Woodland Mountain Institute, USA, and Makalu Barun Conservation Project, Kathmandu.

WECS. 1992. Water and Energy Commission. Kathmandu, Nepal.

World Bank. 1996. World Development Report, Washington, D.C.

Wyatt–Smith, J. 1982. The Agricultural System in the Hills of Nepal: the Ratio of Agriculture to Forest Land and the Problems of Animal Fodder. APROSC Occasional Paper, Kathmandu.

7

Community-based conservation: Experiences from Pakistan

G.M. Khattak

Country profile

Pakistan is a rectangular mass extending north-east to south-west over about 88 million hectares, located between China in the north, India in the east, the Arabian Sea in the south and Iran and Afghanistan in the west. Mountains and foothills on the north and west cover about half its area. The remaining half comprises the Indus Plain towards the east, intersected by the Indus river and its tributaries.

The country is largely arid, with three-fourths receiving an annual precipitation of less than 250 mm, and 20 per cent of it less than 125 mm. Only about 10 per cent of the area in the northern Himalayas and the Karakoram mountain ranges receives rainfall between 500 mm and 1500 mm (GOP 1992).

Of the country's total area of about 88 million hectares, 24 per cent is cultivated, of which about 80 per cent is irrigated and the balance, rain-fed (GOP 1996). Forests and grazing-lands cover about 4 per cent and 34 per cent of the area, respectively, 38 per cent is not fit for agricultural uses, and 2 per cent is under urban and other uses (IUCN 1992; Mian and Mirza 1993).

Pakistan is inhabited by over 132 million people, growing at about 2.8 per cent annually. About 68 per cent of its people are rural and 32 per cent urban. Its civilian labour force is around 36.7 million of whom about 34.9 million are employed. Agriculture employs about half of the employed labour force and contributes 25 per cent to the country's GDP. The country's per capita income at current prices is about $495. Its literacy rate in 1995–96 was estimated at 38 per cent, 50 per cent for males and 25 per cent for females (GOP 1996).

Wild biological diversity

Roberts (1977) (cited in IUCN 1992) distinguishes the following major ecological zones in Pakistan: dry alpine and cold desert, alpine scrub and moist alpine, Himalayan dry coniferous with ilex oak, Himalayan moist temperate forest, subtropical pine forest, subtropical dry mixed deciduous scrub forest, Baluchistan juniper and pistachio scrub forest, dry subtropical and temperate semi-evergreen scrub forest, tropical thorn forest, mangrove and littoral forest, and sand dune desert.

Pakistan has about 5,000 species of wild plants, about 372 of them endemic. It has about 188 species of mammals, 666 species of migratory and resident birds, 174 species of lizards, 14 species of amphibians and fishes and about 20,000 species of insects and other invertebrates (IUCN 1992). According to the Forestry Sector Master Plan (GOP 1992), 'The varied composition of Pakistan's fauna is of outstanding international value and interest. It represents vividly the transitional features between two of the world's six major zoogeographical regions—the Western Palearctic and the Oriental regions.'

Major threats to biodiversity

Destruction of habitat is by far the most important threat to biodiversity. Since the dawn of canal irrigation about a century ago, the tropical thorn forest has almost been eradicated from the Indus Plain. Increasing upstream storage of flood waters for controlled irrigation since the 1960s is threatening the survival of the riverain, mangrove and littoral forests. Woody vegetation in all ecosystems is declining mostly because of its removal for use as fuel and the failure of its regeneration due to heavy uncontrolled grazing; the latter has also degraded the non-woody vegetation.

Heavy pressure of hunting, combined with habitat destruction is partly responsible for the decline of game species:

> Wild animals have been hunted to extinction, among them the Lion [*Panthera leo*], Tiger [*Panthera tigris*], Cheetah [*Acinonyx jubatus*], One-horned rhino [*Rhinoceros unicornis*] and Chausinga [*Tetracerus quadricornis*]. Today the Ibex [*Capra ibex*], Snow leopard [*Panthera uncia*], Wild ass [*Equus hemionus*], and Houbara bustard [*Chlamydotis undulata*] all face extinction due to hunting pressure. Falcons of all types,

not just the peregrine [*Falco peregrinus*], are being trapped to satisfy the Middle Eastern market for hunting birds. (IUCN 1992; scientific names added).

Rural community dependence on wild biological resources

Superimposed on the large ethnic diversity of rural communities in various parts of Pakistan, speaking a variety of languages and dialects, is the diversity of lifestyles adapted to the dictates of their habitat. The communities depending on irrigated farmland in the plains usually own about 3 hectares of farmland on an average and a few heads of cattle, buffaloes and goats per household, and are resident all the year round in their hamlets. Communities living off semi-arid rain-fed farmland in the foothills may own somewhat larger farms and a few more livestock. Such farming communities are also resident, sending their livestock to forage on the field boundaries and the countryside. Inhabitants of the arid zone are generally organized in transhumant pastoral communities owning small herds of goats and sheep, making optimum use of the scanty and unpredictable rain which may fall anywhere in their domain.

Among the communities inhabiting the mountains, there is considerable variation of lifestyle. Those residing at mid-elevation in the northern mountains, with perennial streams for irrigation, are settled small-scale agriculturists. Mountain communities residing at higher elevations are semi-transhumant, shuttling between a lower abode with irrigated or rain-fed agricultural land where they spend the winter, and a higher abode, generally in the alpine zone where they take their livestock for grazing during the summer. Then there are a few entirely migratory pastoral tribes which do not have a permanent abode but range over large areas, moving with the season to take advantage of the availability of forage as well as opportunities for occasional labour.

Woody biomass and fodder/grazing are the most dire needs of the rural communities of Pakistan. About 79 per cent of the households of Pakistan (91 per cent rural, 52 per cent urban) depend on firewood from trees grown along fields, or from the forests and foraging areas, as the sole domestic source of energy (Ouerghi and Heaps 1993). Pakistan has a livestock herd of about 15 million cattle, an equal number of buffaloes, 48 million goats and sheep, and 3 million donkeys and camels (GOP 1990). About half the sustenance of the cattle and the total sustenance of goats, sheep, donkeys and camels is supplied by the natural vegetation on uncultivated lands. Rural communities of Pakistan are therefore heavily dependent on grazing land for food as well as cash income.

Official efforts at conservation

History

Though the whole of Pakistan except the bleakest parts of the deserts can support woody vegetation of varying density, its current forest area is only about 4 per cent of its total geographical area. Massive deforestation has occurred over millennia when the area was occupied by successive civilizations dependent on grazing and cultivation. The plains in the main Indus Valley and in the subsidiary valleys in the mountains were occupied by conquerors and the vanquished were pushed to the mountains where they had to clear forests for subsistence agriculture and grazing. Conservation of hunting species for the elite in game reserves was a common tradition. The riverain forests of Sindh were raised by the local rulers for this purpose (Stebbing 1921).

With the British occupation of the area, the pace of deforestation picked up momentum. Large-sized timber from the mountain forests became saleable in large quantities for the first time in history, for the construction of cantonments and railway tracks. Large-scale canal irrigation in the Indus Plain started the demise of the tropical thorn forest which is now almost complete.

British efforts at forest conservation started towards the middle of the 19th century with the passing of the first Forest Act in 1865. Prior to this, in Hazara, Forest Conservancy Rules were promulgated in 1857. According to these, all forests in Hazara were declared the property of the government. The local villagers and right-holders were allowed to cut trees for their personal use, and break up land in the forests for cultivation with the permission of the Deputy Commissioner. Non-right-holders could also obtain trees but on payment of tax to the government. Half of the tax recovered was paid to the local villagers. Grazing was allowed without restriction, but causing fire was punishable. To ensure a sustainable supply of firewood to the railways and the river steamers, the Changa Manga irrigated plantation was started in 1866. The planting of riverain areas was also started in the 1860s.

Under the amended Hazara Forest Regulation of 1873, the wooded lands of Hazara (NWFP) were divided into two categories: the Government Reserved Forests and the public wastelands which later came to be called the *guzara* forests. The Reserved Forests were handed over to the Forest Department (FD) for management and the *guzara* forests were set aside to meet the domestic requirements of the local people under the general supervision of the Deputy Commissioner. But the government retained the right to conserve and manage them and charged a share on their sale proceeds known as the seigniorage fee (Jan 1965). The *guzara* forests were also transferred to the management of the FD in 1950. The forests of the former princely states of Dir, Swat and Chitral also came under the management of the Department with the abolition of the states in the 1970s. All the efforts at conservation were based on the regulatory approach, without involving the local people in the management of the forests even when these were set aside for meeting the domestic requirements of the local people (as in *guzara* forests). Forests in the Punjab, Sindh and Baluchistan were mainly reserved in the 19th century.

Current conservation policies

Policy pronouncements for the conservation of natural resources are framed by the Government of Pakistan in the Ministry of Environment, Local Government and Rural Development. Their implementation is generally the responsibility of the provincial FDs. The most recent conservation policies of the Government of Pakistan are the following:

The Forest Policy Statement (FPS), 1991, with the following objectives: (*a*) increasing forest area from 5.4 per cent to 7 per cent in 15 years; (*b*) sustainable development of forests, grazing lands and wildlife; (*c*) promoting social forestry programmes on watersheds and farmlands; (*d*) conserving biological diversity and maintaining ecological balance; (*e*) containing environmental degradation in the watersheds; (*f*) promoting income generation and self-employment in the rural areas; (*g*) promoting NGOs and private voluntary organizations to create public awareness; and (*h*) integrated and participatory management of natural resources.

The National Conservation Strategy (NCS), 1992 (IUCN 1992), with the objective (*a*) to maintain essential ecological processes; (*b*) to preserve the biodiversity of natural resources; (*c*) to restore degraded natural resources cost-effectively; (*d*) to ensure the sustainable use of natural resources; (*e*) to ensure balanced and diversified development that maintains, if not increases, the sum of options available to future generations; (*f*) to improve the efficiencies with which natural resources are used; (*g*) to improve the efficiencies with which associated resources and human derived capital (e.g., community infrastructure) are used and managed; (*h*) to give priority to the conservation and improvement of best soils and sweet water; and (*i*) to give priority to preventing deterioration of fragile ecosystems with large downstream effects.

The Forestry Sector Master Plan (FSMP), 1992 (GOP 1992), had the following objectives: (*a*) development of forestry through afforestation, reforestation and natural regeneration of forests and public lands, and through farm and social forestry on private lands; (*b*) harmonizing environmental protection, maintenance of biodiversity and wise economic use of forest resources; and (*c*) increasing the productivity of forests, farmlands, watersheds and ranges to contribute towards meeting the country's requirements of forest produce and services.

All these policies are very ambitious and will require tremendous resources and a long time to implement in full. The implementation of NCS is proposed to be initiated through launching programmes for protecting watersheds, forestry and plantations, restoring rangelands and improving livestock, protecting waterbodies and sustaining fisheries, protecting biodiversity, and supporting institutions for common resources.

Conservation areas

The Forestry Sector Master Plan (1992) estimates the forest area under the control of the FDs of Pakistan at about three million hectares. These comprise coniferous, scrub, mangrove and riverain forests as well as irrigated plantations.

The extent of protection legally available to them varies in accordance with their legal category. For example, in the reserved forests, all acts are prohibited unless they are permitted as rights or through orders in writing. In the protected forests, all practices are permitted unless they have been specifically prohibited by notification. The protection extends to all the resources such as trees and other vegetation, wildlife, and fish in the forests.

In addition, Pakistan has 175 PAs (national parks, wildlife sanctuaries and game reserves) covering about 7 million hectares, mostly managed through the regulatory approach.

There are a wide range of conflicts between the people residing in or around the PAs and the agencies responsible for their management. In certain protected forests, the local people may not accept the ownership of the forests by the government. In forests owned by households or communities, the owners do not readily accept the right of the government to restrict resource utilization. A persistent conflict in forests is between the need to close a part of the forest to regenerate the tree crop, and the continual requirement of the local people for grazing.

In areas protected for wildlife conservation also, there is considerable conflict on grazing, since heavy grazing of domestic livestock depletes the forage available for wildlife. Outside PAs, the conflict is mainly due to the reluctance of the people to accept any controls on the killing of wildlife. In fisheries, the major conflict is between the fisherman's desire to catch the maximum possible number of the fish and the manager's desire to enforce a sustainable level of harvest.

Public attitude towards these efforts

Till the 1970s, the attitude of the local communities to conservation efforts was uniformly negative, due to the overbearing attitude of the managing departments and the failure of the rural communities to see any link between these efforts and the quality of their life. For example, the local people were not allowed to cut trees in the reserved forests and graze livestock in some parts. Every now and then parts of the forest were harvested and carted away to be sold to the highest bidders. But the local people could not even gain employment in this process as they were not trained in timber harvesting.

Even where the local people have a right to 60 to 80 per cent of share in the sale proceeds of timber (as in the forests designated as Protected Forests, and *guzara* forests), there is intense conflict between the forest managers and the local communities around innumerable restrictions on timber harvesting and the long delays that occur between the start of the timber harvest and the payment of the people's share. The situation is further complicated by innumerable conflicts among the local individuals with respect to the extent of their ownership and rights over the resources. Yet another major conflict is between forest managers and illegal woodcutters.

Attempts at involving local communities in conservation

Watershed Management Programme

The first major attempt at involving the local people in the planting of their mountain land was the watershed management programme launched in the 1970s with the assistance of the World Food Programme, still under implementation. This programme seeks to improve the quality of life of the mountain farmers and assist them in planting forest trees and sometimes fruit trees on their mountain lands.

So far the programme has been implemented by the FD securing the cooperation of landowners through individual contact. Attempts have recently been made to form organizations of owners, right-holders and where possible, also the users, to assist in planting and the sustainable management of existing and future plantations.

Kalam Integrated Development Project

A Swiss-assisted Kalam Integrated Development Project was started in 1981, initially for the improvement of forest management, timber harvesting and regeneration of the coniferous mountain Protected Forests. It was soon realized that the project would not make much headway as a traditional forestry venture. It was therefore converted into an integrated development project, adding agriculture, horticulture, human resources development, organization of local communities and village infrastructure development, to secure the participation of the local people. Through the introduction of off-season vegetables in the local farming system, the project has made considerable improvement in the quality of life of the local people who are now assisting actively in the protection of forests in which they are entitled to a 60 per cent share in the proceeds from the sale of timber. The local people were originally not in favour of regenerating the forests for the fear that this would reinforce the claim of the government to the ownership of forests. However, working together on the project for the past two decades has softened their resistance and they are now even allowing the regeneration of the forests. Because of its great success, the project has created a model whose outlines are now being followed by all donor-assisted forestry-related projects in Pakistan.

The project is being implemented with the assistance of 182 community organizations and 17 women's organizations. Since the inception of the project, attempts have been made to improve agriculture in the area through the introduction of off-season vegetables, viticulture, improved horticulture, bee-keeping, seed multiplication, improved livestock management, improved marketing of fruits and vegetables, food preservation and kitchen gardening, poultry farming and raising forest nurseries. Members of Community-Based Organizations (CBOs), and women's organizations are trained by representatives of the concerned government line departments.

An important component of the programme is imparting training to the local people in ecologically sustainable tree-felling. Besides reducing the loss of soil and the breakage of trees in the process of felling, it also increases the income of the trained local people who can now seek employment as skilled forest workers.

As a result of this beneficial programme, the support of the local people has been secured in forest protection. During 1995–96, 10 peoples' checkposts jointly staffed by the community and the FD were maintained to check the smuggling of timber. In addition, 13 village-level FPCs function along the major wooded valleys to ensure the protection of the forests. The FD facilitates the formation of these committees and provides them full support in their work (GONWFP 1996a).

Himalayan Jungle (Palas Conservation) and Development Project

The remote Palas Valley, covering about 1,400 sq. km, is Pakistan's most outstanding remaining tract of West Himalayan forest. It is a global hot-spot for avian diversity and has a rich mammalian and plant diversity. The project was initiated by Bird Life International in 1991 with the partnership of several Pakistani government and non-government agencies. A low-key informal dialogue was started with the local tribesmen through a local facilitator, together with participatory rural appraisal to identify 'entry points for working with the community'.

In 1992, the area was hit by a devastating flood. Since there were no relief agencies in the area, the project got heavily involved in rendering emergency aid to the local people. This was followed by the rehabilitation of the physical infrastructure such as water-mills for milling the staple maize, irrigation channels and small foot-bridges. This greatly enhanced the visibility and credibility of the project among the local people as well as the government.

The project subsequently initiated the formation of pro-active village committees based on the traditional *jirga* and 'written gentlemen's' agreements with them. The project is now concentrating on the economic conditions of local people through the improvement of the maize crop and the introduction of horticulture. A biodiversity profile of the area is being prepared, as also plans to sustainably manage the forests to both benefit local people and foster biodiversity (Duke 1997). An expanded five-year 'Palas Conservation and Development' project to be implemented with the financial assistance of the European Community (EC) started in April 1997. This includes extension of biodiversity conservation and environmental awareness programmes through support to a Kohistan Wildlife Unit; participatory forest management, including setting aside forests of highest biodiversity value from commercial timber harvesting ('core zones'); sustainable use of remaining forests ('sustainable use zones'); conservation of biodiversity and sustainable use of non-timber forest produce (NTFP) in all forests; sustainable agricultural development for improved nutrition and income generation; and improved livestock and rangeland management. The plan also aims to enhance local economic benefits from forests, agriculture and livestock, develop basic infrastructure such as bridges and footpaths, and improve health, nutrition and sanitation conditions. It is hoped that a functional network of CBOs including an All-Palas CBO, village CBOs, special interest CBOs and women's CBOs will be set up. These will facilitate participatory planning, monitoring and evaluation of natural resource use throughout Palas, with locally defined indicators of success and sustainability. The project also hopes to enhance the capacity of WWF–Pakistan and/or other environmental NGOs as support agencies, and strengthen the legal, policy and institutional framework for participatory management.

Malakand Social Forestry Project

Though large areas of private land have been planted under watershed management programmes, no effective institutions have so far been created for their sustainable management. The Dutch-assisted Malakand/Dir Social Forestry Project, started in 1987 as a quest in this direction, seeks to enable the local communities to sustainably manage their forest and grazing lands. An integral part of the project is organizing the local communities and securing their active participation from the planning stage and throughout the implementation. Improvement of grazing lands (so far considered too refractory to handle) is an important component of the project.

At present, 59 village development committees are working in the area. A village development committee is formed as a result of the interaction of the project social organizers and the village council of elders. It usually comprises representatives of major sections and groups interested in the tasks the project proposes to undertake. Sometimes these groups live in harmony, sometimes in conflict. So a pre-condition of starting work in the village is the execution of an agreement specifying the distribution of rights, concessions and benefits between the groups. Often such negotiations have resulted in the reduction of conflict among the communities as a result of mediation of neutral groups. But social unity among communities in the area is still in its infancy. The area is rife with landlord–tenant disputes around which the project has to work. The major focus of the work is on forest tree planting on denuded communal land, and rotational grazing on such areas. The basis of the work is a participatory village land use plan which is developed at the start of the project. About 19,000 hectares have been planted so far, and controlled grazing implemented over about 7,000 hectares. The project has also organized 27 women's organizations around forest nursery development and management and distribution of fuel-efficient stoves (GONWFP 1996b).

Khunjerab National Park

The Khunjerab National Park was established in 1975 with the special objective of protecting the Marco Polo sheep. But little progress could be made towards the attainment of the objective because the local people claimed that they had not been adequately compensated for relinquishing their grazing rights in the area. In 1991, the local people challenged the constitution of the National Park in a court of law. With the mediation of WWF and the local authorities the conflict was resolved in 1992 with the signing of a management agreement by representatives of the local communities, the Park and the local civil administration. The salient features of the agreement are:

- Traditional grazing concessions will be allowed in accordance with scientific range management principles, with a maximum of 100 yaks allowed in the core zone during winter.

- Khunjerab village organization will assist the park administration in protecting wildlife against hunting, killing and disturbance.
- Qualified local graziers will be employed on 80 per cent of the positions in the Park.
- Subject to maintenance of quality, local graziers would be preferred for providing services to Park visitors.
- When essential, for the sustenance of certain wildlife species, the concerned graziers would vacate their grazing areas for the requisite period. For this, they will be compensated through employment in the Park.
- Local graziers would be paid 70 per cent of the net revenue from trophy hunting within the Park, after such hunting is allowed.
- A Park Management Board would be constituted, consisting of high-level government functionaries of various departments, representatives of various local communities, and of NGOs active in the area.
- The Local Administration reserves all the rights to suspend all concessions and benefits allowed in this agreement if the conditions of effective wildlife protection and appropriate rangeland use are infringed.

Since the signing of this agreement there have been no significant problems with the local graziers. Based on the agreement, the WWF has prepared a management plan for the Khunjerab National Park (Ahmed 1996).

The Bar Valley (northern areas)

The Bar Valley is a remote pocket in the northern areas of Pakistan whose 1,100 inhabitants used to hunt ibex for food. Nineteen hunters of the valley would kill the animals and distribute the meat to all the households. When WWF approached them for conserving the ibex, the community came up with the demand for being compensated for the free meat they got from ibex hunting. Though they were promised a share in trophy hunting for which a fee of $3,000 each is levied, they still asked for immediate compensation. WWF therefore advanced them Rs 2,40,000 (about $6,000) through the local administration. In 1995, the government allowed three ibex for trophy hunting for $9,000. The people's share out of this at 75 per cent amounted to Rs 2,70,000 out of which the community paid back the amount advanced by the WWF (A. Ahmed, WWF, personal communication).

Protection of biodiversity in the Suleman Range

The Suleman Range supports the largest area of pure *chilghoza* or *Pinus gerardiana* forests in Pakistan, and also provides habitat to the endemic wild goat, the Suleman markhor *Capra falconeri*. Even though the trees yield edible nuts which are marketable, the local tribesmen were felling the forests through contractors for quicker returns and over-hunting the wild goat for meat.

In 1991, the WWF surveyed the area and diagnosed that the dependence of the tribesmen on felling the forest could be reduced if suitable land could be developed for agriculture and disputes resolved among individuals in the tribe. WWF offered financial assistance for renovation of an abandoned water channel. In addition, 7,000 apple trees, 5,000 of which were donated by the Baluchistan Forest Department, were distributed among the tribesmen. Irrigated agriculture and apple cultivation has considerably improved the economic conditions of the local tribesmen and about 200 forest owners have signed conservation agreements for protecting about 50 sq. km of *chilghoza* forest. Also, the killing of markhor is now reportedly a rare occurrence (A. Ahmed, WWF, personal communication).

Conservation through community participation in Gunyar village

Gunyar is a village of about 200 households in the Malakand Agency of the NWFP. As a response to natural calamities, the local youth have organized themselves into a voluntary association since 1987, rendering free service to the community. The association has 115 members who have agreed to abide by its manifesto. They elect their office bearers annually and work under the guidance of the village elders as well as the village general assembly. The Government of New Zealand has financed the construction of a community centre and a female handicrafts centre for the women who now act as focal points for their work. Since 1992, the World Conservation Union (IUCN), with a grant from the Canadian International Development Agency (CIDA), is assisting them in the rehabilitation of a formerly bare hillside, improvement of their agriculture and livestock husbandry and introduction of improved poultry. Efforts are also made to interest other government agencies as well as donors in the welfare of the village. All the work is done by the village activists free of any remuneration though small honoraria are paid to individuals who subsist on daily wages labour. IUCN mainly provides linkages with competent individuals in government departments, research institutes and universities in the areas where villagers need advice, training and the supply of requisite materials. The major impact of the project has been to increase the competence and confidence of the association to implement complicated tasks which would assist the village in effectively participating in government and donor programmes.

Other projects

Apart from the above, environmental rehabilitation projects assisted by the European Commission and the Asian Development Bank strongly emphasize the sustainable use of natural resources through participation of local communities from the planning stage (GONWFP 1995). A GTZ–KFW-assisted Siran Forest Development Project, started in 1991, seeks to improve timber harvesting and forest management and regeneration in the reserved and *guzara* forests of Siran and Agror–Tanawal Forest Divisions through community participation. A recently started three-year Global Environment Facility (GEF)-financed project, Maintaining Biodiversity in Pakistan with Rural Community Development, seeks to enhance the capacities of rural communities in 15 valleys of northern Pakistan to sustainably manage their wildlife species and habitats (IUCN 1996a).

Local community responses

The response of the local communities to offers of participation in the management of renewable natural resources inside and outside PAs has so far depended entirely on the magnitude and pace of economic returns to them. Where these are sizeable and likely to accrue in a year or two, the response is enthusiastic, as occurred in the Kalam Integrated Development Project, after the communities were taught how to grow and market their off-season vegetables. Difficulties arise in semi-arid areas, without sources of irrigation. In such areas, the only incentive which can be offered to the communities is income from the sale of grass from areas closed to grazing, or sale of firewood from shrubs which regenerate consequent to restrictions on grazing. But it is difficult to make the local communities agree to grazing closures, or even grazing control. Recently, attempts have been made to give communities a share in income derived from controlled harvesting of wildlife, and get their assistance in conservation. The results are not yet known.

Though the FDs have welcomed the participation of local communities in planting on treeless, privately and communally owned land, they have not so far made a major headway in this task in the commercial forests, except in the Kalam Integrated Development Project. The major difficulties have been to evolve the terms of a partnership which can ensure the sustainable management of forests, appropriate and sustained returns to the communities, and the acceptability of the terms of partnership by stakeholders often of widely divergent interests, such as, forest owners, forest right-holders, non-right-holder forest users and the government.

Major weaknesses and strengths in the attempts

The most important weakness in the initiatives taken so far in the participatory approach to environmental rehabilitation in Pakistan is their dependence on donor funding. This makes the spread of the approach conditional on the availability of grant aid, which is rapidly on the decline. The donor-supported projects are not yet sufficiently sustainable to permit the withdrawal of the donor without the collapse of CBOs. The major challenge appears to be to evolve a system of community organization which can be financed entirely from revenues earned from the natural resources without significantly reducing the current returns to the communities and to the government. This may be possible in the commercial forests where more effective protection, training of local people in forest operations, and introducing efficiencies in forest management, timber harvesting and utilization, and forest regeneration, may yield total benefits in excess of those being derived at present. It is difficult, however, to visualize a sustainable participatory approach to the management of natural resources in the semi-arid and arid tracts unless the government has the means to invest the required resources as the cost of environmental rehabilitation.

One conclusion however, appears to be inescapable; community participation in renewable natural resources rehabilitation would only be sustainable if undertaken as a component of a broad integrated rural development programme including as many facets of the lives of the local people as possible. Though forest trees, grasses and wildlife appear the most important considerations to the conservationist, they are often the least important from the viewpoint of the rural households whose predominant concerns are health, food, employment, education.

The major strength of the participatory integrated approach appears to be its promise in the task of the rehabilitation and sustainable management of the renewable natural resources. To most reflective foresters in Pakistan, the traditional authoritarian approach to forest protection is no longer viable. Organizing, training and empowering the local communities to assist in the protection, management and utilization of renewable natural resources appears to offer a viable alternative to the current system of a government agency attempting the task alone under prevailing local antagonism, or at best, total indifference.

Independent community/NGO efforts at conservation

History

In the area now comprising Pakistan, there does not appear to be a strong tradition of community participation in the management of forests, grazing lands, wildlife and fisheries. Rural communities, however, used to close their woodlands and grazing lands to tree-cutting and grazing for specified periods whenever they appeared to be over-exploited. The tradition of such closures to grazing still survives in remote tribal areas. Elsewhere, traditional community interest comes into play when people's shares in the proceeds from sales of timber have to be distributed or when a community transgresses into the grazing domain of another.

The great majority of Pakistani NGOs operate as welfare or relief organizations (Aftab 1994). The first massive application of integrated rural development through community organization was in the northern areas of Pakistan through the Aga Khan Rural Support Programme (AKRSP). Started in December 1982, the programme is now operating in six districts of northern Pakistan covering more than 1,000 villages (AKRSP 1995). Its coverage is comprehensive, encompassing social organization, women's issues, mountain infrastructure and engineering services, natural resource management (agriculture, livestock, forestry), human resource development, enterprise development, and credit and savings. Because of its great success, efforts are being made to replicate the experience in other areas: Sarhad Rural Support Corporation (SRSC) in the North-West Frontier Province, Baluchistan Rural Support Programme (BRSP) in Baluchistan, and the National Rural Support Programme (NRSP) operating in six regions of Pakistan.

Since the approval of the National Conservation Strategy in 1992, several rural development NGOs have been formed around interests in the improvement of irrigation and water resource management and forest tree planting. Almost all rural development NGOs include forest tree planting in their mandates. Several umbrella organizations in Pakistan endeavour to encourage the environmental and rural development NGOs. The World Conservation Union (IUCN), through its NGO/Community Support Unit, is actively involved in strengthening the organizational and environmental capability of NGOs, especially in Sindh and the North-West Frontier (IUCN 1996b).

Donor initiative has also assisted in the emergence of environmental and rural development NGOs. The forestry programmes assisted by the governments of Switzerland, Netherlands, Germany individually and the European Community as a whole include community organization as an essential component. Since the 1990s, the World Bank, the Asian Development Bank and the UNDP have also been encouraging the participation of NGOs in their environmental rehabilitation initiatives.

All these efforts have succeeded marvelously. On a larger scale, the work of umbrella organizations such as the AKRSP, NRSP, SRSC, BRSP, larger NGOs such as the Orangi Pilot Project, Sungi, Baanhn Beli and others, and on a smaller scale, the several small NGOs fostered by donor programmes, have brought about a considerable awakening for environmental rehabilitation among the rural populations.

Official policy towards such efforts

The Government of Pakistan is supportive of such initiatives. It is a member of IUCN and hence a partner in all environmental and rural initiatives taken by it through the National Conservation Strategy, Sarhad Provincial Conservation Strategy and other provincial strategies under formulation. All the other umbrella organizations for participatory rural development (AKRSP, NRSP, SRSC, BRSP) also have its blessings. In addition, it also encourages the involvement of other donor agencies in such ventures.

Major weaknesses and strengths in these efforts

The most important weakness of all these efforts so far is their dependence on donor interest and donor funding. Though all these initiatives claim to foster sustainability, even the most advanced of them is so far unsustainable without donor funding.

The greatest strength of the approach is ensuring the participation of the communities at the grassroots level in the identification of their problems and their resolution with their full participation. This approach provides an opportunity for resolving problems which have defied solution so far, e.g., the replacement of the complex, time-consuming and corruption-riven process of distributing the shares among local people from the proceeds from sale of timber by a simpler, quicker and more transparent system acceptable to the people as well as the government.

Constraints and opportunities for community-based conservation

Existing and potential constraints

Lack of sustainability: Lack of sustainability is the most important constraint for CBC. The conventional hope has been that community participation in conservation would increase the incomes of the local people and that their collective savings would be ploughed back in the development of the area. Even where untapped local resources were available, e.g., land and water in the northern areas, and their utilization through CBC almost doubled the incomes of the local people, it has not so far been possible to plough their savings into the development of the area, for a variety of reasons, mainly the paucity of profitable and secure investment opportunities. Few technologies are available for increasing the productivity of land significantly under a climatic regime dominated by low and highly erratic rainfall. Even under irrigated conditions, small landholdings set the limit on significantly increasing the incomes of the local people.

Under such conditions, it has not so far been possible to persuade the local people to finance the superstructure created for community participation under various initiatives. Sustainability would appear to demand that a country-wide superstructure be available to which CBOs at the grassroots level could latch on for continual support. Though the local government departments of the provinces could perform this function, in their present form, they have neither the will nor the capacity for this task.

Lack of enthusiasm of the line departments of the government: Conservation has so far been the domain of the line departments of the government, all of whom, including extension agencies, are used to operating in the authoritarian mode. Though the FDs have taken a lead in CBC, barely 10 per cent of its officer cadre may so far be convinced of the merit of the approach. Other government line departments concerned with conservation, such as, Agriculture, Livestock and Irrigation, may only have a nodding acquaintance with the approach as part of a donor-assisted project. A massive re-education of the conservation-related government line agencies would be essential to enable CBC to take off in real earnest.

Uncertain land tenure and rights of communities: A drive for universalizing CBC is no doubt hampered by uncertain and inequitable land tenure, and by rights of communities recorded in the 19th century and therefore no longer realistic. But these are highly refractory perennial problems. A start has to be made somewhere. The current approach is to start from areas where such disputes are minimal, and to gradually expand the coverage of the programmes around such foci, while avoiding areas of acute conflict, especially those likely to result in bloodshed. This approach appears to be sound for the future too.

Inadequate finance: Under the present conditions, donor funding is the major source of finance for this endeavour, with local funds being scarce. This is likely to remain so for the foreseeable future too.

Existing and potential opportunities

Dialogue and mediation: The main problems with tenurial disputes between the government and the local communities, and among the communities and individuals in a community, are the inflexible positions taken by the parties to the disputes. CBC provides a mechanism of dialogue and mediation by which the differences between the parties to the disputes can be discussed and progressively narrowed down.

It must be realized that the existence of differences need not obviate the adoption of CBC in all cases. For example, the approach has worked very well in the Protected Forests of Kalam under the Swiss-assisted Kalam Integrated Development Project even though the ownership of the forests is disputed between the government and the local people. Through its emphasis on integrated rural development, the project was able to benefit the local communities so much that they considered it prudent to overlook the differences on chronic issues such as the ownership of forests.

Proposals for the future

Policy and law

Both the Forest Policy of Pakistan (1991) and the National Conservation Strategy (1992) favour CBC, and forestry laws in force over most of the country do not debar the process. Availability of information on policies and laws which have facilitated the process in other countries would enable Pakistan to improve its policies and laws further.

Promoting existing or new mechanisms for fostering CBC

Dialogue: Dialogue is the universal approach used in CBC in Pakistan. The AKRSP and the initiatives it has spawned, mostly use a stylized system of dialogues to create CBOs. Donor-assisted projects are more flexible, accepting the wide diversity of socio-economic conditions in Pakistan.

Creation of stakes for conservation: In Pakistan, where people are generally individualistic, and communities are mostly riven by disputes, it is essential to discover potential stakes for conservation. To appeal to the local people, social organizers must address the immediate problems of the people. Few rural communities are inspired by visions of sustained yield forestry or rotational grazing. They are more concerned with employment, health, education, water supply, farm to market roads, etc. The current trend in Pakistan therefore is to undertake CBC as part of a wide-ranging integrated rural development programme.

Benefit-sharing: Ignorance about benefit-sharing is the most important single factor which has hampered the extension of CBC to the commercial forests of Pakistan. It is easy for the forest departments to allow the use and sale of grass and fuelwood shrubs from previously blank privately owned land to the owners of the land. It is quite a different matter to allow them any significant benefits in commercial forests which they do not enjoy as a legal right. Even if the functionaries of the forest departments are convinced of the need for benefit-sharing with the local people, they may first have to amend their laws and rules and convince their finance departments before they would be allowed to take necessary steps towards benefit-sharing. And while doing so they would always be liable to charges of discrimination from the local people.

To enable Pakistan to devise systems of benefit-sharing in forests of various legal categories, under diverse socio-economic conditions, it would be essential to study the various systems of benefit-sharing in other countries under similar conditions and their successes and failures.

Legal and political empowerment: What has been stated for benefit-sharing, applies equally to legal and political empowerment. Most Pakistani foresters think that they are already practising participatory forestry because they frequently talk to the local people on the benefits of conservation. Not many of them realize that CBC does not come about unless the organized local communities are legally and politically empowered to participate in the management of the forests. But this is easier said than done, particularly in Pakistan which has little experience of this process. Again the first step is to study what other countries in the region have attempted successfully and then to endeavour to initiate steps in this country.

Tenurial certainty: Tenurial uncertainty is often cited as a reason for not proceeding with CBC. Such views do not take into consideration the difficulties in changing tenure in the short run. In Pakistan, tenure has evolved over generations and has not materially changed in spite of several land reforms. If we wait for a significant change in tenure before expanding CBC, we may never be able to make a start. We must therefore attempt to make the best we can of the opportunities and flexibilities available in skirting the prevailing tenure through the process of dialogue, mediation and conflict resolution.

Education and training: For Pakistan, education and training is the most important action item in its quest for CBC. Its formal forestry education focuses almost entirely on physical and biological sciences. The same is true of animal husbandry and irrigation. Though rural sociology courses are taught in agricultural education, their application is hardly covered at all. It is essential to introduce significant contents of basic and applied social sciences in all disciplines relevant to the conservation of renewable natural resources.

Introduction of social sciences as well as their application to resource conservation through community participation must be included in the syllabi of all educational institutions dealing with natural resources and taught competently and creatively. But formal education alone has a long gestation process for changing outlooks. To bridge the gap, it is essential to offer massive short-term training to the entire cadre of the current employees of government line departments dealing with natural resources. Such training can most profitably be organized around the nearest donor-assisted CBC programmes relevant to the concerned departments.

REFERENCES

Aftab, Safia. 1994. *NGOs and the Environment in Pakistan.* Sustainable Policy Institute, Islamabad.

Ahmed, A. 1996. Management Plan, Khunjerab National Park. WWF–Pakistan, Lahore.

AKRSP. 1995. *Thirteenth Annual Review.* Aga Khan Rural Support Programme, Gilgit, Northern Areas, Pakistan.

Duke, Guy. 1997. Extracts from poster presented at the World Conservation Congress Montreal, October 1996. Unpublished.

GOP. 1990. *Census of Agriculture Vol. 1.* Economic Affairs and Statistics Division, Government of Pakistan, Lahore.

———. 1992. *Forestry Sector Master Plan.* Government of Pakistan/Asian Development Bank/UNDP.

———. 1996. *Economic Survey 1995–96.* Finance Division, Economic Advisor's Wing, Government of Pakistan, Islamabad.

GONWFP. 1995. *Palas Conservation and Development Project (PC-I).* Department of Forests, Fisheries and Wildlife, Government of North-West Frontier Province, Peshawar.

———. 1996a. *Pak-Swiss Kalam Integrated Development Project: Highlights.* Department of Forests, Fisheries and Wildlife, Government of North-West Frontier Province, Peshawar.

———. 1996b. *Social Forestry Project Malakand-Dir, Report of Fact-Finding Mission.* Department of Forests, Fisheries and Wildlife, Government of North-West Frontier Province, Peshawar.

IUCN. 1992. *National Conservation Strategy.* IUCN–The World Conservation Union–Pakistan, Karachi.

———. 1996a. Maintaining Biodiversity in Pakistan with Rural Community Development. Annual Report. IUCN–The World Conservation Union–Pakistan, Karachi.

———. 1996b. *The Pakistan Programme.* IUCN–The World Conservation Union–Pakistan, Karachi.

Jan, A. 1965. *Working Plan for Guzara Forests of Haripur Forest Division (1965 to 1974).* Office of the Chief Conservator of Forests, NWFP (Territorial), Peshawar.

Mian, A. and **M.Y.J. Mirza.** 1993. *Pakistan's Soil Resources.* Environment & Urban Affairs Division, Government of Pakistan and IUCN–The World Conservation Union.

Ouerghi, A. and **C. Heaps.** 1993. *House-hold Energy Demand: Consumption Patterns.* Energy Wing, Government of Pakistan.

Roberts, T.J. 1977. *Mammals of Pakistan.* Earnest Benne Ltd., London and Townbridge.

Stebbing, E.B. 1921. *Forests of India.* John Lane, The Bodley Head Ltd., London.

8

The practice of community-based conservation in Sri Lanka

Avanti Jayatilake, Nirmalie Pallewatta and
J. Wickramanayake

Introduction

Sri Lanka, a tropical island with a land extent of 65,610 sq. km, is located close to the south-eastern part of the Indian peninsula. The country possesses rich scenic beauty, extensive waterways, mineral and biotic resources and an ancient cultural heritage.

About 75 per cent of the island is a flat lowland peneplain (average 75 m above sea level) that surrounds the central hills, comprising steeply rising second (75–5,000m) and third peneplains (500–2,500). The northern and eastern parts of the peneplain have isolated hills sometimes rising up to 600m or more (MTEWA 1995). Some of these isolated hills have valuable biological resources unique to these sites. The country's relatively drier areas, agro-climatically described as the dry zone of Sri Lanka, lie exclusively in this first peneplain covering north-western, northern, eastern and south-eastern parts of the island. This area is dotted with about 12,000 human-made traditional irrigation water storage/reservoir systems that collect rainwater during the NE monsoons. Rainfall in the country on an average ranges from less than 1,250 mm in the drier areas to about 2,500–5,000 mm in the central hills. Although the mean annual temperature in the lowlands is around 27ºC, the higher elevations of central hills often record 10º–15ºC of daily variations.

Socio-economic profile

Sri Lanka has a multi-ethnic society. Sinhalese make up 76 per cent of the population while Tamils (15 per cent), Muslims (7 per cent) and other groups (2 per cent) constitute the rest. Buddhism is the predominant religion. Hinduism, Islam and Christianity are the other religions practised. Total population is 18 million, growing at an estimated average of 1.3 per cent annually. With the growth in population, unmistakable signs of environmental stress are appearing in various parts of the island. The projected population of 25 million by the year 2040 will no doubt create unprecedented demands for food, fibre, energy, land and water for settlement, agriculture, trade and other developmental activities.

The country has a literacy rate of 90 per cent, a low infant mortality rate and a high life expectancy. Agriculture accounts for 25 per cent of the country's Gross Domestic Product (GDP), nearly half of the total employment, and 40 per cent of export earnings. More than 90 per cent of rural communities depend on agriculture (paddy and subsidiary crops) and on export-oriented tea, rubber, coconut and other minor export crops. The industrial base is narrow and remains relatively small as an employer compared with the agriculture and service sectors.

The governments are elected democratically since independence from British rule in 1948. There is an executive presidency (voter-elected) and a parliamentary system of people's representation. Constitutional reforms instituted in 1988 have devolved a considerable degree of power to the provinces with elected or appointed leaders at the local level. However, Sri Lanka's economic performance has fallen behind compared to the early 1960s. The economy grew rapidly during the late 1980s with liberalization, when substantial inflow of foreign assistance took place. However, ongoing civil disturbances have undermined the development efforts throughout the country.

Legal ownership of land

Over 95 per cent of the forest lands have belonged to the state since British times, and are under the jurisdiction of the Wildlife Conservation Department, the Forest Department (FD) and the Land Commissioner's Department in current times. Rivers, major reservoirs and tanks that support most of the wild biological diversity belong to the Irrigation Department, the Mahaweli (River) Authority and the Agrarian Services Department.

Biological diversity profile

An array of ecosystems, covering diverse terrain and climatic types, have resulted in high biological diversity in the country. Available information indicates that Sri Lanka has greater biodiversity per unit area than any other country (except Indonesia) in Asia (Baldwin 1991). It is one of the 11 areas identified by the Committee on Research Priorities in Tropical Biology as demanding special attention because of its high biological diversity, endemism and vulnerability to habitat destruction. The lowland rainforests, the forest ecosystem with by far the highest levels of biodiversity, occupy less than 10 per cent of the forest land in the country. The remaining forest cover is concentrated in the dry zone and hill country wet zone.

Community relations to wild biodiversity

Traditional practices

'The key to success of indigenous way of life is sustainability of natural systems' (Medawewa 1994). Community conservation practices are closely linked to Buddhist and Hindu cultures of Sri Lanka. For example, Sri Maha Bodhiya, a seedling from the main *Bo* (*Ficus religiosa*) tree in India, where the Great Buddha attained *Nirvana*, gifted to Sri Lanka about 2,500 years ago and planted at Anuradhapura, is among the world's oldest trees. Some large trees (e.g., *Mesua thwaitsii*) are worshipped as abodes of deities. Moreover, the hydrological civilization that prevailed over several thousands of years has left some very valuable lessons for natural resource use and management. This included an extensive network of irrigation tanks which not only ensured continued supply of water in the dry zone but also were a repository of biodiversity in the area. An interesting tradition in the country was the naming of villages or townships after plant or animal species (Weerawardana 1994). For instance, Colombo may have derived its name from the *Kolon* tree (*Nauclea cordifolia*) (*Kadamba* in Sanskrit and Pali) or a large mango tree that was a prominent landmark. The ancient sea port of northern Sri Lanka was named 'Jambukola' referring to the leaves of the *Jambu* tree (*Szygium jambos*). A classic example of the community's co-existence was *Kurulu Paluwa* (meaning 'bird's area') in the paddy fields. This was a crop protection measure; a separate strip of land near the jungle was sown only for the birds, while various environmentally friendly methods were used to repel them from other field sites. Trees grown along the edge of the fields were also for the birds to rest and eat as the birds helped to control pests. In the case of *chena* (shifting cultivation), large trees were not felled when the forests were cleared. The fruits on the upper branches of tall trees were also left unplucked. This is not merely due to the danger of climbing to reach them but also recognition of the needs of other birds and animals (Weerawardana 1994).

Current interaction

Many communities still directly depend on the natural resources for their basic biomass requirements as well as livelihood, e.g., fodder, fruits, fuelwood and others. Hunting for food (though prohibited in the PAs) also continues for both personal consumption and sale in the local markets. Often destructive means are also employed for these extractions, such as, setting forest fires to chase animals, and use of toxins to poison fishes. Illicit fellings of forests for high-value timber is also not uncommon, but mainly by henchmen of politicians and wealthy private entrepreneurs living in towns. Encroachment, however, is more a result of poor communities trying to live off subsistence agriculture on lands adjacent to forests.

Threats to wild biodiversity

According to 1989 estimates, the closed tree canopy forest cover in Sri Lanka has declined to 20.2 per cent of the land area (1992 forest map by Forest and Land Use Mapping Unit) from 44 per cent in 1956. It is estimated that an average of 42,000 hectares per year of forest land has been lost during the period 1956 to 1983, and thereafter an average of 54,000 hectares per year.

Population growth and heavy dependence on land-based economic activities have posed major threats to the country's biological diversity and wilderness. The wet zone of the country (occupying just 24 per cent of the country) is under the greatest pressure because it is settled by 55 per cent of the population. Until recently, governments only considered these areas of rich biodiversity as resources to be exploited or converted into other socio-economic uses, leading to their steady degradation. There is an added pressure of poaching, illicit extraction of timber, mining, encroachment and habitat destruction in the forest ecosystems.

Other critical habitats such as coral reefs and wetlands are also severely threatened as conservation efforts have so far been directed mainly towards the forests with tree canopy and grasslands. This is evident from the fact that only one wetland ecosystem (Bundala) and one coral reef area (Hikkaduwa, on the southern coast) are protected under conservation laws.

Extraction of turtle eggs is common in areas where turtles are nesting. Turtles entangled in the nets of fishermen are often killed at sea and their shells sold for ornamental products. These activities are now prohibited by law. Coral mining, though decreased to some extent after this law, continues in some stretches of the coastal belt. The very lucrative export industry in ornamental fish is also a very serious threat. Certain natural products such as nests of birds, corals, endangered fish varieties, some endemic plants, plant parts, moths, butterflies, leaf insects, live spiders, and turtle shell products are still being illegally exported.

Institutional and legislative aspects

The state owns 82 per cent of the land in the country (Land Commission 1990). It is interesting to examine whether this state tenure has helped to conserve wild biodiversity or has contributed to its destruction. As in many countries of this region, it has worked both ways. The management authority of this land has been diffused among a large number of state agencies with differing objectives and mandates. The absence of a national land use plan, poor coordination among authorities and the conflicting political priorities of successive governments have resulted in major destruction of land-based resources.

The state agencies in charge of forest ecosystems and wildlife PAs, the Department of Wildlife Conservation (DWLC) and the Forest Department (FD), have been placed under two different ministries in the 1994 institutional reforms. As it is, both these agencies lack adequate resources, skills or capacity to manage areas under their jurisdiction. The placement of these two agencies in two different ministries has put considerable strain on already scarce governmental resources, and affected coordination and exchange of information between them. This fact has also been recognized by the Forestry Sector Master Plan (FSMP) adopted in 1995. The Ministry of Environment (set up in 1988) is entrusted with the task of protecting Sri Lanka's environment and is currently engaged in formulating the Biodiversity Action Plan (BAP).

Both DWLC and FD have so far largely adopted an enforcement attitude towards the management of these lands. Until recently they considered the communities living around the forests and other ecosystems as threats. The only interaction with the community most often has been to investigate poaching or removal of timber without permits. On the other hand, the authorities always blamed the inadequacy of laws for their failure to punish major offenders who violated the laws, often with political patronage. This attitude of the enforcement agencies has led to the development of major distrust among communities.

Aspects related to policies and practices

Lack of policy directions and practices that address biodiversity conservation results in eventual destruction of biological resources. For example, promotion of high yielding varieties; recommendation of higher doses of fertilizer and pesticides without monitoring; trade (especially export); conversion of extensive areas into monoculture under major irrigation schemes or into commercial varieties like tobacco; use of ecologically sensitive areas like hill slopes for cultivation; setting up large-scale plantations in habitats important for threatened species; filling up wetlands for development; destruction of mangroves for aquaculture; and lease of river and tank beds for mining activities, have resulted in serious impacts on the biodiversity of the country.

Until 1990, the country did not have a clearly defined wildlife policy while the forest sector has had many policy statements. Most of these have been issued by the FD. However, these initial statements were more the objectives of the FD than a national policy on forest. In fact, they had very few references to wildlife, watershed, environmental management and community involvement. The emphasis has now changed with the current comprehensive National Forest Policy (NFP) approved in 1995. The new policy has recognized the role of the public sector, local resource users and the role of NGOs in forestry. Table 8.1 gives the proposed distribution of roles between the governmental and non-governmental sectors in forestry.

Table 8.1: Proposed distribution of roles in forestry management between the government and non-governmental sectors

Forestry Partners	*Protected Areas*	*Multiple-use Natural Forests*	*Home Gardens and other Agro-Forestry*	*Forest Plantations*	*Industrial Production*
National authorities	Policy and legislation, Finance and audit	Policy and legislation, Finance and audit	Policy and legislation, Access to Finance	Policy and legislation, Access to Finance	Policy and legislation, Access to Finance
FD, FPU, DWLC	Policy formulation, Macro-level planning, Enforcement, Management, Monitoring and extension	Policy formulation, Macro-level planning, Enforcement, Management, Conservation, Monitoring and Training and Extension	Policy formulation, Macro-level planning, Monitoring, Management and Conservation, Training and Extension	Policy formulation, Macro-level planning, Leasing, Enforcement, Monitoring, Training and Extension, Conservation, Management	Policy formulation, Macro-level planning, Extension, Supply of wood
Wildlife Trust	Management of income-generating activities, Patron of conservation, Education of the public	Support in conservation			
Other state institutions	Law enforcement, Industry licensing, Land-use monitoring, Education, Research, Collaboration in extension	Law enforcement, Industry licensing, Land-use monitoring, Education, Research, Collaboration in extension	Law enforcement, Industry licensing, Land-use monitoring, Education, Research, Collaboration in extension	Law enforcement, Industry licensing, Land-use monitoring, Education, Research, Collaboration in extension, Environmental monitoring	Policy formulation, Law enforcement, Industry licensing, Environmental monitoring, Education, Research, Collaboration in extension

Table 8.1 continued

Table 8.1 continued

Local people	Participation in conservation, Authorized utilization	Participation in management and conservation, Authorized utilization, Protection	Management and utilization, Conservation, Protection	Non-resident cultivators, Hired labour, Protection	Labour services, Supply of wood
NGOs	Extension, Mobilizing and facilitating, Capacity and skill building, Participation in conservation and management, Advocacy of private rights, Law enforcement, Monitoring	Extension, Mobilizing and facilitating, Capacity and skill building, Participation in conservation and management, Advocacy of private rights, Law enforcement, Monitoring	Extension, Mobilizing and facilitating, Capacity and skill building, Advocacy of private rights	Extension, Mobilizing and facilitating, Capacity and skill building, Advocacy of private rights	Monitoring
Other non-state sector (including forest industry and estates)		Support to conservation, Authorized utilization, Harvesting transport utilization	Harvesting and transport utilization	Management, Harvesting and transport utilization	Management, Supply of wood for manufacturers

Source: FSMP 1995.

Official efforts at conservation

Although traditional practices valued conservation, official recognition of the importance of conservation is relatively recent. Changes in traditional attitudes towards conservation and land use practices began with the colonial administration. Until the early 19th century most of the hill country and low country dry zone were forested. Only the extreme south and south-west were cultivated with paddy and coconut plantations. From 1830, vast areas covering the middle altitudes of the central hills were cleared for coffee plantations, which were later replaced with tea in the 1850s. Forest clearance in the dry zone began in the late 1860s and accelerated with the gradual introduction of settlements. Land use changes were accelerated by the Wasteland Ordinance in 1840 whereby all the land that was not under cultivation (or had a private ownership) at that time became the property of the British Crown. This was about 95 per cent of the total land as most areas were under community use and management. This very seriously affected the traditional relationship of the communities with these resources. In 1873, eminent botanists Hooker and Thwaites expressed concern and warned about denuding hills for large-scale plantations. Eventually in 1938, forests above 5,000 feet were decided to be kept as climatic reserves; this was incorporated into the Forest Policy through an amendment. The emphasis on wildlife protection was officially recognized in the Fauna and Flora Ordinance of 1937 whereby Strict Nature Reserves, National Parks and Intermediate Zones (for controlled hunting which was later removed), in the Crown Lands were declared. Lands with private ownership having significant wildlife were declared as sanctuaries. Table 8.2 contains a summary of the policies and laws on forest conservation.

However, felling of natural forests for timber extraction continued long after independence. Finally, in 1972, a sub-committee was appointed to examine the reasons for the public outcry resulting from the decision to fell a prime natural forest, Sinharaja, unique for its biodiversity. The government that came into power in 1977 pledged to stop this practice and by mid-1980s, a Forestry Master Plan was produced. But a strong NGO protest claiming that the Plan does not adequately address forest conservation led to an Environmental Impact Assessment (EIA) on the Forestry Master Plan. This resulted in a major revision of the Plan where the government responded with specific immediate actions, including a moratorium in 1990 on logging

Table 8.2: Chronology of key policies and laws concerning conservation of forest biodiversity

Year	*Policy/Law (Competent Authority)*	*Provisions for Biodiversity Conservation*
1848	Timber Ordinance No. 24	Reservation of forests, largely for timber production
1873		Hooker advocates protection of natural forests above 5,000 feet as climatic reserves
1885	Forest Ordinance No. 10 (Conservator of Forests)	Protection of forests and their products in reserved forests (including stream reservations) and village forests, primarily for sustained production; also, protection of wildlife in sanctuaries
1907	Forest Ordinance No. 16 (Conservator of Forests)	Protection of forests and their products in reserved forests and village forests, primarily to provide for controlled exploitation of timber
1929	First authoritative forest policy statements	Preservation of indigenous flora and fauna
1938	*Amended	*Clearing of forests prohibited above 5,000 feet. Plantations to be gradually converted to indigenous species
1937	Fauna and Flora Protection Ordinance No. 2 (Director of Wildlife)	Protection of wildlife in national reserves (i.e., strict natural reserves, national parks and intermediate zones comprising only Crown Land). In sancturaries, habitat protected only on state land while traditional human activities may continue on privately owned land
	*Amendment Act No. 44 in 1964	*Nature reserve and jungle corridor incorporated as categories of national reserve
	*Amendment Act No. 1 in 1970	*Intermediate zone to provide for controlled hunting removed from ordinance
	*Amendment Act No. 49 in 1993	*Refuge, marine reserve and buffer zone as additional categories of national reserve

Table 8.2 continued

Table 8.2 continued

Year	*Policy/Law (Competent Authority)*	*Provisions for Biodiversity Conservation*
1953	National Forest Policy *Re-stated in 1972 and 1980	Emphasis on conserving forests to preserve and ameliorate the environment, and to protected flora and fauna for aesthetic, scientific, historical and socio-economic reasons
1969	Unesco Biological Programme and 1975 Unesco Man and Biosphere Programme	Arboreta representative of the main bioclimatic zones established and demarcated in forest proposed reserves
1982	Mahaweli Environment Project	Network of protected areas established to mitigate impacts of Mahaweli Development Projection wildlife and to protect catchments in the upper reaches of Mahaweli Ganga
1988	National Heritage Wilderness Areas Act No. 3 (Conservator of Forests)	Protection of state land having unique ecosystems, genetic resources, or outstanding natural features, in national heritage wilderness areas
1990	National policy for wildlife conservation (approved by cabinet)	Objectives include the maintenance of ecological process and preservation of genetic diversity; *ex situ* conservation recognized as important for threatened species (See section 3.4.3)
1995	National Forestry Policy (approved by cabinet)	Over-riding priority given to conservation of biodiversity and protection of watersheds (see section 3.4.3)

in the wet zone until the conservation value of remaining natural forests had been assessed. This moratorium still continues; and a National Conservation Review with inventories of species of plants and animal groups on a geo-specific database has been carried out.

In 1970, the Department established biosphere reserves (BRs) amounting to 120,000 hectares within the lands available to them, to help protect them better from exploitation. Sinharaja (now a World Heritage site) was declared a National Heritage Wilderness area in 1988.

National reserves (NRs) (415,000 hectares) and sanctuaries (205,000 hectares) were established to protect wildlife under the jurisdiction of the DWLC. Sanctuaries have private lands within the declared areas and restricted economic activities are allowed. No private properties are allowed within NRs. There are five types of NRs established under the Fauna and Flora Protection Ordinance (FFO) (1948). They are (in order of legal protection): Strict Nature Reserves (SNRs), National Parks (NPs), Natural Reserves, Jungle Corridors and Intermediate Zones. With regard to wildlife protection, the next major step after the introduction of FFO and the setting up of the DWLC was the Mahaweli Environmental Project (MEP) in 1982. This project, formulated following the EIA carried out on the now famous Mahaweli Project, resulted in the establishment of a network of PAs.

The government also developed a Wildlife Policy in 1990, in response to a recommendation made in the National Conservation Strategy (NCS) (1988). Sri Lanka has also signed several international conventions relating to biodiversity protection. The most significant of them is the Convention on Biological Diversity under which the government is meant to prepare a Biodiversity Action Plan through a participatory process.

The Central Environmental Authority; set up under the National Environmental Act (NEA), 1980, has taken some pioneering steps to conserve biodiversity. The Wetland Conservation programme initiated in the early 1990s to coordinate and direct the activities affecting wetlands, has given a scientific direction to conservation programmes in these areas. The EIA process introduced under NEA makes it mandatory for all proposed major development projects to study and report on significant environmental impacts. However, there is still more potential for the use of EIA.

However, one major reason for all these not being very effective so far has been that there has been little or no involvement of the local communities in either the formulation or implementation of most of these plans. The authorities generally believe that communities lack scientific knowledge, or the skills needed to make useful decisions and therefore it is not worth involving them in the design stages. Once designed, the projects and activities are handled by the implementing agencies through their officials placed in the field. These field officers most often do not have a good relationship with the community, as they have assumed the role of protectors of the resources on behalf of the government. Until recently, many officers did not believe that they needed to communicate or coordinate with the communities to exercise their authority.

Conflicts between people and these agencies are clearly evident at various levels. One of such serious conflicts is that of encroachment (in the hope of subsequent regularization), especially in the forests under the Land Commissioner (Government Agency). However, the encroachments in the FD and DWLC managed areas had reportedly gone down during the period of 1956 to 1983, but no recent estimates are available. The continued use of resources by the people living adjacent to the officially protected areas, in the absence of available alternatives, also creates these conflicts.

There have been some efforts towards involving the people in the recent past but most of these were not carried out with actual appreciation of the concept but under donor (funding agency) or other influences. One such effort was in the period 1980–92 under the Asian Development Bank (ADB)-assisted Community Forestry Project. This, however, fell short of institutionalizing community participation. The programme had limited success because of its top-heavy approach and for not reflecting people's aspirations. With the forest policy now endorsing community involvement (with the qualification 'wherever possible'), the FD is looking at avenues for making it a reality through legal and institutional changes. In this regard the concept of social forestry where lands are leased to communities to grow certain specified species and own the resulting forests, is being tried out. The DWLC is however still hesitant to accept any such programmes, and the local communities continue to be distrustful of them.

An encouraging development in this regard has taken place with the Coast Conservation Department (CCD) as a part of the Coastal Zone Management Plan (CZMP) on an experimental basis. The CZMP (1990) recognizes the need to devolve coastal resource management responsibilities to local authorities and initiated a Special Area Management (SAM) plan in some coastal areas. Hikkaduwa, a developed tourist town on the south-western coast, and Rekawa, a rural community living along a coastal lagoon with ecologically sensitive natural resources, were identified as SAM sites (see the following sections for details).

Independent community/NGO efforts at conservation

In Sri Lanka there have been small farmer development, health and sanitation, housing, irrigation and water management, women's development and adult education issues with community participation as an important component but such efforts in natural resource conservation have been initiated only in recent times.

CBRM projects

The Community-based Resource Management (CBRM) project of the Natural Resources Environment and Policy Project (NAREPP) of the United States Agency International Development (USAID) was launched in late 1991. The starting point for a CBRM is the recognition that communities have a knowledge base gathered from experiences of living with nature, social interaction with other humans and the challenges of livelihood. This stock of knowledge is not readily apparent. One of the initial agreements among those launching a CBRM is that the communities need to be assisted to articulate and systematize their experiences; to share and analyze and, perhaps, improve it by incorporating appropriate new knowledge from outside.

The objectives of CBRM are: Local-level resolution of biodiversity conservation and environmental problems through meaningful participation of affected communities and the collaboration of authorities; public and private sector cooperation and awareness building among policy-makers of the usefulness of the CBRM approach.

The process of CBRM includes a pre-CBRM phase where the major stakeholders are identified, the concept of the CBRM introduced, identification and/or formation of Community-based Organizations (CBOs) takes place, awareness and education is carried out, and major project strategies are selected. The next phase begins by implementation of activities addressing the issues at hand, institutional development of CBOs, promotion of new CBOs, strengthening of existing ones (e.g., through training in financial management, leadership strategies, technical assistance, material support, etc.), studies and research (e.g., on human–elephant conflict, medicinal plants, problems of housing in filled-up wetlands). Finally, there is the formation of CBO apex bodies as instruments for resource management.

Two areas where CBRM is being tried out are the Kahalla Pallekele Sanctuary in the North-Western Province, and the Ritigala Strict Nature Reserve in the North Central Province.

Coastal Resource Management Projects (CRMPs)

As mentioned earlier (in the section on official efforts at conservation), the Coastal Conservation Department has initiated CRMP in some areas, to generate community participation in conservation. Two prominent examples are Hikkaduwa and Rekawa, both on the south-western coast of Sri Lanka. These areas were taken up under the Special Area Management SAM Plan of the CCD, and the CRMP projects launched thereafter. The major features of these efforts are explained below. (See also Ekaratne et al. in this book; DeCosse and Jayawickrama in this book.)

The CCD and the CRMP played the role of catalysts initiating the resource management process in these areas by mobilizing the local communities (RSAMCC 1996). The project was initiated with the formation of the area profile. A coordination committee was created consisting of the major stakeholders. This body oversees the preparation of the profile, helps in the formation of the management plan for the area, and is ultimately involved with the implementation of the plan.

To begin with, both the communities were very sceptical about the programmes given the prevalent distrust of government agencies. However, the attitudes are now slowly changing after having separately influenced certain major decisions on investments planned for the areas through this mechanism. The project authorities ensured that the voice of the community representatives is not suppressed by the outside 'experts'. This has not only given the community more confidence but has also increased their trust on the project staff and the government officials involved. But for these opportunities, the community may not have continued its participation for long and the committee would have become another bureaucratic exercise. However, in either of the areas the communities have not yet assumed the entire opportunity to manage the area as well as the biodiversity.

Two major advantages of this approach have been, the involvement of the local authorities (rather than the central ones), who are closer to the communities as well as a long time and effort taken to elicit actual participation of the people. Both these ensured adequate and prior consultations with the major stakeholders.

Compared to CBRM (predominantly NGO-led), the earlier SAM approach had a heavier emphasis on the government agencies. Therefore, the progress made under the different Divisional Secretaries (DSs) (who are the heads of the coordinating committees) was very much dependent on the interest taken by the individual officers. In addition, there were no established NGOs involved with the process at any given stage. The disadvantage of this aspect was felt when the project at one point experienced severe obstacles due to constraints in funding, and lack of participation from all sections of the local community.

Efforts are now being made to build new or encourage existing CBOs. These efforts have been met with mixed success in the two areas. The Rekawa rural community has reacted positively by taking necessary steps to get together under a common leadership acceptable to the majority of the community. However, this has not happened yet at Hikkaduwa, reportedly because the community is more fragmented and heterogeneous. The less-privileged sections have not accepted leadership of the elite, which to begin with worked with the project staff and volunteered to organize the community into a foundation.

Turtle protection

Another small-scale community-based activity was started at the Rekawa area with the aim of protecting turtles, by a British NGO, along with several national NGOs. This programme, however, did not receive much community acceptance or success as the communities viewed it as an effort largely focused on conservation without offering any alternatives to their economic problems.

Both the above efforts however were initiated with external donor funds and are now finding the implementation difficult as the support for that has not yet come in.

Constraints and opportunities experienced in implementing CBC efforts

It has to be realized that all communities have had some interaction with the government and other agencies, and many have not been very pleasant. Therefore, before commencing the project it is important to understand these past experiences and keep in mind the fears and scepticisms that have arisen from them.

Community members at the SAM site Rekawa had had unpleasant interactions with the government agencies in the past and had been seriously affected by some ill-conceived projects implemented in the area. Therefore, the community reacted by ignoring the presence of the SAM project staff who were in close association with the GOSL officials. However, a marked change in their attitudes resulted when the project staff's intervention led to the coordinating committee offering them an opportunity to influence an important decision on aquaculture in the area. After studying the proposals and EIAs, the community rejected the proposal to safeguard both the local environment and livelihoods. The mistrust of the community towards the authorities and the project was eventually partly overcome because of such rapport-building efforts.

The Hikkaduwa experience has shown that in any community-based effort it is very important, first, to take into account the disparities and differences within the community so that all sections' interests are adequately addressed. Second, to build the project such that it aims at ultimately improving or enhancing the existing capacity of the community to equitably share the benefits and responsibilities of the implementation, as also to identify with the process itself. In Hikkaduwa, the coordinating committee (under the chairmanship of the DS), faces the danger of not surviving beyond the project as there is no organized community representation responsible for follow-up.

Compared to the above two coastal sites, the initiative at the Kahalla Pallekele and Ritigala sites was by NGOs, therefore, there was greater acceptance from the communities. It was only after the community expressed interest in resolving the problems in consultation with the government agencies that meetings were organized and facilitated by the NGOs. However, unlike the CRMP–SAM project, the NGOs have not so far been able to establish a permanent body/committee to coordinate the matters with the authorities on a regular basis. However, strategies are now being drawn up to bring the government agencies into the scenario at a regular basis. One of the main advantages of this process is that it has no direct project personnel in the field, and the community does not depend on outsiders to carry out the tasks. In Ritigala, for instance, activities have progressed enough to allow the community leaders to take responsibility of managing and making decisions collectively and the facilitating NGOs have withdrawn to take an advisory role if required. It is clear that the degree of sustainability of the project is considerably higher under such circumstances.

The way ahead

Though there are many such efforts in the country, one major hindrance they face is the lack of policy support and commitment from the government. The policies and laws endorsing government authority and affecting community rights to influence decisions and manage natural resources need to be seriously reviewed and amended. It could start at the provincial levels as it is easier to bring about changes at such levels.

The government should facilitate the formation of partnerships among local authorities, community organizations and private sector entities. This will require a change in attitude such that the capabilities of these partnerships is recognized and power and responsibilities are devolved to the institutions thus developed.

Programmes that are currently being implemented need to give greater attention to the issues of community ownership as well as the social, economic and ecological sustainability of the process.

The definition of 'participation' also needs to be clear in this process, considering that it could range from mere consultation to actually taking part in the planning, and implementation (see Pimbert and Pretty in this book).

There is still some resistance among the government officials towards people's involvement. Re-orientation programmes for the staff of the agencies handling the natural resource conservation to introduce them to the social dimensions of conservation and make them aware of people's aspirations, need to be taken up.

REFERENCES

Baldwin, M. (ed.). 1991. *Natural Resources of Sri Lanka: Conditions and Trends*. Natural Resources, Energy, and Science Authority and the United States Agency for International Development, Colombo.

FSMP. 1995. *Forestry Sector Master Plan*. Forestry Planning Unit, Ministry of Agriculture, Lands and Forestry, Colombo.

Land Commission. 1990. Report of the Land Commission, 1987. Department of Government Printing, Colombo.

Ministry of Transport, Environment and Women's Affairs (MTEWA). 1995. *Strategy for the Preparation of a BAP*. Colombo.

Medawewa, S. 1994. Sri Lankan Indigenous Knowledge and Biodiversity Conservation. *Soba* v(3), November: 48–52.

RSAMCC. 1996. *Special Area Management Plan for Rekawa Lagoon, Sri Lanka*. Rekawa Special Area Management Coordination Committee, Coastal Resource Management Project, Coast Conservation Department and National Aquatic Resources Agency, Colombo.

Weerawardana, W.P.W. 1994. Religious and Other Socio-cultural Factors in the Conservation of Biological Diversity. *Soba* v(3), November: 36–39.

PART 3

emerging issues

9

Issues and opportunities in co-management: Lessons from Sri Lanka

Philip J. DeCosse and Sherine S. Jayawickrama

As Sri Lanka's environmental resources come under increasing pressure from an expanding population and economy, one option for improving resource management is to engage those people who have a stake in resources to manage them better. These efforts, described as collaborative resource management, or co-management for short, have gained considerable attention inside and outside Sri Lanka in recent decades.[1] They also yield a number of significant lessons, analyzed in this paper.

Co-management—terms and concepts

There is considerable difference of opinion in Sri Lanka about what constitutes 'co-management'. The differences stem from a lack of clarity concerning the resource and the rights associated with it, the community and other stakeholders in the resource-management process, and the relationship between communities and resources.

What is a 'resource'?

Although most resource types are defined by land use (e.g., forestry, fisheries, wildlife management), they can also be defined by their physical features (water resources, soils, watersheds), their biological features (plant, animal species), and by the ecosystems which are present in them (wetlands, grasslands, coral reefs). Whatever the resource type, it can signify a potential threat to a community as well as a potential benefit. For instance, the elephants in Kahalla Pallekele may be a benefit to the urban inhabitants of Colombo who want to preserve them, but they are a distinct threat to the people of the area.

Resources can also be characterized on the basis of the rights associated with them. In any common property management system, the major rights can be identified as: 1) rights of direct use; 2) rights of indirect economic gain; 3) rights of control; 4) rights of transfer; 5) residual rights; and 6) symbolic rights (Crocombe 1971 cited in Lynch 1991). To this list can be added rights of exclusion, which allow outsiders to be excluded from use of the resource. When 'use rights' over land are altered, it need not imply a change in ownership, but rather the bundle of rights associated with it. Management strategies for resources to which a community has rights of direct use, control, and transfer will be different from strategies for resources for which a community has only symbolic rights. Communities living next to a Strict Nature Reserve (SNR) may have symbolic rights—this applies to resources which act as symbols that define a community—and rights of indirect economic gain, but their rights of direct use are by definition very limited. A clear understanding of this diversity of rights that a community has with respect to a given resource is a prerequisite to designing co-management strategies for the resource.[2]

Finally, resources can be described on the basis of ownership categories: (1) state-owned and managed, (2) state-owned under leasehold, (3) privately owned by individuals, and (4) privately owned by communities. Legally, most forests, wetlands, waters and protected areas (PAs) in Sri Lanka are state-owned and managed, although lease agreements are being explored for management of forests. Although private community ownership of resources occurs in other parts of the world 'today...community ownership in Sri Lanka is almost non-existent' (Nanayakkara 1996). Even in ancient times, informal ownership and management of resources by communities was common, but most resources were legally owned by the monarch.

Although most of Sri Lanka's natural resources are under the legal ownership and management of the state, their rapid degradation is testimony to the fact that the state's management is not effective. Panayotou and Ashton (1992: 201) argue that 'most tropical forests are *de jure* state property, but *de facto* open access with an undefined but large number of non-exclusive claimants'. Most other natural resources are under similarly open-access regimes. Co-management is one means of trying to introduce elements of sustainable common property management systems into the management of open-access resources.

What is a 'community'?

Initially, exploration of the co-management concept in Sri Lanka focused on the community, hence the use of the term 'community-based resource management'. Because the term 'community' suggests the existence of a single, cohesive social organization residing in a well-demarcated area, it can become a source of confusion in the conceptualization and analysis of co-management. In fact, most co-management projects do not work with a single community but with a set of communities or even with an artificial grouping of disparate people and organizations who may be united because of having a stake in the same resource. In the case of The Asia Foundation's (TAF) Ritigala Community Resources Management Project, 14 villages comprising Muslims, Christians and Buddhists have now been moulded into a single community with a perceived common stake in the SNR and in improving their livelihood, but it was not so when the project began. The greatest danger of using the term 'community' is that it may give a misleading suggestion of the potential of the group of stakeholders to come together to manage a resource in common.

The use of the term 'community-based resource management' may also diminish the importance of other stakeholders in the co-management process. To ensure that all co-management stakeholders are identified and included in the process, the major categories of co-management stakeholders are divided into five groups: the community itself, local support institutions, outside local beneficiaries, central resource institutions and external stakeholders (Figure 9.1).

The 'community' includes those who live in the immediate vicinity of the resource and who receive direct benefits or suffer direct costs from it. The community may have existed prior to the co-management effort, or it may have to be created out of common concern for management of the resource. The community may or may not have a strong relationship with the resource (see discussion of the community–resource relationship later). Under 'local support institutions' are included those NGOs, government officials, or other organizations whose objective is to improve management of the resource or to improve the livelihood of the primary stakeholders. 'Outside local beneficiaries' include those who may benefit from direct interaction with or use of the resource, but who do not live in the vicinity of the resource and are likely to have little stake in its

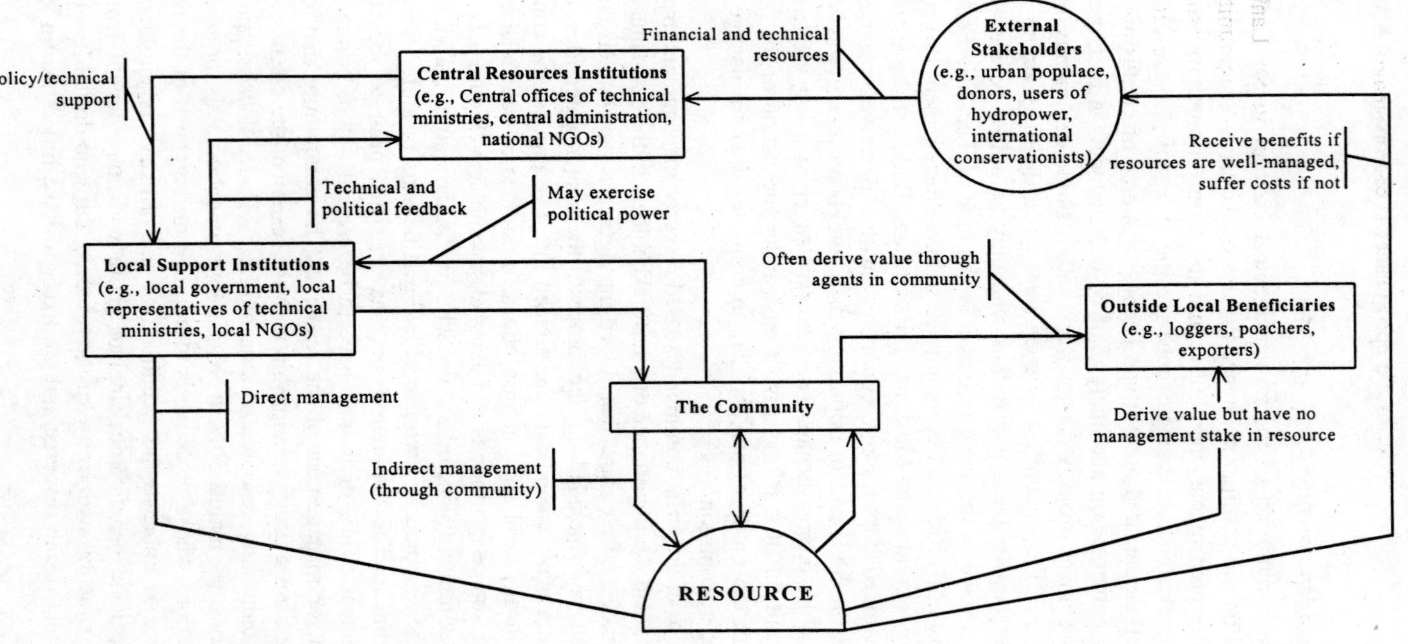

Figure 9.1: Co-management stakeholders and their relationships

sustainable management. This group can include traders of products from the resource (e.g., loggers, poachers) and others who directly consume the resource. 'Central resource institutions' are those government and non-government actors who constitute a source of expertise and funds from which other stakeholders can draw in the management of the resource. Finally, 'external stakeholders' are defined to include those people who benefit from improved management of a resource, but have no direct interaction with it. These could be urban dwellers who place value on the continued existence of a resource or beneficiaries of the power generated from water captured in well-managed watersheds. The relationships between the five groups and the resource are shown in Figure 9.1.

The community–resource relationship

It is often implicitly assumed that a community living in the vicinity of a resource will have a stake in the sustainable management of that resource. This is not always the case. A careful understanding of the relationship between communities and the resources of interest is thus an essential prerequisite to the planning of co-management projects, particularly if they are to make the community the primary focus of resource-management activities.

To attempt a clarification of the relationships between communities and resources, four primary types of relationships can be distinguished, and their implications for co-management analyzed. In the first type of community–resource relationship, the community realizes little or no value from the resource, in spite of living next to it. In such cases, co-management is not an appropriate approach since the community has little incentive, apart from perhaps emotional attachment to or reverence for the resource, to ensure that it is managed sustainably. To the extent that resource degradation is occurring, it is the likely result of actions by outside local beneficiaries, and resource-management improvements should, therefore, focus on this group rather than the community. An example of this relationship in Sri Lanka includes the communities surrounding the Attidiya Sanctuary who are predominantly engaged in wage activities in Colombo and the surrounding areas, and realize virtually no benefits from the Sanctuary. Although it is possible that a stake could be 'created' within the community, this is not likely to be sufficient incentive for the community to actively engage in resource management.

In the second type of community–resource relationship, a significant benefit or cost of the resource accrues to only a few members of the community, while the rest of the community has little interaction with the resource. In such a case, co-management is not likely to be appropriate, since it generally implies involvement by all or most members of the community. If the members of the community who benefit from the resource can be identified, then it may be sufficient to train or educate these few in sustainable resource management methods to ensure better management of the resource. An example of this type of relationship in Sri Lanka would be the small groups of specialized turtle egg poachers who operate at many points along the southern coast. The rest of the communities in which the poachers live have very little influence over their resource use actions, and thus to improve management, it will probably not be helpful to engage the broader community.

In the third type of relationship, the benefits or costs derived from the resource by the community are small, but they are widely distributed across the community. Community-based co-management has the potential for success in these communities, since virtually the entire community has a stake, however small. Communities such as those surrounding the Ritigala Forest would be an example of this relationship, as they enter the forest to harvest a small number of medicinal plants like wild cardamom which cannot be found in the market. Although the value of the cardamom and other non-timber forest products (NTFP) is not great in comparison to their total income, the community recognizes that the forest gives them something they cannot get elsewhere.

In the fourth type of relationship, a significant benefit or cost of the resource accrues to a broad cross-section of the community. Here the opportunity for community-based co-management is very good, since the entire community has a large stake. When the community has a strong sense of its relationship to the resource, including some established social rules for resource management, then the opportunity to establish private communal ownership may be good. Introduction of alternative income generation activities should be done only with a careful prior assessment of the opportunity costs to the community of giving up use of the resource. Although there are not many examples of entire communities that derive significant benefits from resources in Sri Lanka (for instance, there are very few forest-dependent communities), there are numerous examples where a resource is a significant threat or cost to the community at large. An example of this relationship are the communities menaced by elephants in Kahalla Pallekele.

The community–resource relationship types described above make it apparent that a community must have a strong *incentive* if it is to be engaged in the management of a resource. The community must also have the *capacity* and knowledge to manage resources (Ascher 1994). Typically, such capacity relies upon indigenous knowledge. If communities have little interaction with resources, and if traditional management practices and systems have been eroded by decades of exclusion by the authorities, then it cannot be assumed that they still have the capacity to manage resources. Honadle and VanSant (1985) argue that the assumption that communities have the know-how to manage resources is one of the fundamental myths in development. The capacity and know-how of rural Sri Lankans to manage resources is limited in comparison to many countries, as a result of the historical and cultural context. Therefore, any design of co-management efforts in Sri Lanka must pay special attention to assessing whether communities have the incentive and the capacity to manage the resource concerned.

The four preceding types of community–resource relationships assess the level of ongoing interaction between community and the resources in their vicinity. A fifth type of relationship is based on a desire of community members to own the resource. In the context of extreme land shortage for settlements, many rural inhabitants are interested in land-based resources only for the land, not for any of the resources. This often results in encroachment.

'Co-management' defined

Using these terms and concepts, therefore, co-management is defined as: The active engagement of communities and outside local beneficiaries in the collaborative management of *de facto* open-access resources by local support institutions and central resource institutions. This definition highlights two points in particular. First, the use of the term 'active engagement' should make it clear that co-management is not appropriate for those communities which have no relationship with resources of concern. The second point is that both communities and outside local beneficiaries should be engaged in the resource-management process, with no presumption that either of the two is of greater importance.

A note on participation

One other idea merits discussion. Common to most work on resource management is an emphasis on the importance of participation of communities in the process, which is alternatively described as participatory or bottom-up planning and implementation. Although not new to development planning, the importance of participation has gained an increasing acceptance over the years. Virtually all of the USAID-funded co-management efforts in Sri Lanka have made participation an essential element in their approaches (see Wijayaratna 1994; White 1996; The Asia Foundation 1995).

Framework for co-management in Sri Lanka

For most of the post-independence era, state agencies mandated to manage natural resources have traditionally perceived their role not only as 'protectors' but also as 'policemen' and have perceived local communities either as passive observers that can be ignored or as potential threats that can be controlled. The main thrust of resource management has, therefore, been to restrict and control peoples' interaction with resources. The adoption of this 'command and control' approach has been based on the notion that the country's natural resources are the property of the state. This mentality has been reinforced by the fact that the state does, indeed, own more than 82 per cent of the nation's land area and its land-based resources. Over 28 per cent of the country is administered by the Forest Department (FD) and the Department of Wildlife Conservation (DWLC).

Historical and cultural context

In the pre-colonial era, a system of traditional service tenure, called *rajakariya*, prevailed in Sri Lanka. The king owned all the country's resources and would bestow on citizens—individuals and communities—the right to use and manage tracts of land and other resources in return for service to the monarchy. Although citizens did not legally own land, almost every family was provided a tract of land and could decide how it would be used. These rights could be revoked if the monarch was not served as required. Resources such as forests and irrigation works were managed collectively and communities had accepted methods of controlling and allocating their use. It was generally recognized that families or communities which had held tracts of land for long periods of time had customary ownership. Two aspects of this ancient system are important to note. First, community-based tenurial rights were recognized. Second, systems for joint management of resources existed.

Colonial land laws changed this system of communal resource management irreversibly. The Crown Lands Encroachment Ordinance (CLEO) of 1840 declared that all lands for which title was not registered belonged to the Crown. Since customary rights were impossible to establish, vast amounts of land were vested in the state and subsequently transferred to British planters for coffee cultivation. This land included large areas of fallow *chena* (shifting cultivation) land.[3] The Waste Lands Ordinance (WLO) of 1897 consolidated this transfer of lands. Although the Land Reform Laws of the 1970s were enacted with the stated goal of redistributing land to the landless, they resulted in adding to the state's already excessive inheritance of land.

The landownership issue colours the historical and cultural context for natural resource management. The CLEO made thousands of people landless overnight. The problem of landlessness has grown more severe over the years; more than 19 per cent of rural workers are landless (FAO 1985 in Land Commission 1990: 109) and there has been no significant progress in dealing with this problem. As a result, *poverty and landlessness* have become the root cause of many resource-management problems in Sri Lanka. Parcels of land that have been passed down through generations are highly fragmented now and are impossible to cultivate efficiently. Consequently, expansion to accommodate new generations is sometimes possible only through encroachment. The *Report of the Land Commission*, 1987 (Land Commission 1990) puts encroachment at 6 per cent of the nation's total land area.

The outcome of the CLEO, the WLO, and the Land Reform Laws was simply that too much land was vested in the state. Even after major attempts to transfer state land, the state still owns some 82.3 per cent of the country's land area (Land Commission 1990). The state has not been able to manage this vast amount of land effectively. Although access to these lands is legally restricted, the lack of enforcement and absence of any semblance of management has fostered the impression that it is 'no one's land' (De Silva 1993). As a result, large amounts of land have became *de facto* open-access areas which have been encroached upon and unsustainably used.

The self-sufficient, sustainable lifestyle of ancient Sri Lankan communities was eroded not only because of the sudden loss of lands but also because of the pervasive nature of the commercial economy that took firm roots in the British period. In Sri Lanka, most rural communities have easy access to nearby towns and interact closely with the commercial economy. While many still use forests and other natural resources, the extent of their reliance is very small and, in many cases, the knowledge and use of traditional management practices has all but disappeared.

The legal framework

The Fauna and Flora Protection Ordinance (FFPO) of 1937 (amended by the Fauna and Flora Amendment Act of 1993), the key law that provides for the protection of wildlife and flora in PAs, is administered by the DWLC. The main thrust of the FFPO is *regulation* and *restriction.* A large part of the ordinance lists out prohibited acts, permit requirements and fines. This ordinance does not recognize that local communities or other stakeholders can play a role in the management of PAs or in the protection of wildlife. Moreover, with the exception of the provision recognizing communities' customary fishing rights in PAs, it does not acknowledge the need to reconcile PA management with the needs of surrounding communities. The Forest Ordinance administered by the FD, provides the legal framework for the management and conservation of forest land other than PAs. When this ordinance was enacted in 1907, its primary thrust was revenue collection from timber production. Over the years, the ordinance has been amended to reflect conservation concerns. On the whole, however, the Forest Ordinance remains largely regulation oriented. Rural communities are allowed to use certain types of forest products only if they have obtained the required permits. As a result, much of the forest officers' time is spent issuing permits, monitoring their use, and appearing in court against violators (Forestry Planning Unit 1995a). The ordinance contains little recognition of the role of rural communities in forest management. Although the primary emphasis is on keeping people out of forests, there is one provision that could be used for co-management. This provision enables the Minister of Lands to 'constitute any portion of a forest [as] a *village forest* for the benefit of any village or group of village communities' and to 'make regulations for the management of village forests'.

The Coast Conservation Act of 1981, administered by the Coast Conservation Department (CCD), seeks to regulate development activities in the coastal zone.[4] The Act emphasizes the importance of establishing a scientific basis for coastal zone management and the need to reconcile the socio-economic needs of local communities with coast conservation. Although this Act is regulation-oriented to a large extent, it provides a basic framework for collaboration among stakeholders.

Another piece of legislation that could potentially have significant implications for the introduction of co-management efforts on a broad scale is the 13th Amendment to the Constitution. Enacted in 1987, this Amendment provides for the devolution of a wide range of legislative and executive powers to the provincial government level. These powers cover environmental protection, public lands, agriculture, irrigation and many other areas linked to natural resource management. There is currently confusion over the division of responsibilities between the central government and provincial governments and a marked lack of implementation capability at the provincial level. As a result, provincial government institutions currently do not play a significant role in co-management efforts. (See the section on institutional framework for a more detailed discussion.)

The issue of community-based property rights can be an important factor in co-management arrangements. In Sri Lanka, apart from legislations such as the CLEO of 1840, several Supreme Court judgements have refused to recognize traditional communal ownership of natural resources and have clearly demonstrated an aversion to excluding 'outsiders' from using these resources (Nanayakkara 1996). The fact that communities do not have the legal right to exclude outsiders is a major constraint to the sustainability of co-management efforts. The Land Settlement Ordinance (LSO) contains a rare acknowledgement of communal land rights in Sri Lankan law. The LSO has a special provision to allow settlement officers to set apart state land as a 'communal *chena*' reserved for the use of inhabitants of a certain village (De Silva 1993). Once this land has been declared, outsiders can use the land only with the consent of the villagers. This is unique in that it gives the community the right to exclude outsiders.

There are areas in existing legislation that could be used to strengthen the framework for co-management. For instance, the State Lands Ordinance and the Land Development Ordinance can be used to transfer ownership of state land and, therefore, can strengthen incentives for community management of land-based resources. Likewise, the Agrarian Services Act and the Irrigation Amendment Act, which grant legal recognition to resource user groups (i.e., Farmer Organizations) can lend enormous credibility to these groups and facilitate their acceptance as equal partners in co-management efforts. Currently, Farmer Organizations have the right to manage *de facto* open-access resources such as irrigation canals, in perpetuity.

The policy and planning framework

Policy-makers at institutions such as the DWLC and the FD have been slow to incorporate the 'bottom-up' approach into their policies and plans, though they are beginning to recognize its importance.

The National Forest Policy (NFP) and the Forestry Sector Master Plan (FSMP), both adopted in 1995, constitute the first coherent, long-term framework for forest development in Sri Lanka, and are a far cry from the production and regulation-oriented 'keep people out' approach reflected in previous forest laws and policies. The NFP declares that, in the protection and management of natural forests and forest plantations, 'the state will, where appropriate, form *partnerships* with local people, rural communities, and other stakeholders, and introduce appropriate tenurial arrangements'. It acknowledges that the government has not been effective in managing all forest lands and that local communities do not have the rights and incentives to use forests sustainably (Forestry Planning Unit 1995b). It also recognizes that deforestation will continue unless the problem of landlessness is addressed. The FSMP identifies security of tenure as one of the most important incentives for sustainable forest management.

Although both the NFP and the FSMP provide a very supportive framework for co-management, they include little detail on how these policies and plans are to be implemented. This would require a complete re-orientation of the FD bureaucracy.

In the area of wildlife and PA management, there have been no recent attempts to comprehensively review and revise the existing policy and planning framework. Existing DWLC policies emphasize enforcement of regulations to exclude people from PAs and provide little encouragement for collaborating with local communities.

The lack of scientific research and national-level planning has been felt most acutely in the area of elephant management. Much of the landmark elephant research conducted in Sri Lanka is approximately 20 years old and only covers elephant populations in small areas of the country. Although the human–elephant conflict has increased rapidly over the past two decades, there has been no coherent policy or strategy response to this. Instead, the DWLC has tended to deal with each trouble spot on a case-by-case basis with ad hoc responses.

In addition to national-level policies, there have been several management plans that have been developed for specific resources like forests and PAs. For instance, the FD has developed management plans for nine conservation forests in the wet zone including the Sinharaja and Knuckles forests. Seven of these management plans include detailed strategies for engaging local communities in resource-management activities. The DWLC has also developed management plans for several wetland sites under its management and for a few PAs declared by the Mahaweli Authority. None of these management plans have been implemented with any degree of success.

The revised Coastal Zone Management Plan (CZMP), 1997, is an update of the CZMP adopted in 1990. The plan recognizes that the regulatory approach used by CCD in its first 10 years is not sufficient to achieve effective coastal zone management. This approach focused primarily on issuing permits for relatively large development projects. While this helped to prevent adverse impacts on coastal resources that might have been caused by these projects, it was not able to deal with the degradation caused by the cumulative effects of the continued use of coastal resources by individuals or communities. The CZMP strongly advocates the concept of Special Area Management (SAM) which is a co-management approach (Coast Conservation Department 1996).

The Biodiversity Action Plan (BAP), also to be finalized, will advocate the involvement of communities in biodiversity management. The Strategy Document for the preparation of the BAP clearly states that biodiversity conservation should be 'centred at the grassroots level through community participation' (M/TEWA 1995). Networks of NGOs dealing with biodiversity issues have been established to provide input into the preparation of the BAP and ensure that local-level concerns are addressed.

Although the state has authority over most natural resources, a relatively supportive policy environment exists for community participation in the conservation of these resources. The problems, however, arise in the implementation stage for two major reasons. First, threats to effective resource management frequently arise as an outcome of policies of other sectors. Although the M/TEWA is charged with coordinating among other ministries to mitigate adverse environmental impacts, it is typically the sectoral ministry's policy that prevails. Second, involving other stakeholders may require a complete change in attitudes of agency bureaucracies.

The institutional framework

Although the DWLC's approach is almost completely enforcement oriented, there is growing recognition among field officers, in particular, that their work would be much more effective if they had a less confrontational relationship with local communities. With experience in places like Ritigala (see later sections), DWLC has begun to recognize that communities can play a role in PA management.

The FD, in spite of the fact that the FSMP and NFP provide a strong framework for co-management and that it has three years of experience with participatory forestry, is still not equipped to adopt co-management on a broad basis.

The CCD is a leader in terms of its endorsement and adoption of the co-management approach. It has identified the active engagement of local communities and other stakeholders as a prerequisite for sustainable resource management and has led the development of policies, plans and legislation to facilitate use of the co-management approach on a broad scale.

At the provincial level, Provincial Councils (PCs) possess legislative and executive powers over many areas including natural resource management, public lands, irrigation and agriculture. Although the 13th Amendment on this was adopted in 1987, implementation has been weak because of confusion over the division of power between the central and provincial governments and because of the lack of technical capability and staff resources at the provincial level (De Silva 1993). Moreover, with the exception of the Northwestern Province, no PC has even attempted to actively manage the natural resources of its province.[5] PCs serve as the link between the central government and Pradeshiya Sabhas, which future co-management efforts should take advantage of.

At the local level, governance and administration is complicated by the fact that two parallel institutions (the Divisional Secretariat and the Pradeshiya Sabha) function with a poor definition of roles and very limited coordination. The Pradeshiya Sabha (PS) is a locally elected body which reports to the PC. The Divisional Secretary (DS) is appointed by and reports to the Ministry of Home Affairs and Public Administration. PSs and DSs often administer the same geographical areas and engage in very similar activities. Of the two institutions, the Divisional Secretariat is better funded and has a higher level of technical capacity. The DS has the authority to coordinate the activities of field officers of government agencies and historically has a good rapport with the community. On the other hand, PSs are elected and more accountable to local communities and have close links with the provincial government. Co-management initiatives should involve both these institutions.

NGOs are important players because they typically have a better rapport with local communities. NGOs also have specific areas of expertise—for instance, Sarvodaya in community empowerment, March for Conservation in scientific knowledge, and Wayamba Govi Sanwardana Padanama (WGSP) in rural development.

Review of co-management approaches in Sri Lanka

In this section, we will examine several approaches that have been used in Sri Lanka to share responsibility for resource management among local communities, government agencies, NGOs and others.

Shared Control of Natural Resources (SCOR)

The SCOR project, managed by the International Irrigation Management Institute (IIMI), works in the Huruluwewa and Upper Nilwala watersheds to pilot test a participatory approach to sustainable resource management in watersheds. SCOR focuses primarily on 'increasing the sustainable productivity of land and water resources' by integrating conservation concerns with production goals (Wijayaratna 1995: 1). SCOR's strategy is to first organize and strengthen user groups in the project areas and to then facilitate the establishment of formal state-user agreements in order to increase users' control over the relevant land and water resources. An integral part of the SCOR philosophy is that security of tenure enhances the incentive to engage in sound production practices that have long cost-recovery periods. The type of tenure security that SCOR advocates is shared control (i.e., some degree of 'communal' ownership) rather than exclusive individual property rights.

Coastal Resources Management Project (CRMP)

The CRMP focuses its field activities on Hikkaduwa and Rekawa on the southern coast. CRMP uses these two sites to demonstrate the potential of the Special Area Management (SAM) concept. SAM is a co-management approach in which communities work with the local and national government to develop and implement management plans for the sustainable use of resources within a defined geographic setting. Very early on in CRMP's work, the CCD took 'ownership' of the SAM concept and has since championed this approach to integrated coastal management.

At both Hikkaduwa and Rekawa, there are entire ecosystems which are *de facto* open-access resources used by several groups of local communities. In Hikkaduwa, the Marine Sanctuary is being rapidly degraded by over-use and poor management. Much of the threat to the Sanctuary's famous coral reef comes as a result of glass-bottom boats, fishing boats, untreated waste discharged by hotels and restaurants, and tourists walking on the corals. The SAM process in Hikkaduwa brings the stakeholders—hoteliers, restaurant owners, glass-bottom boat owners, and fishermen—together with the local government, CCD, DWLC, and other relevant parties to jointly develop strategies to manage the Sanctuary. In Rekawa, the lagoon and surrounding lands are gradually being degraded. Through the SAM efforts, organization of the community for sustainable resource management has begun (see Ekaratne et al. in this book).

The Asia Foundation's special projects on community-based resource management

(i) Kahalla Pallakele Human–Elephant Conflict project: The Human–Elephant Conflict (HEC) project started in 1993 and is implemented by an NGO coalition named the Wana Jana Mithuro Sanvidanaya (WJMS) in 45 villages in the divisions of Galgamuwa and Giribawa in the Northwestern Province. High levels of human–elephant conflict in this area result not only in regular elephant injuries and deaths but also in severe damage to communities (in terms of death, injury and property/crop damage). Villagers residing in this area are poor and heavily dependent on *chena* cultivation and cannot withstand the losses inflicted by elephants. The basic thrust of the WJMS strategy is to address the elephant-related issues by first helping villagers to address their socio-economic problems. The assumption underlying this approach is that better socio-economic conditions and improved governance will reduce villagers' vulnerability to elephant-related damage and consequently reduce the pressure on the elephant population which is the local resource that the HEC project aims to protect with the help of the local communities. This is different from the other resources discussed in this study for two reasons. First, it generates only costs for the primary stakeholders. Second, the resource interacts with many other communities and resources in a fairly large geographic area.

(ii) Ritigala Community-based Resource Management project: The Ritigala CBRM project, started in 1995, covers the regions bordering on and including the Ritigala SNR in the Anuradhapura district. Established in 1941 under the authority of the FFPO, the SNR is a unique cultural and biological heritage, in particular with respect to medicinal plants. Threats to the SNR have included harvest and sale of hardwood, cattle grazing, poaching, collection of plants for food and medicine and firewood collection.

The CBRM project facilitated the establishment of the Ritigala Community-based Development and Environmental Management Foundation (RITICOE), which now runs project activities. The project works on three fronts to increase the economic opportunities of the community and ensure that the SNR is more sustainably managed: 1) education and awareness-raising; 2) promoting liaison between actors currently or potentially involved with the SNR; and 3) introducing income-generating opportunities. Technical support for the medicinal plants work has been provided by the Bandaranaike Memorial Ayurvedic Research Institute.

Other co-management efforts in Sri Lanka

Forest Department: The FD's first formal social forestry initiative, funded by the Asian Development Bank (ADB), is now widely considered a failure both in terms of expected outputs and community participation. This initiative, launched in 1982, addressed fuelwood scarcity. The extent of community participation was that farmers were contracted to plant seedlings provided by the FD.

Learning from this experience, the ADB-funded Participatory Forestry Project was launched in 1993. This project works in almost all parts of the country to provide farmers with more 'ownership' over the afforestation process. The FD conducts a Participatory Resource Assessment (PRA) prior to site selection and introduces participatory forestry activities only if local communities are capable of and interested in taking an active role in the afforestation process and if their participation has the potential to decrease current pressure on forested areas. The FD uses four agro-forestry models to encourage the conversion of non-forest lands to forests.

There is another co-management approach that the FD has begun using recently: 'informal agreements' with communities. In areas where communities do not have a close interaction with forests, the FD tries to create an incentive for them to help protect forests from threats of felling and clearing. The FD makes a 'deal' with the community—that it will bear the capital cost of a school, tank, road, clinic, etc. if the community agrees to protect the forests (i.e., to stop being agents for illicit timber fellers, to report illicit felling to forest officers).

The World Conservation Union (IUCN/Sri Lanka): With GTZ funding, IUCN is pilot testing the co-management approach in five villages adjacent to the Knuckles forest. The thrust of the initiative is to 'weane communities off the forest' by providing assistance to help villagers improve their socio-economic conditions, which, it is assumed, will reduce the need to use the forest and will, therefore, reduce pressure on the forests.

Discussion

Sri Lanka's experience demonstrates that there is no single formula for co-management. In fact, the significant feature of the co-management approach is its flexibility and its ability to bring together many sets of actors with divergent interests. However, the community is almost always at the centre of co-management efforts while outside local beneficiaries often do not come into the picture at all. For instance, TAF's projects in Kahalla Pallekele and Ritigala focus very heavily on mobilizing the community to engage in project activities even though considerable damage may be caused by poachers, etc. The heavy emphasis on the community should not be considered a weakness in these approaches. The community is a visible and often cohesive group of actors that are relatively easy to define and, more importantly, easy to reach. Outside local beneficiaries—for example, in the case of a forest resource, illicit fellers—are often not easily defined or reached and, therefore, hard to incorporate into a project approach. It is unclear exactly how this group can be engaged in resource management. It is clear, however, that they should be involved because they have as much capability to damage or restore the resource as any other group.

Most of these initiatives offer actors no opportunity to be involved in the actual management of the resource. For example, in Ritigala, the local communities cannot even legally step into the SNR let alone help to manage it. In Hikkaduwa, although stakeholders can use the Sanctuary, actual management is solely the responsibility of the DWLC. The IUCN approach is to divert the community from using the forest resources. In this way, many co-management efforts have emphasized participation over management. This is not to deny the value of participation in itself. In many cases, community mobilization and awareness-creation have achieved a remarkable change of the community's attitude towards resource management. Before the TAF projects in Kahalla and Ritigala started, the community felt that elephant management and preservation of the SNR was the government's responsibility. The community now feels 'ownership' over the resource issues and believes it has a significant role to play in resolving them.

Although collective effort is one of the vital features of co-management, many Sri Lankan projects focus mostly on individual activities. The FD's Farmers' Woodlots model, although it is the centrepiece of their Participatory Forestry Project, provides plots of land to individual farmers who do their tree planting and cultivation independent of the larger community. This is quite unlike social forestry projects in many other parts of the world where entire communities obtain rights to parts of the forest which are then 'communally' managed.

It is interesting to look at why these three features of co-management (multi-stakeholder participation, management responsibility and collective action) are so weakly fulfilled when the country has a relatively strong framework for collaborative management. The nature of the *community–resource relationship* explains this partially. In Sri Lanka today, the commercial economy has penetrated almost every corner of the country and, with the exception of fuelwood, communities fulfill most of their material needs through market transactions. Many communities, therefore, do not have the capacity or the incentive to engage actively in conserving their resource. Some co-management initiators in Sri Lanka have recognized this. In the FD's 'informal agreements' approach, the community makes a deal to keep outsiders away from the forest in return for a road, school, etc. This shows that even when the community–resource relationship is weak, communities can be important actors in co-management merely because of their proximity to the resource and their subsequent ability to act as 'watchdogs'.

The manner in which the community–resource relationship is addressed is often the key to a sustainable co-management arrangement. If the resource is vital to a community, then it has an incentive to manage this resource efficiently and sustainably. This incentive can be strengthened by ensuring that the community will be able to: 1) enjoy the benefits of sound management in the future; and 2) exclude outsiders from enjoying these benefits. In Hikkaduwa, glass-bottom boat owners understand that the reef damage their boats cause will reduce their future income. They, therefore, have the incentive to limit and improve their use of the Sanctuary. However, the fact that they cannot prevent new boat owners from obtaining permits erodes this incentive.

Landlessness associated with poverty is one of the major causes of resource degradation in Sri Lanka. As successful as participatory activities may be, as a community's population expands, the ultimate need is land. For this reason, every co-management effort that seeks to be effective and sustainable must address this issue and the issue of poverty. Many co-management efforts link conservation and development objectives. Underlying this approach is the assumption that the socio-economic improvements generated by development activities will reduce pressure on the resource. In the cases of Kahalla Pallekele, Ritigala, and the villages adjacent to the Knuckles, co-management projects have provided employment opportunities, improved access to services, and increased incomes. However, there is no evidence that this has resulted in improvements in the quality of the resources concerned. This suggests that even though socio-economic improvement does not always result in improved resource quality, it is almost always a strong incentive for stakeholders to participate and stay engaged in co-management activities.

Another lesson to be learned is that central resource institutions can play a vital role in establishing a larger policy, legal, and technical framework for co-management and ensuring the sustainability and replicability of the effort. In the cases of Hikkaduwa and Rekawa, NARA has established a sound technical framework for SAM planning and the CCD has established a supportive policy and legal framework for SAM implementation. This has lent a great deal of weight to the co-management effort. Moreover, since the CCD has established firm 'ownership' of the approach, it is committed to replicating it at other locations. On the other hand, in the absence of a supportive DWLC framework for elephant management, TAF's success in reducing the human–elephant conflict in the Kahalla Pallekele area cannot have a positive impact on the overall problem of elephant management.

It is important also to keep in mind that policies do not always originate from above. TAF seeks to create the demand for policy change from below. There has been a marked change in the DWLC's attitude toward co-management since TAF's projects began, as it now sees the community as a mature, demanding and potentially useful group.

A noticeable feature of many Sri Lankan co-management efforts is that they are based more on assumptions about the potential of community involvement than on solid information and good resource assessment. Two of the common assumptions made are that: 1) the community knows better than anyone else how to manage their resources; and 2) damage done to the resource by outsiders is done with the collusion of the community. Too much faith in these assumptions can lead to efforts that put 'too many eggs' in the community 'basket' and fail to consider and address the underlying causes of resource degradation.

Conclusions and recommendations

Co-management is both viable and necessary for Sri Lanka

Since the state has neither the funds nor the staff to effectively manage the 82.3 per cent of the land area it owns, engaging resource users at the local level may be the only effective means of ensuring the sustainable management of resources. It is clear that co-management is not only a viable but also a necessary option for achieving more sustainable management of environmental resources.

In spite of its potential benefits, co-management should not be blithely considered a panacea for resource-management problems.

Although local communities should always be consulted about the management of resources in their vicinity, it is not always appropriate for them to be 'actively' engaged in a formal co-management process. This might be true in the case of a resource which is of considerable importance to the nation and which justifies direct management by the relevant government institution. As Panayotou and Ashton argue, these resources should 'be accorded full protection and effective enforcement of ownership by the state. This does not preclude a role for the private sector and local communities, but such a role needs to be strictly regulated and closely monitored' (1992: 211). The decision about whether or not to employ a co-management approach should be linked to a careful understanding of the community–resource relationship. In most cases, co-management activities should be coupled with enforcement.

A clear understanding of the community–resource relationship is essential

Co-management project design must include a careful assessment of the relationship between communities and their resources, because it can have a significant bearing on the success of the co-management effort. In cases where a resource generates minimal benefits for a community, it may be more effective to try a direct incentive approach. The greater the community's interaction with the resource and the higher the proportion of the community that gains or loses from that interaction, the more likely is the success of co-management projects.

Evidence from Sri Lanka makes it clear that rural communities' relationship with land-based resources has more to do with a simple desire to *own land* than with a reliance on the resource. Where resources are under threat from encroachment, the response should include a host of policy measures designed to resolve the land question.

Resource assessments and monitoring systems must be included in project design

Without an understanding of the conditions and trends of the resources to be managed, it is not possible to know whether co-management projects are effective. Experiences from outside Sri Lanka have shown that most co-management activities are launched without such resource assessments. Many Sri Lankan experiences suggest that adequately broad *resource assessments* have not been conducted, because implementors did not have the financial and technical resources to carry out such assessments.

Resource assessments can indeed be expensive if they are exhaustive, but they need be neither. Cost-effective means of carrying them out can be found (Valadez and Bamberger 1994; World Bank 1996; Marks 1996). Three options are worthy of consideration. First, national technical institutions can develop economies of scale if they are engaged to carry out similar assessments at different sites. Because of its experiences in SAM planning, NARA has now developed expertise in certain coastal resource assessments. Second, 'sectoral' resource assessments can generate much of the basic knowledge required to understand a resource problem, leaving limited data collection to be done at a particular site. Third, resource assessments can be made more cost-effective by making them a training ground for Sri Lankan graduate students in the natural sciences.

Development of resource assessments must be linked to simple and cost-effective *monitoring* systems. Since resource changes take a longer time to be visible, many co-management projects have used 'level of participation' as a performance indicator. This does not effectively indicate changes in the quality of the resource.

The impact of 'outside local beneficiaries' has been underestimated

Most co-management approaches have focused on the community as the primary stakeholder. Yet, in many cases of resource degradation, the group defined as 'outside local beneficiaries' are the cause of degradation rather than the community. There is an unwillingness to recognize the importance of these outside stakeholders in Sri Lanka. In keeping with the language used by the CCD, the term 'collaborative' rather than 'community-based' resource management should be consistently used in Sri Lanka. This allows for a *broadening of the co-management concept*. There is also an urgent need to design and test various approaches to bring this set of actors into co-management.

The link between alternative income generation and resource management is unclear

There is little evidence from Sri Lanka, or indeed from elsewhere around the world, to show that introduction of alternative income-generation activities results in long-term *reduction of pressure* on resources. Although it is assumed that such activities will be a sufficient incentive for community members to stop over-using resources, such assumptions have often underestimated the impact of outsiders on the resource and the true opportunity cost to villagers of giving up use of the resource.

In conserving PA resources, regional poles of economic development have a greater likelihood of reducing resource pressure than do the localized income-generation activities of co-management projects. When families who once survived off resource consumption from PAs are offered significant alternative income sources (e.g., full-time jobs in factories), their consumption of resources from PAs will likely decline. Co-management efforts must therefore be incorporated into larger national and regional policy and development initiatives.

In addition, a resource is of critical national importance and the likelihood of local resource users being 'lured' away from resource degradation through alternative income-generating activities is unclear, especially where the *direct incentive* model ought to be further explored on an experimental basis.

Community participation is necessary but not sufficient for sustainable resource management

Whether *communities* have a large stake in the sustained management of resources in their vicinity or not, they *must be engaged* in the management process. Considerable attention has been paid to encouraging these participatory processes in Sri Lanka, and they have met with a great deal of success. In both Ritigala and Kahalla Pallekele, for example, the communities now have a clearer idea of their potential for resolving their own resource-management problems. Although improved participation is essential, co-management planners must recognize that it is *not sufficient* for ensuring sustainable resource management. Along with participation must go the 'negative' incentive of enforcement and penalties.

The institutional and policy framework for co-management must be further improved

It is time that the government prepared a more comprehensive policy and technical framework for collaborative resource management.

The FSMP, the NFP, and the soon-to-be-released CZMP include strong policy support for co-management. It is clear that the absence of a policy of involving communities and other stakeholders in the management of PAs and wildlife is a serious constraint. The DWLC must incorporate co-management into its policies and plans.

Supportive policies are dead-letters if implementing agencies do not have the capacity and the commitment to put them into action. Although the FSMP provides an excellent framework for co-management, the FD is unlikely to implement these plans. Institutional capacity of national agencies to support co-management has reached the most advanced stage in the management of coastal resources (by the CCD, NARA and other collaborating institutions). Other central resource institutions have much to gain from building on their successes.

In addition, much more attention has to be paid in the future to building the capacity of provincial and local governments and of NGOs.

The legal framework for co-management is inadequate and demands priority attention

Although future co-management projects will focus on the land base and all of its inland and coastal water bodies, virtually all *legal ownership* and *use rights* over these resources are in the hands of the state. If co-management is to succeed, the package of rights accruing to communities should be formally modified in cases where the community–resource relationship is strong. Communities must be granted more extensive use rights over them, perhaps by legally recognizing them as corporate bodies. In the near term, efforts should be made, perhaps under special permission of the responsible resource management institution, to grant and test stronger use rights to resources for communities, including the right to exclude outsiders from using the resource.

Collaborative partnerships between NGOs and government institutions are often effective

Projects implemented by NGOs have the advantage of being more sensitive to the needs of local communities. NGOs do not, however, have at their disposal either the technical expertise required to design co-management projects or the resources to conduct these projects on a significantly large scale. To be effective, therefore, partnerships must be formed between NGOs and government institutions, so that the NGOs can provide the link to the community and government agencies can provide the link to funding and can facilitate replicability and sustainability.

Lessons learned from pilot activities in Sri Lanka show that the sequencing of involving NGOs and government institutions is an important determinant of sustained community involvement. For instance, if the government gets involved too early, then the community assumes that the government will do all the work. If the NGO begins its work prior to government involvement (as TAF did in Ritigala and Kahalla Pallekele) and if the community stakeholder can develop a clear perception of itself and its goals, the prospects for more active participation in the co-management process are greater.

Selection of sites should use pre-determined rather than ad hoc criteria

The co-management identification process should be sectoral in scale and should select *pre-identified criteria* which would contribute to project success. If communities are to be engaged in the resource management process, they must 'still retain a sense of community of ownership' and those 'protected areas where effective management is already in place' should be given high priority (Nanayakkara 1996: 39–40). To gain economies of scale in this identification process, national technical ministries should take the lead in identifying the criteria and the resulting high priority sites. Site selection should also be preceded by an analysis of the costs and benefits not only of past attempts at co-management, but also of more traditional resource-management options and direct incentive agreements.

NOTES

1. For details on history of conservation and co-management in Sri Lanka and on some of the examples mentioned in this article, see Jayatilake et al. in this book.
2. See Bruce et al. (1985) for a thorough discussion of the bundle of rights that may be held by resource managers.
3. In ancient Sri Lanka, *chena* cultivation was not an unsustainable agricultural practice. Farmers had a system of rotation for land use. Fallow land was left for some years to regenerate and then re-used.

4. The coastal zone is defined as the area lying within 300 metres landward of the mean high water mark and 2 kilometres seaward of the mean low water mark.
5. The Northwestern Province adopted the first Provincial Environmental Act in Sri Lanka in 1990.

REFERENCES

Ascher, William. 1994. *Communities and Sustainable Forestry in Developing Countries.* The Center for Tropical Conservation, Durham.

Bruce, John J., L. Fortmann and **J. Riddell.** 1985. 'Trees and Tenure: An Introduction', in L. Fortmann and J. Riddel (eds). *Trees and Tenure: An Annotated Bibliography for Agroforesters and Others*. ICRAF, Nairobi.

Coast Conservation Department. 1996. Revised Coastal Zone Management Plan, Sri Lanka (Draft), Coast Conservation Department, Colombo.

De Silva, Lalanath. 1993. *Economic Development Projects: An Analysis of Legal Processes and Institutional Responses*. Natural Resources and Environmental Policy Project, Colombo.

Forestry Planning Unit. 1995a, *Sri Lanka Forestry Sector Master Plan*, Ministry of Agriculture, Lands & Forestry, Colombo.

———. 1995b, *National Forestry Policy and Executive Summary*, Ministry of Agriculture, Lands & Forestry, Colombo.

Honadle, George and **Jerry VanSant.** 1985. *Implementation for Sustainability: Lessons from Integrated Rural Development*. Kumarian Press, West Hartford.

Land Commission. 1990. *Report of the Land Commission, 1987*. Department of Government Printing, Colombo.

Lynch, Owen J. 1991. Community-based Tenurial Strategies for Promoting Forest Conservation and Development in South and Southeast Asia. Paper prepared for a USAID conference on environmental and agricultural issues, Colombo, Sri Lanka, 10–13 September.

Marks, Malcom K. 1996. Monitoring and Evaluation Toolkit. Report prepared for the Agriculture Sector Development Grant, Phase II, A project of the USAID. International Resources Group Ltd, Prime Contractor, May.

M/TEWA. 1995. Strategy for the Preparation of a Biodiversity Action Plan. Ministry of Transport, Environment and Women's Affairs, Colombo.

Nanayakkara, G.L. Anandalal. 1996. Common Property Utilisation and Development. *Soba Environmental Publication* 6 (1): 36–42. October.

Panayotou, Theodore and **Peter S. Ashton.** 1992. *Not By Timber Alone: Economics and Ecology for Sustaining Tropical Forests*. Island Press, Washington, D.C.

The Asia Foundation. 1995. Annual Report, Colombo.

Valadez, Joseph and **Michael Bamberger.** 1994. *Monitoring and Evaluating Social Programs in Developing Countries: A Handbook for Policymakers, Managers and Researchers*. EDI Development Series, The World Bank, Washington, D.C.

White, Alan T. 1996. Collaborative and Community-based Management of Coral Reef Resources: Lessons from Sri Lanka and the Philippines. Paper presented at the National Workshop on Integrated Reef Resources Management, Malé, Maldives, March 16–20. Working Paper No. 2/1996 of the Coastal Resources Management Project, Sri Lanka.

Wijayaratna, C.M. 1994. Shared Control of Natural Resources (SCOR): An Integrated Watershed Management Approach to Optimise Production and Protection. *Sri Lanka Journal of Agricultural Economics* 2 (1): 60–97.

———. 1995. *A Participatory Holistic Approach to Land and Water Management in Watersheds*. International Irrigation Management Institute, Colombo.

World Bank. 1996. Staff Appraisal Report: India Ecodevelopment Project. South Asia Department, Agriculture and Water Division, Report No. 14914-IN. August.

10

Grassroots conservation practices: Revitalizing the traditions*

Madhav Gadgil

Introduction

Thirty years ago, Slobodkin (1968) asked if there were many prudent predators in the animal world. Prudent behaviour, in his definition, would involve deliberate restraints on present levels of predation in the interest of long-term sustenance of prey populations. He came to the conclusion that there was no evidence of prudence amongst animals; if animals left some prey alone it was only because it was more expensive to hunt it, in terms of energy or risks of injury, than alternative sources that would meet their needs. Humans then are the only prudent species known, for humans often do leave alone resources which could be harvested at lower energetic costs or risks than other exploited sources. The epic *Ramayana* starts with the famous lines of the sage Valmiki reprimanding a hunter for killing one of a pair of copulating storks. In South India, pelicans, storks and cormorants often nest smack in the middle of a village, with villagers leaving them well alone, although they may hunt them in other seasons and places. These traditions are carried over in modern times too with full protection to breeding birds in protected areas (PAs) like Ranganthittu Bird Sanctuary near Mysore, Karnataka, or Keoladeo National Park, Rajasthan.

Social context of conservation

Despite being the most destructive of the animal species then, humans are also the only species to observe deliberate restraints in

* The author is grateful to the Ministry of Environment and Forests, Government of India, for financial support.

resource harvests, to take pains to protect and nurture living resources, as women of Khejadai village near Jodhpur did in the 1680s to prevent the Maharaja's men from cutting down their sacred *Prosopis cinerarea* trees, or the women of the Garhwal Himalaya did in the 1970s to prevent the contractor's axemen from cutting down their oak and birch trees. Forms of such conservation measures have of course changed over time, in step with changes in the ways human societies relate to their base of natural resources.

Till 10,000 years ago, all human societies subsisted through hunting and gathering. This involved collection of resources from a relatively restricted resource catchment, of a few hundred square kilometres at a time, by a hunting band consisting of 50–60 individuals. Such bands were nomadic, shifting with time, or with resource exhaustion. Australian aborigines were at this stage of economy when first contacted by Europeans a few centuries ago. They evidently did display some elements of resource conservation behaviour, such as observance of some sacred sites where life might be left immune from human interference. But elsewhere there is evidence that hunter-gatherers, when colonizing continents such as the Americas, may have indulged in excessive hunting and been responsible for extinction of favoured prey species of larger birds and mammals (Gadgil 1995).

Cultivation of crops and husbanding of livestock radically changed the relationship of humans to the natural world. People now became more sedentary, their groups increased in size and they developed stronger forms of territoriality. In the more primitive stages, such societies, often termed horticultural societies, practice very low-intensity shifting cultivation which produces little surplus for exchange (Lenski and Lenski 1978). Under these conditions, human groups became more firmly tied to particular localities. Of course when newly colonizing a region, such people would have the option of moving on to newer localities as their populations grew because of the advantage of numbers and availability of food reserves. But as a region becomes saturated by primitive subsistence agriculturists, people would come to be increasingly dependent on the living resource base of their particular, relatively restricted territories. Overuse of the resources of these territories may have serious adverse consequences for such societies, weakening them, especially in occasional years of poor crop production or severe winters, and subjecting them to defeat, expulsion from territory, even massacre by neighbouring groups (Gadgil and Guha 1992).

It thus appears plausible that small-scale, homogeneous societies with limited resource catchments, would tend to develop community-level institutions for regulating harvesting behaviour, promoting sustainability (Gadgil et al. in press). Gadgil and Berkes (1991) identified four kinds of 'rules of thumb' as social restraints underlying such indigenous biological conservation practices:

1. Provide total protection to some biological communities or habitat patches: These may include pools along river courses, sacred ponds, sacred mountains, meadows and forests. For example, sacred groves were once widely protected from Africa to China (Gadgil 1991; Yu 1991), in fact, throughout the Old World. They continue to be so protected even after the population's conversion to Christianity in the tribal state of Mizoram in north-eastern India, now being called 'safety forest', while the village woodlot from which regulated harvests are made is called the 'supply forest' (Malhotra 1990). Ecological theory suggests that providing such absolute protection in 'refugia' can be a very effective way of ensuring persistence of biological populations (Joshi and Gadgil 1991).

2. Provide total protection to certain selected species: Trees of all species of the genus *Ficus* are protected in many parts of the Old World. It is notable that *Ficus* is now considered a keystone resource significant to the conservation of overall biodiversity (Terborgh 1986). Local people seem to be often aware of the importance of *Ficus* as affording food and shelter for a wide range of birds, bats and primates, and it is not difficult to imagine that such understanding was converted into widespread protection of *Ficus* trees at some point in the distant past. Taboos with apparent functional significance may also be placed on some less obvious species within the ecological community. For example, some Amazon fish species considered important for folk medicine are taboo and are avoided as food, as statistically shown by Begossi and de Souza Braga (1992).

3. Protect critical life history stages: In South India, fruit bats may be hunted when away foraging, but not at daytime roosts on trees that may be in the midst of villages. Many waders are hunted outside the breeding season, not at heronries, which may again be on trees lining village streets. Cree Indians of James Bay in the subarctic are avid hunters of the Canada goose, a major subsistence resource, but never kill or even disturb nesting geese (Berkes 1982). The danger of overharvest and depletion of a population is clearly far greater if they are hunted in these vulnerable stages, and the protection afforded to them seems a clear case of ecological prudence (Slobodkin 1968; Gadgil and Guha 1992).

4. Organize resource harvests under the supervision of a local expert: Many traditional resource harvesting systems rely on the guidance of a traditional expert to organize the harvest, control access, supervise local rules and generally act as a 'steward' (Feit 1986). Examples may be found in diverse geographical areas such as the Canadian North, Central Africa and Oceania (Berkes 1989). This practice ensures the proper use and transmission of practical ecological expertise. Further, in some societies, major events of resource harvest are carried out as a short-term, prescribed group effort. Thus, many tribal groups engage once a year in a large-scale communal hunt. Such a group exercise may have served the purpose of group-level assessment of the status of prey populations, and their habitats. This in turn may have helped in continually adjusting resource harvest practices so as to sustain yields and conserve diversity.

It may be noted that even today's scientific prescriptions for conservation of biodiversity are little more than such 'rules of thumb', as can be seen from the debates such as SLOSS (Some Large or Several Small) concerning the feasibility of varying sizes of PAs. Indeed, as Slobodkin (1988) argues, current ecological theory helps us little in arriving at practical prescriptions for resource use and conservation. Such prescriptions are best derived from long-term observations of a particular ecosystem—very much the forte of indigenous knowledge (Ehrlich 1987).

Many such practices thus lead to conservation of biological diversity coupled with sustainable harvests of biological resources. But these societies did not pursue conservation or sustainable use as explicitly stated objectives. Rather the stated purposes tend to be to evade divine displeasure or to adhere to social norms. Indeed individual members of such small-scale societies seem to be highly motivated to keep the gods happy and to abide by social conventions, which ensures compliance.

Over much of the world today, small-scale horticultural societies have given way to larger-scale agrarian or industrial societies (including pastoral communities). These changes have been triggered by technological advances permitting higher levels of surplus production, initially through cultivation and later through manufacture. This permits movements of foodgrains and other commodities over larger distances expanding societal resource catchments. This expansion has twofold implications: people are no longer intimately tied to their own localities, depending largely on the resources they gather or produce with their own labour. Societal well-being is then less firmly linked to the well-being of the local ecosystems, weakening the feedback from the over-use of resource base. At the same time, local communities are no longer in firm control of their own resource catchments, as resources are processed, transported, and traded on larger spatial scales.

Under these circumstances, small-scale homogeneous societies make way for large-scale, stratified societies with stronger centralized state apparati assuming greater and greater control over the resource base at the cost of local communities. With much weaker motivation for sustainable use, such societies often indulge in over-harvests focusing on those resources which are at any given moment most profitable to use. As these are exhausted, the pressure shifts in a sequence to those next most profitable. In such societies, technological change also gathers pace, so that newer and newer kinds of resources, less and less accessible resources can substitute for the more usable, more accessible resources that are exhausted. It is not that these large-scale societies do not become aware of the possible dangers of resource over-use. Indeed when they do so, they tend to explicitly state resource conservation as a societal objective and attempt to implement it through centralized regulation by the state machinery (Figure 10.1). These measures of large-scale societies are however not necessarily any more effective than those of small-scale societies. Indeed, given the difficulties that large, often ineffective bureaucracies have of dealing with complex situations, such resource conservation measures of large-scale societies may in fact be far less effective.

It is these difficulties that the large-scale societies face which have today led to a revival of interest in community-based conservation (CBC) practices so characteristic of small-scale societies. This revival is of particular relevance in the Indian society, which is today an intricate mosaic of hunter-gatherers (e.g., Sentinelese islanders in Andamans, who remain entirely isolated and self-sufficient to this day), shifting cultivators (e.g., many groups of north-eastern India), subsistence cultivators (e.g., small holders of rainfed lands in semi-arid tracts of Karnataka and Andhra Pradesh), nomadic herders (e.g., Gujjars of Himachal Pradesh or Bakarwals of Kashmir), practitioners of intensive, irrigated, chemicalized agriculture (e.g., many

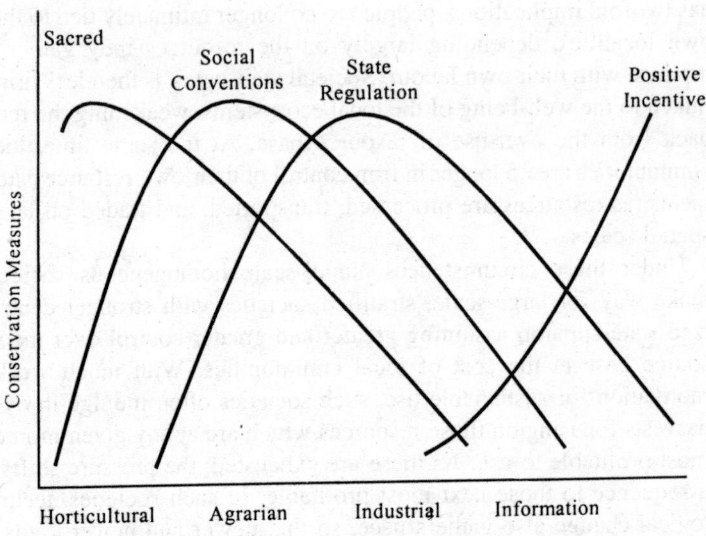

Source: Gadgil et al. (in press)

Figure 10.1: Dominant forms of social institutions involved in implementing restraints in the use of biological resources at different stages of societal evolution

larger farmers of Punjab), and those in organized services-industries-sectors (e.g., industrial labour of Mumbai). As Irawati Karve (1961) suggests, the Indian society is still not so much a single large-scale society, as an agglomeration of thousands upon thousands of caste, tribe, community groups, each with its own distinctive way of life. In this complex society many resource use practices, including conservation practices, characteristic of older small-scale societies still persist, albeit invariably in an attenuated form.

Forces of change

Four kinds of forces—social, cultural, economic and political—are responsible for erosion of the CBC practices deriving from traditions of small-scale societies. Social sanctions for violating accepted norms were the basis of the implementation of the prescribed restraints. Such sanctions depend on the authority of institutions like joint families, tribal or caste councils. These social institutions are rapidly dissolving as people are displaced, move around and accept new occupations. In Siddapur taluka of Uttara Kannada district in Karnataka, for instance, our field investigations showed that sacred groves have been felled in villages like Golgodu by one or two individuals who are now quite willing to defy the edicts of the local caste council, edicts that they would have found impossible to violate in earlier times.

Equally potent are cultural changes, especially changes in religious belief systems. Religious practices of the horticultural societies, often termed animist, are rooted in a view of the world as a 'community of beings'. In this view humans are members of a larger natural community which embraces other animals, plants, even inanimate entities such as rocks, streams or mountain peaks. Many of these other non-human community members are respected, revered and treated as sacred due to which individual trees or groves or stretches along rivers or mountain peaks are regarded as sacred and given protection, such protection being the basis of practices promoting conservation of biological diversity. Religious characteristics of agricultural societies such as Islam and Christianity tend to substitute beliefs in a supreme God detached from the concrete natural world for such nature worship. This desacralization of nature is not so complete in the eastern religious systems such as Hinduism and Buddhism which incorporate some elements of nature worship in their practices. Nevertheless, even in these religions, monumental temples come to replace great forests as objects of worship. Indeed, Hindu priests today often perform rituals to propitiate deities resident in sacred *peepal* trees or sacred groves to permit their being cut down, often to be replaced by an idol and a temple. Christian priests tend to be more aggressive and demand destruction of such sacred animist objects.

Another set of beliefs, those of modern science, also consider attribution of sacred qualities to nature as superstitious and therefore to be rejected. In fact, several rationalists have condemned any suggestions that these traditions have played a worthwhile role in the conservation of biodiversity and deserve to be appropriately incorporated in modern conservation practices.

Thus, social and cultural changes attendant on modernization go against these community-based traditions of nature conservation. Even more potent are modern economic and political forces. These traditions had their origin amongst subsistence societies which largely depended on local resources. The plants, animals, forests, freshwater ecosystems, lagoons in coral reefs being protected by these societies, had little market value. Today these biological resources often have considerable market value, and since money can now procure a diversity of commodities, people are motivated to convert local diversity of biological resources into the common currency of money, and use this money to procure a diversity of other, often man-made objects, that markets can supply. Everywhere, therefore, people are tempted to liquidate sacred trees and forests to make money. Similarly, they are now willing to eliminate monkeys once regarded as sacred, if they are hindrances in the way of making money, as for instance by causing damage to apple orchards.

Furthermore, there is an overall change in the power structure in modern times, with local communities losing control over resources in favour of centralized bureaucracies or industrial corporations. These agencies are also motivated to liquidate the biological resources protected as sacred to dedicate them to commercial use. In fact, state bureaucracies and commercial enterprises generally work together to take control of sacred groves or sacred tanks from local communities and clearfell them or convert them to carp culture, respectively.

Community revivals

Thus, a variety of strong forces are today eroding CBC practices. Yet these practices have by no means disappeared. Indeed some persist on a widespread scale; hundreds of thousands of sacred *peepal* (*Ficus bengalensis*), banyan (*Ficus religiosa*) and other fig trees still dot the Indian countryside. Other practices are even being revived. Thus, in many parts of north-east India sacred groves were liquidated in the 1950s following widespread penetration of market forces and conversion to Christianity. But in this region of extensive slash and burn cultivation the sacred groves used to serve as firebreaks; in their absence fires began to consume villages. Many groups have therefore reinstated total protection to forest belts surrounding their villages. These are protected through the same system of social sanctions as was used earlier to enforce protection to sacred groves. However, having embraced Christianity, the tribals no longer regard these as sacred, instead they are termed safety forests or forest reserves (Malhotra 1990; Gadgil et al. in press). This revival has been facilitated by the fact that most of the north-eastern tribal groups concerned still retain their traditional social organization. Moreover, the land in this region is still largely in private hands, owned by tribal chiefs, or by communities.

There is also now a greater awareness of the imperatives for biodiversity conservation in both the large-scale societies of India and elsewhere, as well as amongst the smaller social groups. Over the last 50 years there have been other assertions of such CBC. Thus, in the Udupi taluka of coastal Karnataka, villagers have set up a large sacred grove called Pilar kaan. In Siddapur taluka, also in Karnataka, individual farmers have established small sacred groves on their own farms. In the same state there has also been a parallel programme of establishment of sacred groves, called *pavitravanas*, by the Forest Department (FD).

State-sponsored conservation

India has a vigorous programme of nature conservation sponsored by the centralized bureaucratic apparatus. The scale of the size of nature reserves in this system is of course much larger than the sacred spaces mentioned above (Figure 10.2). These are protected primarily through state-sponsored regulation on the basis of codified laws (Figure 10.1). The state machinery responsible is strongly motivated to accumulate regulatory powers; it is also preoccupied with narrow sectoral interests. In consequence, it views the basic subsistence demands of local people as diametrically opposed to interests of conservation. In addition, the present model of development is seen by the state and larger society to be achievable only in opposition to conservation.

Such an attitude implies that there are very large opportunity costs attached to any conservation effort. Consequently, such efforts enjoy the support of a relatively narrow segment of society, mostly from amongst urban middle classes. In particular, there is little support to state-sponsored conservation efforts by the local people. Conservation is therefore forced to rely on guns and guards, making it a very expensive proposition. Furthermore, given its lack of accountability, the state apparatus in charge of conservation tends to squander much of the resources made available to it on salaries and perks, on buildings and vehicles. The Indian state, in common with many other countries, goes about the business of conservation in a highly wasteful and inefficient manner.

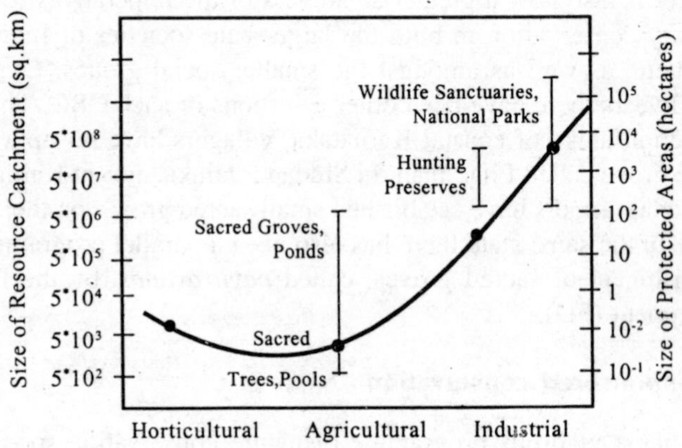

Pure hunter-gatherer societies have catchments of the order of 5×10^3 sq. km. It is not certain if they maintained any PAs. Slash and burn and low input settled cultivators have somewhat smaller resource catchments around 5×10^2 sq. km and maintained PAs ranging in size from around 0.01 to 100 hectares. More advanced agrarian societies have larger resource catchments of the order of 5×10^5 sq. km, and they maintain hunting preserves of the order of 100 to 1,000 hectares in size. The modern industrial societies have resource catchments spanning the whole biosphere and concentrate on PAs of a thousand to a million hectares.

Figure 10.2: Relation between sizes of resources catchments and sizes of protected areas

On top of this, conservation is practised in a simply wrong-headed fashion. The science of ecology has no broad generalizations that can be applied to make detailed management decisions at the field level (Ludwig et al. 1993). Rather, such management is best based on long-range, locality-specific observations of the behaviour of particular ecosystems, adjusting human interventions in light of the observed effects. The state apparatus has little such information available, and most conservation prescriptions therefore tend to be arbitrary, and often result in wholly undesired consequences.

The deficiencies of this approach are evident in what has happened at the Keoladeo Ghana National Park (previously the Bharatpur Bird Sanctuary), Rajasthan, over the last 12 years. This is a wetland of several hundred hectares created by bunding a tributary of the river Yamuna some 150 years ago. The impounded waters have always been used for irrigation in the dry months of the summer, and the wetlands themselves have provided excellent grazing for cattle and especially water buffaloes. These wetlands have been attracting enormous numbers of aquatic birds. The locality served as a hunting preserve of the Maharaja of Bharatpur in the pre-independence days, supporting shoots of tens of thousands of ducks and teals in a single day.

Following independence, the Bharatpur wetlands came to constitute one of the first wildlife sanctuaries of the country. This of course led to the suspension of winter shoots. But the use of water for irrigation and grazing by livestock continued. In the 1960s some scientific studies were initiated at Bharatpur, but these focused on migration patterns, and did not provide information on the functioning of the wetland. In particular, there was no scientific information pertaining to the impact of grazing on the ecosystem and birds. It was nevertheless assumed by scientists as well as forest managers that elimination of grazing would be highly desirable. Bharatpur used to support the endangered Siberian crane in the winter and scientists of the US-based Crane Foundation wrote to the then Prime Minister, Mrs Indira Gandhi, urging her to ban grazing in the Sanctuary.

As a result of these demands, the government imposed a ban on grazing in 1982. The local villagers, several hundreds of whose cattle and buffaloes grazed in the Sanctuary were never consulted; no provisions were made for alternative fodder supply for these animals. As a result the villagers protested against the ban, there was police firing in which several people were killed and the ban was made to stick (Prasad and Dhawan 1982). But it turned out that the ban was counter-productive from the perspective of the birds, especially the wintering waterfowl. In the absence of grazing, a grass, Paspalum, grew unchecked, choking out the wetland. Money was then spent on bulldozers to remove the grass. But these were nowhere as efficient as the buffaloes, and Keoladeo National Park continued to deteriorate as a bird habitat (Vijayan 1987). The villagers are now allowed to harvest grass by hand.

The Bharatpur incident has several lessons. The assumption that all human use is detrimental to conservation, held by respected scientists and managers, was evidently invalid. The local villagers need not have been forced to pay the opportunity cost of desisting from grazing in the Sanctuary. The additional expenses in protection and use of bulldozers that this policy entailed is a wasteful use of scarce resources.

While conservationists have thus succeeded in lobbying for elimination of subsistence demands from PAs, they have had less success in resisting the demands of the industrial sector. Consider for example, the case of Dandeli Wildlife Sanctuary, once amongst the most extensive of India's PAs encompassing over 5,000 sq. km of the hill tracts of Western Ghats in the Uttara Kannada district of Karnataka. In areas once within the Sanctuary, exhaustive harvesting of bamboo was permitted to supply the West Coast Paper Mill, incidentally at rates less than 0.1 per cent of the market price. Also permitted was mining for manganese, and construction of a series of reservoirs on the Kali river and its tributaries. The area of the Sanctuary is now reduced to 800 sq. km.

An alternative

Efficiency and equity

Any alternative approach must aim at greater efficiency by keeping the costs of conservation as low as possible. It should also serve the interests of equity by passing on the costs to those who would benefit from the conservation effort, and those who are wealthier and therefore in a better position to bear such costs to promote long-term, socially desirable objectives. The costs of conservation may be reduced in five ways: (*a*) by promoting those economic pursuits that are most compatible with the conservation objectives; (*b*) by assigning the role of custodians of biodiversity to those involved in such economic pursuits, so that the custodians are willing to perform for relatively modest levels of compensation; (*c*) by assigning the role of custodians to those in intimate contact with the biological communities being managed, so that the custodians can perform their role more effectively; (*d*) by assigning the role of custodians to the economically underprivileged, so that the custodians are willing to perform for relatively modest levels of compensation; and last, but not the least, (*e*) by establishing a firm link between conservation performance and the compensation paid to the custodians.

The interests of equity would be served in ways compatible with promoting efficiency by preferentially involving the economically weaker sections of the society (*a*) in economic activities deemed compatible with maintenance of high levels of biodiversity; (*b*) as custodians of biodiversity who may receive some compensation for their role.

Empowering local communities

We must then stand the current system of managing biodiversity on its head, and assign the role of custodians of biodiversity to local communities of people who live close to the earth; to the tribals, peasants, herders, fishermen and rural artisans. These are the people who depend for their day-to-day survival on the biological resources of their immediate surroundings. They have for centuries obtained livelihoods without destroying the natural diversity; its rapid erosion has largely followed large-scale commodification of nature in the last two centuries. Their traditions such as sacred groves have permitted survival of species like *Kunstleria keralensis* in the coastal plains of Kerala with population densities exceeding a thousand per sq. km. They have an intimate knowledge of natural resources, albeit often in very limited localities. They are amongst the poorest of Indian people and would be willing to perform for relatively low levels of rewards.

Experience of recent years has shown that communities of such people, also called the ecosystem people by Dasmann (1988), are often willing, even eager to take on the role of custodians of local natural resources when permitted to do so. The most notable instance of this is the experiment of joint forest management initiated some 20 years ago in the predominantly tribal Midnapore district of West Bengal (Deb and Malhotra 1993; Poffenberger and McGean 1996; see also, Sarin et al., and Raju in this book).

Focusing on biodiversity

There is, however, a distinct limitation to the range of diversity that the villagers would thus be motivated to protect on grounds of utility. The larger interest in protection of the entire spectrum of biodiversity would obviously encompass many additional elements of no immediate utility to the local communities. Thus dead trees left standing for years are an important habitat for a whole range of wood-eating insects, wood-rotting fungi and hole-nesting birds that may play no role in village economy. Villagers would then be inclined to quickly harvest such dead trees and use them as timber or fuelwood. If we wish to motivate the villagers to conserve such habitats, we need to offer them additional incentives. Indeed it is clear that if the rationale for conservation of biological diversity includes retaining options for future economic use, or ethical or aesthetic grounds, then simply assuring access to villagers for immediate use would lead to socially sub-optimal levels of biodiversity conservation. The larger Indian (or global) society must then mobilize additional resources to raise the level of conservation efforts towards socially desirable levels. Today such additional resources are indeed being made available, and used in a highly wasteful manner by the state bureaucracy. It is my contention that they should instead flow to local communities who would use them in a far more effective fashion.

Building institutions

Such a flow of resources from the state to local communities is clearly analogous to the subsidies farmers in some countries receive to adopt land use practices in the broader interests of soil conservation, or payments made by Nature Conservancy in the US to landowners to adopt land use practices compatible with conservation of biodiversity. Our focus is on communities, rather than on private parties, for the proposal concerns public lands and waters. To operationalize such a proposal would require answers to a whole series of questions:

- How are local communities to be bounded? How many households should they involve? How homogeneous must such communities be?
- How should parcels of land and water be assigned to particular local communities to manage in the interests of biodiversity?
- How should the national or global society go about assigning conservation value to different elements of biodiversity? How should this information be shared with local communities?
- How can local communities be organized on different spatial scales, to enable them to manage local natural resources effectively in the interests of biodiversity?

- How should we arrive at the levels of financial or other incentives to be awarded to local communities to conserve biodiversity?
- How must we ensure that the conservation performance of local communities is firmly linked to the level of rewards flowing to them?
- How should the funds used to promote biodiversity be generated?

Evidently, we need to design and build a series of alternative institutions to manage the country's biodiversity, while dismantling part of the current wasteful, inequitable machinery. Suggested below is a broad approach to such a task; it is important that there be sufficient flexibility to permit the emergence of institutions appropriate to the local conditions.

Linking land and people

A likely unit for a CBC system in India would be a group of people in daily contact, somewhere between 20 to 200 households living in a hamlet or a small village. Such a group of people would be relatively homogeneous, economically and culturally, and likely to act in a cohesive fashion. It should be authorized to manage all the public lands and waters within a defined territory, and to coordinate actions of private landowners within the bounds of such a territory. This management would of course operate within a broader socially acceptable framework which may specify that no public lands or wetlands should be brought under cultivation, or that no trees belonging to genus *Ficus* should be felled. The local communities should be organized into larger and larger groups within a nested hierarchy with the larger groups serving to coordinate the activities of component neighbouring groups and to resolve disputes. The larger resource management groups should form appropriate links with political institutions at corresponding levels: *panchayat* or village cluster level (population = 10,000), taluka or county level (population = 100,000), and district level (population = 1 million).

A number of difficulties would of course arise pertaining to the complexities of social organization and to the overlap of areas of resource use by neighbouring communities. Special cases of overlap would involve nomadic herder communities. It would be best to create institutions for finding locality-specific solutions to such difficulties. We have elsewhere provided much more detailed suggestions on such an institutional framework (Rao and Gadgil 1995).

Valuing biodiversity

The local natural resource management groups would each have their own system of valuing biodiversity, based on utility, culture, religion. The values may change with time, as when allopathic drugs supplement the use of herbal remedies. These local systems would in the long run be inadequate guides to organizing a national, or a global effort at conservation of biodiversity. It is therefore also necessary to set priorities at a larger level and to ensure that these provide the framework for the local conservation efforts. Such priorities may be set at many different scales; of specific genes, of individuals of specific species, of specific types of ecological habitats and so on. In general, elements representing more isolated evolutionary lineages, with restricted geographical ranges, under greater threats of extermination, of greater economic utility, would tend to be valued more. Establishment of such priorities is a technical exercise and might have to be organized by a technical body such as the Subsidiary Body on Scientific, Technical and Technological Advice to the Conference of Parties of the international Convention on Biological Diversity. It is however important that the relevant information from such an exercise of setting conservation priorities should ultimately reach local communities. Educational institutions at various levels ranging from universities to village primary schools could play a vital role in this process. Indeed, information on the conservation value of different species of living organisms or different types of habitats could constitute an important component of the environmental education curriculum at all levels.

Rewarding conservation efforts

Armed with information on conservation values of different elements of local ecology, and with the authority to manage public lands and waters in their immediate neighbourhoods, local communities could design management strategies that would preserve, even enhance the total conservation value of biological communities within their territories. Human communities in continual touch with the local biological communities would be best equipped to do so. They could then continually adjust their resource use strategies in the interest of biodiversity, provided that they are adequately rewarded. The perception of local people as to the adequacy of such external rewards would to a great extent depend on their ability to directly benefit from biodiversity. An important component of rewards for conservation efforts could then be building the capacity of local communities to add value to local biodiversity. This may, for instance, involve preparing alcohol extracts of ingredients used in the pharmaceutical industry, or organizing eco-tourism. Over and above that the local communities may need to be compensated for foregoing some opportunities for economically more rewarding uses of land or water within their territory. Exactly what level of rewards the local communities receive and what level of conservation effort they put in would depend on the demand for conservation generated at the national or global level, and the supply of conservation effort offered at the local levels. However, we believe that such a system would deliver conservation effort in a far more cost-effective fashion than the current system of investing in regulatory efforts by a bureaucracy which is not accountable for what it ultimately delivers.

Ensuring accountability

Ensuring accountability in the proposed alternative system would depend on periodic monitoring of the biological communities within the territories of the various local communities. Such monitoring would have to be designed at the national or global level, and then adapted to local conditions. Its implementation in the field could best involve local educational institutions guided by appropriate higher level institutions. Indeed, such a programme could form a valuable component of teaching of ecology at all levels. It would of course be necessary to guard against local schools overestimating local levels of biodiversity to attract higher levels of rewards to their own territory. This could be ensured by organizing exchanges of students and teachers across districts or states to serve as independent auditors of the monitoring process. Such visits too would have considerable educational value.

Generating finances

India as a nation state is already investing substantial amounts of state revenue in conservation efforts. Its population is further contributing by accepting costs of conservation, whether it be through foregoing some economically more attractive development options, or tolerating crop damage by elephants. What we propose is that these financial inputs be organized as a national biodiversity conservation fund, perhaps with contributions from international sources such as the Global Environmental Facility. These funds could then be rationally allocated to the various local communities in relation to the levels of conservation value of biological communities in their respective territories. We have suggested elsewhere fuller details of how this system may be organized (Rao and Gadgil 1995).

Conclusions and policy implications

1. Biodiversity elements of value are by no means confined to extensive tracts of pristine ecosystems; they occur even in the midst of extensively humanized landscapes, as with wild relatives of paddy and climax rainforest species in sacred groves on the densely populated coastal plains of Kerala. Conservation of biodiversity must therefore be made a people's movement, as forest protection has been made with the joint forest management programmes. Times are particularly opportune for such an initiative since the 73rd Amendment to the Indian Constitution, dealing with village *panchayat* bodies, has set the stage for a decentralized system of governance, including management of natural resources, throughout the country (see Kothari et al. in this book).

2. The focus of biodiversity conservation efforts must shift from a small number of PAs, to the entire countryside.

3. Local communities should be conferred much greater control over public lands and waters in their own localities, and encouraged to manage these and private lands in a biodiversity friendly fashion by enhancing their capacity to add value to biological resources.

4. Local communities should be further encouraged to maintain biodiversity through appropriate financial rewards.

5. The funds being deployed towards conservation efforts today in the form of salaries and perks of bureaucrats and technocrats, their jeeps and guns and buildings, should be redeployed over a period of time to provide positive incentives to local communities.

6. Technical inputs from the national or global level could help to assign conservation value to specific elements of biodiversity and to organize a reliable, transparent system of monitoring biodiversity levels. Educational institutions at all levels could play an important role in this effort.

7. In the long run a very lean bureaucratic apparatus should be retained to play a coordinating, facilitatory role and to ensure that local communities can effectively enforce a desired system of protection and management of the natural resource base.

8. Such a system would create a very efficient market for conservation performance so that funds earmarked to promote biodiversity would flow to localities and local communities endowed with capabilities of conserving high levels of biodiversity.

9. This system would channelize rewards for conservation action to relatively poorer communities living close to the earth, thereby also serving the ends of social justice.

REFERENCES

Begossi, A. and **F.M. de Souza Braga.** 1992. Food taboos and folk medicine among fishermen from the Tocantins river (Brazil). *Amazoniana* 12: 101–18.

Berkes, F. 1982. Waterfowl management and northern native peoples with reference to Cree hunters of James Bay. *Musk-Ox* 30: 23–35.

———. 1989. *Common Property Resources: Ecology and Community-Based Sustainable Development*. Belhaven Press, London.

Bhat, K.G. 1996. Personal Communication with Shri K.G. Bhat.

Dasmann, R.F. 1988. Toward a Biosphere Consciousness, in D. Worster (ed.). *The Ends of the Earth*. Cambridge University Press, Cambridge.

Deb, D. and **K.C. Malhotra.** 1993. People's Participation: The Evolution of Joint Forest Management in South-West Bengal, in S.B. Roy and A.K. Ghosh (eds). *People of India: Biocultural Dimensions*. Inter India Publications, New Delhi.

Ehrlich, P.R. 1987. Population Biology, Conservation Biology, and the Future of Humanity. *BioScience* 37: 757–63.

Feit, H.A. 1986. James Bay Cree Indian Management and Moral Consideration of Fur-bearers, in Alberta Society of Professional Biologists. *Native People and Renewable Resource Management*. Alberta Society of Professional Biologists, Edmonton.

Gadgil, M. 1991. Conserving India's Biodiversity: The Societal Context. *Evolutionary Trends in Plants* 5 (1): 3–8.

———. 1995. Traditional Conservation Practices, in *Encyclopedia of Environmental Biology*. pp. 423–25. Academic Press.

Gadgil, M. and **F. Berkes.** 1991. Traditional Resource Management Systems. *Resource Management and Optimization* 18 (3–4): 127–41.

Gadgil, M. and **R. Guha.** 1992. *This Fissured Land: An Ecological History of India*. Oxford University Press, New Delhi and University of California Press, Berkeley.

Gadgil, M., N.S. Hemam and **B.M. Reddy.** (in press). People, Refugia and Resilience, in F. Berkes and C. Folke (eds). *Linking Social and Ecological Systems: Institutional Learning for Resilience*. Cambridge University Press, Cambridge.

Joshi, N.V. and **M. Gadgil.** 1991. On the Role of Refugia in Promoting Prudent Use of Biological Resources. *Theoretical Population Biology* 40: 211–29.

Karve, I. 1961. *Hindu Society—An Interpretation*. Deccan College, Poona.

Lenski, G. and **J. Lenski.** 1978. *Human Societies: An Introduction to Macrosociology*. McGraw-Hill, New York.

Ludwig, D., R. Hilborn and **C. Walters.** 1993. Uncertainty, Resource Exploitation and Conservation: Lessons from History. *Science* 260: 17–36.

Malhotra, K.C. 1990. Village Supply and Safety Forest in Mizoram: A Traditional Practice of Protecting Ecosystem, in Abstracts of the Plenary, Symposium Papers and Posters presented at the V International Congress of Ecology, Yokohama, Japan.

Poffenberger, M. and **B. McGean** (eds)! 1996. *Village Voices, Forest Choices: Joint Forest Management in India.* Oxford University Press, New Delhi.

Prasad, A. and **H. Dhawan.** 1982. A Sanctuary for Birds Only? Kalpavriksh, New Delhi. Mimeo.

Rao, P.R.S. and **M. Gadgil.** 1995. People's Nature, Health and Education Bill. *Economic and Political Weekly*, 7 October, 40: 2501–12.

Slobodkin, L.B. 1968. How to be a Predator. *American Zoologist* 8: 43–51.

———. 1988. Intellectual Problems of Applied Ecology. *BioScience* 38: 337–42.

Terborgh, J. 1986. Keystone Plant Resources in the Tropical Forest, in M.E. Soule (ed.). *Conservation Biology*. Sinauer Associates, Inc., Sunderland, Massachusetts.

Vijayan, V.S. 1987. *Keoladeo National Park.* Bombay Natural History Society, Bombay.

Yu, K.J. 1991. Translate the Philosophical Ideal into Reality: Feng-Shui as Applied Human Ecology. Paper presented at the International Conference of Human Ecology, Goteborg.

11

Conserving the sacred: Ecological and policy implications

P.S. Ramakrishnan

Introduction

The concept of the sacred grove in India has its roots in antiquity, stretching into pre-historic times, even before the Vedic age, the Vedas representing the only recorded remains of the thoughts of the ancient Aryans who migrated into this subcontinent. In their migration from the steppes of Central Asia through Balkh in Khorassan (Dandekar 1979) to the Indian subcontinent, the ancient Vedic people of pre-historic times assimilated new environmental values; they also incorporated into their value system the concept of the 'sacred grove' from the original inhabitants of the Indian subcontinent (Vannucci 1994). Though many traditional societies value a large number of plant species from the wild for a variety of reasons, for food or for medicine, sacredness attached to species is perhaps more recent, being part of the post-Vedic Hindu ritualism. Thus the already existing sacred grove concept of the original pre-Vedic inhabitants of India was extended by the Vedic migrants down to the species level.

Buddhism and Jainism, initially branching out as revivalistic religious offshoots of Hinduism led to revivalism of conservation practices too. On one extreme they led to a sect of Jains, the 'Digambara Jains', dead-set against the killing of all living organisms; and at the other end of the spectrum was the sacred landscape of the Sikkimese Buddhists, further refined and concretized from post-Vedic conceptualization, based on a holistic ecological philosophy.

This paper looks at the concept of the sacred from varied spatial scales and considers policy implications for conservation of biodiversity in India, with peoples' participation.

The concept of sacred species

It is reasonable to assume that traditional Hindu society recognized individual species as objects of worship, based on accumulated empirical knowledge and their identified value for one reason or the other. Thus, *Ficus religiosa* (the *peepal* tree) and other species of the same genus form components of a variety of ecosystem types and support a variety of plant and animal biodiversity. The sacred basil or *tulsi* (*Ocimum sanctum*) is worshipped in all traditional homes as a goddess, and indeed is a multipurpose medicinal plant according to traditional Indian pharmacopoeia.

Others may not be worshipped in a religious sense, but form part of the socio-cultural traditions. The socially valued multipurpose *Quercus* (oak) species of Kumaon and Garhwal region of the western Himalayas are important for fodder and fuelwood and serve a variety of functions as an important component of the mountain forest ecosystem (Ramakrishnan et al. 1994b). They support a rich biodiversity in the ecosystem by improving soil fertility through efficient nutrient cycling and conserving soil moisture partly through humus build-up in the soil and partly through a deeply placed root system which has root biomass uniformly distributed throughout the soil profile. It may be noted here that oaks are the climax species in the mid-elevation regions in the western Himalayas. Being valuable timber species, there has been large-scale extraction of trees during the past few decades, and the subsequent large-scale conversion of the climax oak forests to early successional pine forests; pines are favoured by the foresters since they are fast growing and in demand for resins. This has done considerable ecological damage in the region, making the soil more acidic, adversely affecting nutrient cycling and soil fertility, almost forming a monoculture with little undergrowth and low biodiversity. No wonder that oaks are socially valued and are the focal point of much of the folk music and dance of the Kumaonis and the Garhwalis. The genesis behind the now well-known Chipko movement (saving forests by literally 'hugging' the trees) could be traced to the hardships felt by the local people as a consequence of large-scale conversion of oak to pine forests.

Working in north-east India (Ramakrishnan 1992), we soon realized that species which perform key functions in the ecosystem and thereby contribute to support/enhance biodiversity, are also species that are socially valued by the local community, for cultural or religious reasons. Thus the Nepalese alder (*Alnus nepalensis*) coming up in the shifting agriculture (*jhum*) fallow plots is protected from slashing and burning during *jhum* operations. Based on intuitive experience, the shifting agriculture farmer realizes that this species does good to his cropping system; indeed our studies suggest that this species may conserve up to about 120 kilograms of nitrogen per hectare per year through nitrogen fixation (Ramakrishnan 1992). Similarly, the bamboo species *Dendrocalamus hamiltonii, Bambusa tulda* and *B. khasiana*, which are important early successional species and which conserve nitrogen, phosphorus or potassium in the system are also highly valued by all communities in north-east India; they are often grown along the margin of agricultural plots and as part of the home garden system, or used for a variety of traditional activities, such as house construction, making household utensils, tubing for water transportation, thatching material, etc. These and other ecologically important keystone species are also socially selected keystone species. A broader survey done by this author in the Indian context also suggests a close parallelism existing between ecological and social processes (e.g., the locally valued *Prosopis cinerarea* in the arid state of Rajasthan and *Azadirachta indica* throughout the subcontinent, and many others discussed in this paper). The implication is that linking up ecological and social processes is significant for enhancing biodiversity in the ecosystem, building up of biodiversity based on the ecological integrity of the system, and indeed for ensuring rural peoples' participation as an integral part of the ecosystem function, rather than being a manipulator from outside (Ramakrishnan 1995b). This last point is extremely important for rehabilitation of degraded ecosystems and for rural development (Ramakrishnan et al. 1994a). Being able to identify themselves with a value system that they cherish through the socially selected keystone species, the local communities would be able to participate in the rehabilitation programme.

The sacred grove concept

Sacred groves are small patches of native vegetation type traditionally protected by the local communities. They may range in size from less than a hectare to a few square kilometres. Distributed throughout the country (Ramachandran and Mohanan 1991; Ramakrishnan 1992; WWF-India 1996), the current level of protection and conservation of these ecosystem types in India is hampered by erosion in traditional value systems. It is gratifying to note, however, that these traditional methods of social fencing of ecosystem types as conservation patches are being rediscovered.

Whilst there are many reports on general aspects of sacred groves (Box 11.1), the ecological value of the species within them and the ecosystem function of most of these are yet to be fully evaluated. The following example, therefore, is illustrative.

Mawsmai sacred grove: A relict rainforest ecosystem of Cherrapunji in Meghalaya

The climax vegetation at higher elevations in Meghalaya, at Cherrapunji (1,300m altitude), about 50 kilometres south of Shillong, is represented by the sacred grove. It is strictly protected for religious and cultural reasons. Cherrapunji represents a unique ecological situation in that it receives an exceptionally high rainfall of 10,372 mm, which may go up to 24,555 mm in an exceptional year such as 1974.

Though the vegetation is generally stunted because of unbalanced soil and extreme climatic conditions, the plant species diversity is very high. Among the trees in the grove, *Engelhardtia spicata* is the dominant species followed by *Syzygium cumini* and *Echinocarpus dasycarpus*, and important shrubs are *Clerodendron nutans* and *Phlogocanthus* spp.

Operating under highly stressed environmental conditions (including high leaching of nutrients and thin soils), this delicately balanced sacred grove forest ecosystem is highly fragile. It is not surprising then that the landscape of Cherrapunji, except for the relict sacred grove, is largely balded with arrested seral grassland types (Ramakrishnan and Ram 1988). The Cherrapunji ecosystem that stands desertified due to deforestation inflicted sometime in the distant past, now refuses to recover to its original state, as represented by the relic sacred grove. The fact that shifting agriculture, which is the predominant land use, remains banned by the village council for quite some time now, is suggestive of the part played by this land use in creating the present landscape. The sharp boundary between the sacred grove and the balded landscape indicates that the system will never recover through natural processes of re-vegetation.

Linked with the drastic loss in biodiversity is the human suffering which is now immense (Ramakrishnan 1992). Ironically, water is a scarce commodity during the dry months in this high rainfall spot, a

Box 11.1
Distribution of Sacred Groves in India

In India, sacred groves occur under a variety of ecological situations. Many have evolved under resource-rich situations such as in Meghalaya in north-eastern India, Western Ghats region in southern India, or in the Bastar region in the state of Madhya Pradesh in central India. In that sense, this contradicts the viewpoint that forest conservation measures always follow from some perception of resource scarcity, because all the above examples are from resource-rich situations, unlike the situation elsewhere, as in arid regions of Rajasthan or elsewhere in East Africa (Rodgers 1994). Religion and culture are the over-riding considerations in the Indian context.

– According to a report by the Centre for Earth System Studies in Kerala, at present there are 240 groves in the Western region of Kerala, in southern India. The largest of these (20 hectares) is the famous Iringole Kavu. Many of them are linked to temple premises. These *kavus* contain many of the rare and endangered species of the Western Ghats and rare medicinal plants of the region.

– Elsewhere in the Maharashtra region of the Western Ghats further north, Gadgil and Vartak (1974) have recorded a widespread network of sacred groves.

– Variously called *vanis, kenkris, oraans* or *shamlet dehs*, the groves of the arid region of Rajasthan in north-western India act as refugia of biodiversity for the people of the desert (Anonymous 1994).

– The sect of Bishnois, founded about 500 years ago in the Rajasthan desert, has given absolute protection not only to the *khejadi* tree (*Prosopis cinerarea*), a multipurpose legume tree valued by the villagers, but also promoted plant and animal biodiversity within their village ecosystem boundary. Such trees are valued by the local people for pods for food, leaves for fodder and manure and branches as construction material. It is said that some 350 years ago, many Bishnois even laid down their lives by hugging the trees, when the Prince of Jodhpur tried to fell *khejadi* trees for his lime kilns.

– In Meghalaya, in the north-eastern hill region, many sacred groves are still well protected, in spite of a rapid decline in the traditional value system with the advent of Christianity (Boojh and Ramakrishnan 1983; Khiewtam and Ramakrishnan 1989). The traditional religious belief is that the gods and the spirits of the ancestors live in these groves. The Mawphlang grove close to Shillong town is one of the best preserved, set in a degraded landscape all around. Indeed, the Mawsmai grove in Cherrapunji of about 6 sq.km of protected mixed broad-leaved rainforest, is an island in a bleak desertified landscape. Ceremonies which used to be performed regularly in these groves, are no longer common; rituals have been stopped in many of them for the last few years. Removal of plants or plant parts is considered to offend the ruling deity, leading to local calamities. We have recorded in the Cherrapunji region, 21 sacred groves with varied degrees of human disturbance (Khiewtam and Ramakrishnan 1989).

– Elsewhere in the north-eastern state of Mizoram, there are community woodlots called 'supply forests' from which only regulated harvests are permitted; then there are sacred groves which are 'safety forests' from which removal of biomass is strictly prohibited.

contender for being the wettest spot on earth. All the water flows away into the plains down below, as there is no vegetation to hold it and the soil, and there is no soil to recover the forest. For fuelwood, the tribal villager has to trek long distances. The ruins of villages all around remind one of yesteryears and migrations of the past. The tribal, who is traditionally bound to the land and forests, has been forced to seek other avenues for survival.

It is in this context that an understanding of the sacred ecosystem function (Ramakrishnan 1992; Khiewtam and Ramakrishnan 1993) is significant for designing a rehabilitation strategy of the degraded landscape, with peoples' participation (Ramakrishnan et al. 1994a). The plant biodiversity contained therein could be used as a resource for ecosystem rehabilitation based on successional concepts (Ramakrishnan 1992).

The role of keystone species in conserving and enhancing biodiversity, and indeed in manipulating ecosystem function is an important area which has not been adequately explored (Ramakrishnan 1992). Keystone species play critical functions in an ecosystem, and could be used for not only managing pristine ecosystems but also for building up biodiversity in both natural and managed ecosystems, through appropriately conceived rehabilitation strategies (Ramakrishnan et al. 1994a).

The four dominant species in the sacred grove at Cherrapunji, namely *Englehardtia spicata*, *Echinocarpus dasycarpus*, *Syzygium cumini* and *Drimycarpus racemosus* are keystone species in the sense that their leaf litter contains high levels of nitrogen, phosphorus and potassium compared to other species in this ecosystem (Khiewtam and Ramakrishnan 1993). Their function of nutrient conservation in the ecosystem is crucial for rehabilitating degraded ecosystems especially when soil fertility is a constraining factor. Indeed, these four species and others needed for the rehabilitation of Cherrapunji are exclusive to the sacred grove. These and other keystone species are also socially selected keystone species, with implications for rehabilitation and biodiversity conservation with peoples' participation, as already discussed earlier.

The concept of sacred landscape

Until now, the concept of sacred landscape is only somewhat vaguely understood. One of the best examples of a sacred landscape is that represented all along the course of the sacred river Ganga, originating from the higher reaches of the Garhwal Himalaya of the north-west, tracing through the plains of Uttar Pradesh, Bihar and West Bengal, before the river drains into the Bay of Bengal. The sacred land, all along the course of the river, the human habitation and the land-based activities, the temples dating back to antiquity, the sacred cities such as Badrinath, Kedarnath, Rishikesh and Haridwar in the Himalayan and sub-Himalayan tracts, Allahabad and Varanasi in the Gangetic alluvial plains, all together represent a set of interconnected ecosystem types bound together by the sacred river itself. The variety of natural ecosystem types ranging from the alpine vegetation above the timberline, through the temperate oak and pine forests down below, the subtropical moist-deciduous to dry-deciduous forests in the plains and a variety of human-altered ecosystems such as terraced agriculture and valley land agriculture, all along are tightly linked together and controlled by the sacred river and its tributaries in a variety of ways through flooding and silt deposition. The sacred groves of the Bishnois in Rajasthan (Box 11.1), viewed as units around each Bishnoi village, could indeed be enlarged to encompass a cluster of villages forming a landscape unit—to include interacting ecosystem types such as agriculture, animal husbandry and domestic units of the village ecosystem, natural water bodies, and the protected natural ecosystem.

The concept of sacred landscape has found further holistic expression under the Buddhist philosophy of non-violence and kindness to all living beings. One example is cited below.

Demojong in West Sikkim district: A unique example of a sacred landscape

Sikkim has a long tradition of Buddhist religion. Buddhism is practised by about 25 per cent of the local population and the majority religion is Hinduism (70 per cent). Because of the rule by the Chogyal dynasty, since the time the first Chogyal (king) of Sikkim was crowned in 1642 in Norbugang in Yuksom, Buddhist traditions are deeply ingrained into the psyche of the Sikkimese people. This is evident in all walks of life—a rich tapestry woven with Buddhist symbolisms, legends, myths, rituals and festivals, the typical Sikkimese building architecture, and the large number of monasteries and stupas dotting the landscape throughout the state. It is important to note that these traditions are shared by all the three communities; the culture being a blend of the Buddhism of the Lepchas and the Bhutias and the Hinduism of the majority Nepalis.

Whilst Sikkim as a whole is considered to be sacred by the Sikkimese Buddhists, according to the sacred text Nay Sol, the area below Mount Khangchendzonga in West Sikkim, referred to as Demojong, is the most sacred of all, being the abode of Sikkim's deities (Box 11.2).

This region has a number of glacial lakes in the higher reaches. These are sacred lakes. The Rathong Chu, itself a sacred river, is said to have its source in nine holy lakes of the higher elevation, closer

Box 11.2

'Demojong', The Land of the Hidden Treasures!

Padmasambhava, who is highly revered and worshipped by the Sikkimese Buddhists, is considered to have blessed Yoksum and the surrounding landscape represented by 'Demojong' in the West Sikkim district of Sikkim, by placing a large number of hidden treasures (*ter*). Many of these sacred treasures were hidden by Lhabstsun Namkha Jigme in the Yoksum region. It is believed that these treasures are being discovered slowly and will be revealed only to enlightened Lamas, at appropriate times. Conserving these treasures, protecting them from polluting influences is critical.

The area below Mount Khangchendzonga in West Sikkim, referred to as 'Demojong' is the core of the sacred land of Sikkim. Yoksum is considered to be a 'Lhakhang' (altar) and 'Mandala' where the protective deities are made offerings to. No meaningful performance of Buddhist rituals is possible if this land and water is desecrated. Any large-scale human-induced perturbation in the land of the holy Yoksum region would destroy the hidden treasures, the *ters*, in such a manner that the chances of recovering them some time in the future by a visionary will diminish (it is said that the last such discovery was made by Terton Padma Lingpa, when he lived 540 years ago). Any major perturbation to the river system would disturb the ruling deities of the 109 hidden lakes of the river, thus leading to serious calamities.

Indeed, the very cultural fabric of the Sikkimese society is obviously dependent upon the conservation of the whole sacred landscape of interacting ecosystems, as was evident during discussions this author had with a cross-section of the Sikkimese society, cutting across religious, cultural and professional backgrounds of the people. The issue here is not merely a question of protecting a few physical structures or ruins. The uniqueness of this heritage site is that the value system here is interpreted in a more holistic sense—the soil, the water, the biota, the visible water bodies, the river and the less obvious notional lakes on the river bed, are all to be taken together with the physical monuments.

to the mountain peaks. Besides, the river in the Yoksum region itself is considered to have 109 hidden lakes. Both the visible and the notional lakes identified by the religious visionaries are said to have presiding deities, representing both the good and the evil. Propitiating these deities through various religious ceremonies is considered important for the welfare of the Sikkimese people. Indeed, conserving this rich tradition is considered to be significant for peace, harmony and welfare of not only this area but also of the region as a whole.

It is no wonder that Rathong Chu is the focus of religious rituals. During the Bum Chu ritual, considered the holiest of all festivals, held annually at Tashiding, Rathong Chu is said to turn white and start singing, and this is the water to be collected at the point where Rathong Chu meets Ringnya Chu. Attracting thousands of devotees from the state and the neighbouring region, the Bom Chu ritual is predictive in nature, in that it is suggested to be indicative of coming events—possible calamities and prosperity for the people of Sikkim. The water is kept in vases; if it overflows it is indicative of prosperity. Decline in water level is indicative of bad events such as drought, diseases, etc. Turbid water is indicative of unrest and conflicts.

More generalized rituals, such as the one done throughout Sikkim by the Buddhists during Pang-Lhabsol to propitiate the various ruling deities of the mountain peak of the Khangchendzonga, the midlands represented by the Yoksum region, and the lowlands down below, is indicative of the widespread respect with which this sacred landscape region is worshipped by the people.

Of the total catchment area of 328,000 hectares of Rathong Khola, 28,510 hectares is under snow cover. The vegetation is varied, ranging from alpine scrub vegetation at higher reaches to subtropical moist evergreen forests down below. Afforestation is essential over 4,290 hectares of catchment area of the Rathong Chu. This task is crucial for conservation of the sacred landscape as a whole, in particular for controlling erosion and flash floods. According to the data collated by the Himalayan Nature & Adventure Foundation, Siliguri, the Ouglthang and Rathong glaciers are retreating rapidly. Retreating glaciers create several moraine dams, containing sizeable quantities of water. Increased melting of snow and exceptionally high rainfall in a given year could result in dam bursts and flash floods in the lower regions. The 1988 dam burst is an example that could recur, if adequate catchment treatment measures are not initiated immediately.

The Sikkim Himalaya are richly endowed with biological resources spread over a variety of ecosystem types in a range of altitudes, from the alpine *Rhododendron* dominated scrub forest through conifer forests with *Abies densa* and *Tsuga demosa* getting down to mixed evergreen forests dominated by species such as *Castanopsis* spp. The region under consideration here has all these types over a very short transect running down from the alpine to the subtropical zone. Orchids are abundant. With rich wildlife represented by Himalayan black bear, Musk deer, Fishing cat, Leopard cat, Black-capped langur and a rich bird life, this unique landscape unit should be protected.

The region is rich in medicinal plants of value to traditional Tibetan pharmacopoeia, nurtured in Sikkim by the Buddhist monasteries (Tsarong 1995). Conserving these plants and their cultivation would ensure the survival of one of the oldest systems of medicine stretching back to more than 2,500 years. With recent attempts to revive traditional medicine, possibilities of providing sustainable livelihood to local communities through cultivation of medicinal plants are immense.

The entire region right from the Kangchendzonga to the Yoksum lowlands (the sacred region of Demojong) is most appropriate to be declared a National Heritage Site, with all its people, ecological and cultural heritage—the land and the land use systems (the traditional terraced agricultural system included), all the water bodies (the obvious and notional lakes), the Yoksum Chu, the monasteries, the historical sites, and the rich biodiversity—for conservation in a truly holistic spirit.

Conclusion

Ecological implications

At one level, the sacred species occurring in a variety of ecosystem types in the subcontinent, with all the sub-specific variations that they exhibit, should be conserved for human welfare. Whilst many species are of economic value as a source for medicine (*neem* and *tulsi*, for example) or as food (e.g., bananas or mangos amongst plants and cows among the animals), the ecological values could be diverse. As discussed, many of the socially valued species are ecologically significant keystone species having key roles in ecosystem structure and function, in space and time.

Sacred groves are repositories of wild germplasm of value not only for conservation biology but also as a source of germplasm for the rehabilitation of degraded rural landscape (Ramakrishnan et al. 1994a).

The 'sacred landscape' with a variety of natural and traditionally managed ecosystem types is a storehouse of traditional wisdom based on an ecologically meaningful management strategy, which is now being rediscovered in the context of a decentralized network of protected patches of biodiversity in a heterogeneous landscape.

Sustainable development, in my opinion, involves a series of compromises (Ramakrishnan 1995a; Ramakrishnan et al. 1994a). Such a compromise should not only be based on biophysical considerations, but also on integrating the human dimension, viewing ecology in an all-encompassing holistic sense.

Policy implications

With rapid and continued decline occurring in the quality and the number of sacred groves, because of changes in value system and pressure on land due to increasing population and dwindling natural resources, there is an urgent need to document and monitor the existing groves, analyze the scientific basis of these relict ecosystem functional units, and evaluate their value for biodiversity conservation. The research efforts should particularly be directed towards evaluating keystone species in these ecosystems and to exploit them for nature reserve management and for rehabilitation of degraded landscapes. The efforts in this direction should be well-coordinated between governmental agencies such as the Ministry of Environment and Forests, the Wasteland Development Board, other governmental and non-governmental agencies and the scientific community at large. The research and management of these groves should be part of a larger effort to develop a meaningful action plan for natural resource management based on traditional knowledge and technology in regions where traditional societies live. In doing this, ecological perceptions should be closely knit with socio-economic and cultural concerns of local communities so as to ensure their participation both in conserving and indeed enhancing the quality of these groves. In this context, the locally recognized and accepted land tenure system is a case in point. Joint management committees that may be appropriate for improving the quality of these groves and to use them for local benefit to the extent feasible should operate within the local traditional set-up.

Identifying and conserving selected large groves that are well preserved and which are ecologically significant systems should be taken up as an urgent priority. Indeed, the governments concerned should initiate steps to declare a few selected groves representing different ecological zones in the country as National Heritage Sites, for effective conservation and use wherever feasible (e.g.; eco-tourism). The economically valuable germplasm, such as medicinal plants, could then be exploited through cultivation outside the sacred grove, as part of a larger rural development plan.

All identifiable sacred landscapes (I have identified three of them here—the Ganga system, the Bishnoi landscape of Rajasthan, and the 'Demojong' of West Sikkim) should be declared as National Heritage Sites, with a view to their conservation and sustainable management on the basis of a well-developed plan of action. These cultural landscapes with traditional societies living within, the natural and managed biodiversity within them forming an interconnected and interacting ecological landscape, should be rigorously researched upon with a view to link conservation with sustainable livelihood/ development.

The Sikkimese sacred landscape is a unique case where ecological considerations cannot be separated from historical, social, cultural and religious dimensions of the problem. Here is a sacred landscape where the people living in the region are truly integrated within the landscape unit itself, in a socio-economic sense. Therefore, one has to consider sustainable development of the region as an integrated issue with vernacular conservation. Declaring the sacred landscapes as National Heritage Sites and their eventual recognition as World Heritage Sites of UNESCO, would be a step in the right direction not only for conservation but also for evolving and implementing a meaningful sustainable development action plan, with peoples' participation.

REFERENCES

Anonymous. 1994. The Spirit of the Sanctuary. *Down to Earth* 4: 21–36.

Boojh, R. and **P.S. Ramakrishnan.** 1983. Sacred Groves and Their Role in Environmental Conservation, in *Strategies for Environmental Management.* Souvenir Volume, Department of Science and Environment of Uttar Pradesh, Lucknow.

Dandekar, R.N. 1979. *Vedic Mythological Tracts. Select Writings*, I & II. Ajanta Publishers, Delhi.

Gadgil, M. and **V.D. Vartak.** 1976. The Sacred Groves of the Western Ghats in India. *Economic Botany* 30: 152–60.

Khiewtam, R.S. and **P.S. Ramakrishnan.** 1989. Socio-cultural Studies of the Sacred Groves at Cherrapunji and Adjoining Areas in North-eastern India. *Man in India* 69: 64–71.

———. 1993. Litter and Fine Root Dynamics of a Relict Sacred Grove Forest at Cherrapunji in North-eastern India. *Forest Ecological Manage.* 60: 327–44.

Ramachandran, K.K. and **C.N. Mohanan.** 1991. Studies on the Sacred Groves of Kerala. Centre for Earth Sciences, Trivandrum. (mimeo).

Ramakrishnan, P.S. 1992. *Shifting Agriculture and Sustainable Development of North-Eastern India.* UNESCO-MAB Series, Paris, Parthenon Publishers, Carnforth, Lancs. U.K. (republished by Oxford University Press, New Delhi, 1993).

———. 1995a. Sustainable Rural Development with People's Participation in the Asian Context, in S. Samad, T. Watanabe and S. Kim (eds). *People's Initiatives for Sustainable Development: Lessons of Experience.* Asia & Pacific Development Centre, Kuala Lumpur.

———. 1995b. Biodiversity and Ecosystem Function: The Human Dimension, in F. di Castri and T. Younes. *Biodiversity, Science and Development: Towards a New Partnership.* CAB International, London, 1996.

Ramakrishnan, P.S. and **S.C. Ram.** 1988. Vegetation, Biomass and Productivity of Seral Grasslands of Cherrapunji in North-east India. *Vegetatio* 74: 47–53.

Ramakrishnan, P.S., A.N. Purohit, K.G. Saxena and **K.S. Rao.** 1994. *Himalayan Environment and Sustainable Development.* Diamond Jubilee Publications, Indian National Science Academy, New Delhi.

Ramakrishnan, P.S., J. Campbell, L. Demierre, A. Gyi, K.C. Malhotra, S. Mehndiratta, S.N. Rai, and **E.M. Sashidharan.** 1994. *Ecosystem Rehabilitation of the Rural Landscape in South and Central Asia: An Analysis of Issues.* Special Publication, UNESCO (ROSTCA), New Delhi.

Rodgers, W.A. 1994. The Sacred Groves of Meghalaya. *Man in India* 74: 339–48.

Tsarong, T.J. 1995. Tibetan Medicinal Plants: An Agenda for Cultivation, in R.C. Sundriyal and E. Sharma (eds). *Cultivation of Medicinal Plants and Orchids in Sikkim Himalaya.* G.B. Pant Institute of Himalayan Environment & Development, Kosi, Almora.

Vannucci, M. 1994. *Ecological Readings in the Veda.* D.K. Printword (P) Ltd., New Delhi.

WWF-India. 1996. *Sacred and Protected Groves of Andhra Pradesh.* Andhra Pradesh State Office, World Wide Fund for Nature—India.

12

Tribal communities and conservation in India

Amita Baviskar

Introduction

Tribal communities are often understood to be key actors in the field of biodiversity conservation. At the same time, they are among India's poorest and politically most powerless people. Resolving this paradox requires that two sets of concerns—ecological and equity-related—be simultaneously addressed. Drawing largely on experiences from Jhabua, a tribal district in Madhya Pradesh, central India, this paper looks at tribal attempts at conservation against the backdrop of larger processes of development. The roles of the state and a local trade union of peasants and labourers—the Khedut Mazdoor Chetna Sangath (hereafter called the Sangath) in shaping natural resource use and management practices are critically examined. The paper argues that there is no easy distinction between the goals of safeguarding ecology and equity in tribal societies, and suggests that policies and practices need to be more sensitive to wider historical changes in the community–nature relationship and to the internal dynamics of tribal society. Through the analysis of one political struggle by a tribal community, the paper attempts to go beyond the somewhat simplistic assumptions made about tribals and conservation and help refine our general understanding of these issues.

Who are we talking about?

The term 'tribal' is widely used in India and evokes a set of standard images of forest-dwellers, hunter-gatherers, small egalitarian bands, nature-worshippers, drinkers and dancers. However, as is true of all stereotypes, many tribal people in India bear no resemblance to their representation in the popular imagination. There is the official list of Scheduled Tribes, a Constitutional designation which entitles certain tribes to special protection and positive discrimination. However, this list includes not only forest-dwellers, but also farmers, pastoralists, and increasingly, industrial labourers. Certain tribes are not as impoverished as the rest: the case of Meenas in Rajasthan or Chaudhris in Gujarat (Shah 1991) comes to mind. Most significantly, tribal cultural beliefs and practices often merge with those of low-caste Hindus and the boundary between tribe and caste has historically been a porous one. In addition to the Constitutional, occupational, economic, cultural and other aspects, there is also the political; calling oneself tribal may be part of a strategy of political assertion and action, as has been the case in Ladakh in the trans-Himalayan part of India. The wide variation among tribal communities means that one cannot make general statements about nature–culture relationships or assume the existence of 'indigenous knowledge' as an unproblematic category. Since it is hard to construct a generally valid description of the category 'tribe', let me focus on the specific instance of the Bhils and Bhilalas of Alirajpur in Jhabua district, Madhya Pradesh.

Alirajpur: From the past to the present

Alirajpur *tehsil* (administrative sub-division) in Jhabua district lies in the south-west corner of Madhya Pradesh, along the Narmada river. The area is mountainous and sparsely settled; the hills are covered by dry deciduous mixed teak forests, which have been largely cleared for cultivation. The appearance of isolation is misleading, for the Bhil and Bhilala tribal population of the region has dealt with successive waves of Rajput, Maratha and British invaders and settlers. For over 400 years, *adivasis* (tribals) have struggled against the inroads made by states and markets which have transformed their relationship with their environment and with non-*adivasis* (Baviskar 1995).

After India's independence in 1947, the Bhils became citizens and subjects of an independent nation state. One of the first acts of the new government was to conduct a land settlement, formally issuing titles to lands that were then being cultivated, and classifying most of the remaining lands as the property of the Forest Department

(FD). The region was designated a Scheduled Area in 1950, and the state was charged by the Indian Constitution with the responsibility of 'promoting with special care the educational and economic interests of the weaker sections (the Scheduled Tribes) and protecting them from social injustice and all forms of exploitation' (GOI 1978). In practice, however, the state has performed its role as guardian only fitfully, initiating half-hearted welfare schemes like the Integrated Tribal Development Programme.[1] The state's refusal to recognize tribal rights to forest lands and allow tribal communities control over their productive economy has subverted the project of tribal welfare from the very start. The most grievous blow to the tribal cause comes from the state's enthusiastic pursuit of a strategy of national development based on industrialization. Mineral-rich forested areas and the upper reaches of rivers in the hills, prime landscapes for resource extraction, have been gradually acquired by the state 'in the national interest' by pushing out the resident tribal population. Physical displacement in some cases, resource displacement in others, has impoverished the lives of the majority of Bhils.

However, a tiny section among the Bhils—those who were economically and politically more powerful, whose ancestors as village headmen had turned to their advantage their mediating role between the villagers and the state—were in a position to seize the opportunities offered by tribal development programmes and get access to jobs on the lower rungs of the government bureaucracy. Over time, with the spread of state education, this section is growing. While development has brought economic security and lower middle-class respectability to this section, its movement away from the land and into a class dominated by non-tribals has engendered a strong desire to shed the stigmas of tribal identity and adopt caste-Hindu practices. At the same time, the same initial conditions of relative privilege which enabled this section of petty officials to come forth also gave rise to the current tribal leadership.

Forty-five years after independence, rural Alirajpur presents a dismal picture in terms of various human development indicators. According to the *District Census Handbook*, just 4.6 per cent of the population is literate; only 2 per cent of rural women can read and write. Of the *tehsil*'s total rural population of 196,000, only 14 per cent has access to government medical services. A mere 55, or 16 per cent of the 339 villages in the *tehsil* are electrified. Most villages have no source of safe drinking water (GOI 1981). A part of the Bhil belt which stretches from western Rajasthan through Gujarat and Madhya Pradesh to Maharashtra, the population of Alirajpur is overwhelmingly tribal; almost 89 per cent of rural people belong to the Bhil and Bhilala tribes. Another 6.7 per cent belong to various Scheduled Castes. Trade and commerce and the entire state administration in Alirajpur is dominated by non-*adivasis*, with *adivasis* relegated to petty posts. There is a sharp social divide between *adivasis* and non-*adivasis*, with the latter regarding the former as savage, backward and contemptible.

Local political economy: Land, livestock and forests

Jhabua district is one of India's poorest districts. Almost all the people of Alirajpur make their living from the land. About 83 per cent of adult workers cultivate their own land, and another 8 per cent work as agricultural labourers. People grow maize, *jowar* (sorghum) and *bajra* (pearl millet) and several kinds of pulses mainly for self-consumption. Oilseeds such as groundnut and sesame are primarily grown for sale. Although people grow as many as 20 different crops, agricultural productivity is low due to thin soils and unpredictable rains. Only 8 per cent of the total cultivated area of Jhabua district is irrigated. Out of the district's total of 97,674 agricultural holdings, 91,507 holdings (93.6 per cent) are classified by the government as uneconomic (DRDA 1993). With the growing population, more and more people are dependent on increasingly tiny, partitioned plots.

With marginal legal landholdings, people rely heavily on livestock and the forest to sustain themselves. Access to the forest enables *adivasis* to own large herds of livestock—draught animals, a few cows and several goats. The better-off *adivasis* keep buffaloes and large herds of goats. While there is little significant differentiation in the size of legal landholdings among *adivasis*, the size of livestock herds tends to vary quite a bit.

With crown densities of 40 per cent or less, the forests of Alirajpur are classified as highly degraded. This degradation can be traced to a combination of causes such as rapid deforestation in the early part of this century by the Alirajpur princely state for revenue generation (continued till the early 1980s, but on a lesser scale), and a growing *adivasi* population's need for cultivable land. Yet, despite being degraded, the forests are as central to the tribal economy as legal landholdings and livestock. Through the seasons of the year, the cycle of collecting various forest produce marches along with the cycle of agriculture (Baviskar 1995). *Adivasi* homes, constructed entirely of teak, bamboo and *anjan (Hardwickia binata)*, with floors of packed mud and cowdung, are built almost entirely of forest produce. Besides providing house-building material, the forest also yields fodder, fuel, fibre, fruit, medicines, edible gums and numerous other items. The continuity between the forest, animal husbandry and cultivation is reiterated through rituals and taboos that seek to control and manage nature. The dependence on the forest is also expressed in the *gayana*, the Bhilala myth of creation, in which teak and *khakhra (Butea monosperma)* are accorded as much importance as *jowar* (sorghum).

Access to the forest enables *adivasis* of Alirajpur to hold their own economically. Besides self-consumption needs, they trade forest produce along with some of their agricultural produce for merchandise such as cloth, jewellery, iron implements and salt. And in years when the rains fail, people fall back on the forest for survival by selling wood. As one *adivasi* observed, 'during a drought, the forest is our moneylender.' Whereas people from other drought-stricken areas are forced to migrate in search of a livelihood, the *adivasis* of Alirajpur who still have access to the forest manage to stave off starvation and avoid migration by selling forest produce. During a particularly lean summer, when asked why he and his fellow villagers had not migrated for *mazdoori* (wage labour), one Bhilala man replied, 'we are doing *cheek ki mazdoori* right here'; villagers survived by collecting and selling the aromatic resin (*cheek*) exuded from the *halai* (*Boswellia serrata*) tree.

Nevad: Fields in the forest

The most significant contribution of the Reserved Forest to the local subsistence economy is *nevad*. The word *nevad* literally means 'new field', a place cleared for cultivation. However, now it refers only to fields that encroach on forest lands. The statistics about *nevad* speak for themselves: more than a thousand claims for regularization were submitted on one day in May 1994 in Sondwa block alone. These revealed that each household cultivated *nevad* holdings that were three to 10 times larger than its legal holdings. Since the size of legal holdings is too small for subsistence, it becomes imperative to supplement them with *nevad* cultivation. *Nevad* fields tend to be tiny patches of cleared forest, usually discreetly tucked away in the high hills. The land is generally sloping, with friable soils. But despite its drawbacks, this land keeps the *adivasi* economy of Alirajpur on its feet. While there are no precise records about the extent of encroachments in Jhabua, former administrator M.N. Buch estimates that for Madhya Pradesh as a whole, out of a total of 15.5 million hectares of forest land, 1.6 million hectares (over 10 per cent) was lost to encroachment (Buch 1991).[2] There are other interpretations of this: the Sangath, a trade union of peasants and labourers which has been mobilizing *adivasis* in about 90 villages in Alirajpur over the last 15 years for rights to the forest, argues that forests belonged to *adivasis* before they were privatized by the state and felled extensively in the 1930s. It also argues that the process of land settlement in 1949 was simply land alienation since it left out many holdings deep in the forest. However, the rights of *adivasis* to forest encroachments are continually contested by the state.

Since *nevad* is crucial for *adivasi* survival, every attempt by the FD to repossess *nevad* fields has met with widespread resistance. *Adivasis* have made frequent representations to the administration to settle their claims through negotiation, but have been rebuffed. The FD has rarely tried to find an amicable solution to this conflict, preferring to enforce its claim with the unilateral use of force. Most confrontations have been sparked off when FD parties have suddenly descended on an area and started digging Cattle-proof Trenches (CPTs) preliminary to plantation, attempting to cordon off an area that may include customary grazing lands as well as *nevad* fields. A particularly charged encounter occurred in March 1991 in Kiti village when the police fired to disperse assembled villagers (both men and women) from Kiti, Keldi and Vakner, who had been peacefully resisting CPT work. Very often, the FD provokes a confrontation by setting its labourers to digging holes in the middle of *nevad* fields, ostensibly for planting trees, as was done in Pujara ki Chauki village. In the instances when the FD's attempts to reclaim *nevad* have led to retaliation in the form of villagers throwing stones, the crisis has been precipitated mainly because of the state's use of force at the very outset.

Besides occasional violent mass confrontations, the conflict over the forest has also been a ceaseless war of attrition in which the FD has used legal weapons like litigation and confiscation. The registration of thousands of cases, many of them on trumped up charges, against *adivasis* for violating forest laws has trapped people on a treadmill of visits to the police lock-up and the court. People also risk getting their livestock impounded and have to pay to get them back.[3] If a pair of bullocks is confiscated at the time of ploughing during the crucial agricultural season, it can cripple an *adivasi* household's economic prospects for the entire year.

The state's legal actions against *nevad* were supplemented by illegal ones. Villagers recount the beatings they suffered at the hands of forest guards. They remember times when the *nakedar* or deputy ranger would enter their village and order that a feast of chicken and *pannia* (bread cooked between leaves) and *mahua (Madhuca indica)* liquor be served for the pleasure of the officers. Earlier, forest guards would simply demand and receive a bottle of *ghee* or a bag of groundnuts; no one dared resist. When called upon, an *adivasi* had to put aside his work and escort the forest official to the next village, carrying his bag for him. A constant accompaniment to these demands were the monetary bribes that villagers had to pay to persuade the FD to look the other way.

Organizing for sustainable development: The Sangath at work

Since 1982, the FD's ability to get away with such blatant abuse of power has been sharply curtailed by the formation of the Sangath. The Sangath started work in Sondwa block as a sub-centre of the Social Work and Research Centre (SWRC), an organization with its headquarters at Tilonia, Rajasthan. The SWRC is famous as a rural NGO engaged in grassroots development through the use of appropriate technology, handicraft cooperatives, and improvements in community health and education. However, the Sangath activists chose to abandon SWRC's community development model of social work, and instead engaged in direct political action through collective organization and mobilization of *adivasis* against the state and the market.

The Sangath now works in about 90 villages in Sondwa and Sorwa blocks of Alirajpur *tehsil*. Besides the issue of land rights and forest management, the Sangath also tries to ensure that public development funds and welfare services actually reach the villages and are not siphoned off by a corrupt administration. The union has worked in the area of education, using local history, music and myths to revive a sense of pride in *adivasi* heritage. It has tried to revitalize customary modes of dispute resolution within the community, such as arbitration by elders, avoiding recourse to the police and courts. The union runs a cooperative shop at its headquarters in Attha village. Although the number of full-time workers keeps changing, the Sangath has a core of three full-time activists and three tribal leaders. It has now started paying special attention to training *adivasi* youth for continuing political activism. In terms of its origins, organizational structure and ideology, the Sangath is similar to other mass-based organizations—the Adivasi Mukti Sangathan, the Kashtakari Sangathana, the Shramik Sangathana—in the region.

The centrality of *nevad* in the lives of *adivasis* makes it a key political issue for the Sangath. The administration's refusal to negotiate drew the Sangath towards more aggressive tactics like mass demonstrations, hunger strikes, and the obstruction of forest-related work. These have ended much of the petty violence and corruption of the FD. The Sangath claims that its work is not limited to defending *nevad*; according to its pamphlet printed for publicity and fund-raising, 'its objective is the wider cause of empowering *adivasis* in their struggle to live with dignity, without being exploited and cheated, with control over the resources and processes that so vitally affect them'. To this end, the Sangath has tried to work on several fronts—organizing a cooperative shop, teaching in Bhili and Bhilali, initiating a soil and water conservation programme, and generally strengthening *adivasis*' ability to stand up for their rights. For its efforts, the Sangath has been frequently called a 'naxalite'[4] organization, inciting simple *adivasis* to violence. For the most part, the local administration has chosen to treat the mobilization around the forest rights issue as a law and order problem, to be suppressed with state violence.

From the beginning, the Sangath's organizing against exploitation has been necessarily directed against the state, including its agents in the village. The traditional village *patel* (headman), a hereditary office established more than a century ago but which continues to be vested with political power, was usually complicit in exploitation by the state. The Sangath had to challenge the authority of corrupt *patels* and establish alternative structures of power. This task could well have created cleavages within the village community with different factions supporting the *patel's* group versus the Sangath. However, in many villages, the Sangath's rallying cry of defending *nevad* was powerful enough to bring most villagers together and to marginalize the *patels*. The Sangath's relationship with other local government such as the *panchayats* has been more mixed. In 1989, Sangath members contested the local *panchayat* elections and won overwhelmingly. However, once elected, their efforts to implement the Sangath charter were hamstrung because financial and other controls over the *panchayats* were still vested with the state. It was hard to get projects sanctioned without bribing officials or undertaking to purchase materials from their associates. In many cases, villagers who had worked on building contour bunds or schools did not receive the wages to which they were entitled because the state delayed or refused payments. Frustrated by the experience of trying to work with the state, the Sangath decided that it would not participate in the *panchayats*. Individual Sangath members have, since then, contested *panchayat* elections and won, but the organization as a whole has stayed away from the process, believing it to be full of compromises and pitfalls.

The *adivasis* of Alirajpur are forced to cultivate *nevad* fields on fragile hill slopes and collect forest produce in order to avoid migration in search of work. Because *adivasis* have no security of tenure and live under the constant threat of eviction, they cannot invest in improving their land. Their poverty prevents them from planting tree crops which have long gestation periods, and the illegality of their position precludes their receiving loans from the government for making agriculture more productive. In this situation, the first step towards sustainable development has to be recognition of *adivasis*' rights to the forest. The Sangath argues that getting only access to forests is not enough; as long as control is vested with the FD, the forests will continue to be destroyed due to illicit felling by contractors, and expensive afforestation programmes will be launched without any lasting benefits. At the same time, forests are also being destroyed by increasing *adivasi* numbers and their land hunger. The Sangath argues that the solution lies in local control. Only when local communities have the power to decide how they should best manage the forest to meet their needs, can they choose to set aside some areas for protection. With security of tenure for *nevad*, they can invest in land improvement measures.

With the intervention of the Sangath, as *adivasis* have secured partial control over the forest for the last 15 years, they have initiated some conservation measures. The villages of Attha, Gendra and

Umrath have tried to improve *nevad* fields and manage the rest of the forest for sustainable fodder and fuel yields. Labour collectives have worked on bunding and gully plugging to prevent soil erosion from the fields.[5] One year, villagers ran a nursery to grow saplings for planting. These efforts have shown a marked improvement in forest regeneration and in checking soil erosion, but they remain limited to a handful of villages. Their future is also highly uncertain, partly because of the possibility of fresh state initiatives to remove encroachments, and partly because of the organizational crises that periodically affect the Sangath. The few full-time activists bear the burden of coordinating activities in 95 villages; there are occasions when critical institutional support from the Sangath is not forthcoming because the activists are dealing with something else. Yet, *nevad* fields need inputs on a scale that the Sangath cannot provide. Only the state can ensure that a comprehensive soil and water conservation programme covers the entire watershed. But the government is unwilling to make this long-term investment in land resources since it feels that to do so on *nevad* lands would be construed as recognizing *adivasi* rights to the forest. The mutual hostility and suspicion that mark the relationship between the state and the Sangath seem to rule out the possibility of finding a middle ground of cooperation.

The small experiments undertaken by the Sangath indicate people's preference for an approach where they have control over the conservation programme and where their rights are secured. However, a small financial base prevents them from expanding the scale of activity, and from undertaking works that require more bought materials (for water harvesting, for instance) or paying workers for wages foregone. Unlike most NGOs, the Sangath raises most of its funds from its members, impoverished though they are. All households pay union dues and contribute in kind for running the union. Most of the work is done voluntarily. Only the six full-time activists receive small stipends. Specific projects are funded by small grants from Indian NGOs and by donations from supporters; the organization recently received a grant of Rs 2.4 lakhs (approximately $6,900) from the Council for the Advancement of People's Action and Rural Technology (CAPART), a government institution, for soil and water conservation—its largest grant so far. The financial straitjacket within which the Sangath operates is a direct consequence of the organization's decision to chart a path of political confrontation with the state. The kind of political goodwill required to secure funds from donor agencies has never been the Sangath's happy lot. At the same time, it has ideological reservations about accepting foreign funding or any kind of work with a large budget, feeling that it makes the organization less accountable to its membership. This choice has ruled out the possibility of undertaking the kind of resource-intensive land development work that many NGOs have been able to do.

At present, people are being forced to choose between the land or the forest, and necessity compels them to choose the former. If forests are to survive, it is essential that existing agricultural lands be made more productive through protective irrigation, soil fertility improvement and checking soil erosion. A strategy of this sort requires enormous decentralized planning, sustained resources and stable structures of management. In the present context of the political uncertainty about people's rights to land, such a strategy remains a distant dream. Even if the political issue of land rights were to be somehow resolved, there would need to be a sea change in the state's pattern of relating to tribal people, a movement towards recognizing their right to have greater control over development planning and practice. More concretely, state-initiated conservation schemes such as Joint Forest Management (JFM) would stand a greater chance of success if they acknowledged that, in this case, people want to grow food first and trees later. Greater flexibility in state programmes would help people to combine food and tree crops in ways that best suited their priorities.

Though the Sangath has secured a degree of temporary stability for the production economy of Alirajpur's *adivasis*, its achievements have been miniscule in terms of the overall vision of sustainable development. Local access has not been converted into local control and management in most of the member villages, despite persistent efforts by the activists to set up village-level committees to manage natural resources. In some Sangath-controlled areas, the virtual absence of the FD has created a free-for-all with extensive clearing of forests for *nevad*. Certain valuable gum-yielding trees such as *halai* are being tapped so intensively that villagers say that no trees of this species will be left after some years. Recourse to seasonal migration increases every year in the region. The prescription of local access and control over forests is, by itself, not enough to bring about sustainable forest management; rather, the current situation demands multiple interlinked strategies.

The Sangath's vision of ecology and equity

It would be useful at this point to delineate the vision of sustainable development that drives the Sangath's work. This ideology is not available as an articulated doctrine but emerges from an analysis of the Sangath's practices and from activists' statements on specific issues. This model envisages an economy and polity that is largely autonomous from external control, that is internally equitable, and that allows natural resources to be replenished and expanded in a way that first satisfies the subsistence needs of the local population. If basic requirements cannot be met locally, it becomes incumbent upon the state to provide them. Control over decision-making processes as well as over the management of resources would rest primarily with the *gram sabha* (all the adults of a village or hamlet), the body that is best informed about local specificities. The focus on subsistence in the sustainable development model is a way of emphasizing that resource use and management remain limited to that which is necessary for meeting basic needs, without entering into the ever-rising spiral of demand for consumption which would be ecologically hard to sustain. Thus, control brings with it the need to exercise responsibility—towards all the members of the community and deprived people everywhere, and towards maintaining ecological productivity for the future. According to the Sangath, the decentralization of power also entails recognizing the legitimacy of local concerns and knowledge, which are undervalued and ignored by centralized systems.

Whereas other attempts to bring about sustainable development have tried to create model villages like Ralegan Siddhi or Sukhomajri (Pangare and Pangare 1992; Chambers et al. 1989), the Sangath has chosen to mobilize politically in a region-wide fight for rights to natural resources, supplemented by a revival of the tribal history of resistance, customs of dispute resolution, traditional knowledge about herbal medicine and forest use.[6] The Sangath has also participated in nation-wide mobilizations on issues such as the proposed Forest Bill, against liberalization (supporting Azadi Bachao Andolan, a nation-wide federation of organizations opposing the entry of multinational corporations and consumerism), against state repression of the Chhattisgarh Mukti Morcha (a powerful trade union in eastern Madhya Pradesh), and against displacement (through its participation in the Narmada Bachao Andolan, the movement against the Sardar Sarovar dam on the Narmada river). Despite these efforts, the Sangath's project of sustainable development faces an even more serious threat today in Madhya Pradesh.

The state and development: The engine of industrialization

In 1994, the Madhya Pradesh government launched an aggressive programme to shed the state's industrially backward image by courting private industry with incentives such as attractive tax breaks and fast clearances to fully exploit the state's rich endowment of mineral and agricultural resources. Proposed investment since 1991 exceeds Rs 36,000 crore, next only to Gujarat, Maharashtra and Uttar Pradesh (Chakravarti 1995). Its location in the heart of the country and good connections by road and rail make Madhya Pradesh ideal for sourcing and distribution of manufactured goods. Industrialists are also drawn by the availability of relatively peaceful and cheap labour, and to the promises of the state government to build substantial additional power capacity. For a state with more than 60 per cent of its population living below the poverty line, the slogan of 'Let's Talk Business' signifies the government's resolve to prosper by capitalizing on the opportunities thrown up by the new climate of liberalization.

The new liberalization policies[7] have been welcomed by Madhya Pradesh's business community. They have been accorded a more mixed response by the urban middle-class which looks forward to improved transport and power infrastructure but is apprehensive about the worsening of the already severe water scarcity. In the fertile plains of the Malwa region, most farmers are upbeat about the increased demand for cash crops such as soyabean, cotton, oilseeds and sugarcane. Opposition to liberalization has emerged only in tribal areas in the form of protest against particular projects which threaten to displace people. For instance, Bastar is a heavily forested, mineral-rich, predominantly tribal region in south-east Madhya Pradesh, with major heavy industry to its north. In this district, which has experienced resistance in the past to proposed dams, a demand now frequently voiced by tribal leaders and activists is that half the shares of any private industry that is set up in the region should be distributed among the tribal people, the real owners of the region's resources. While this demand seeks to decentralize the liberalization process and secure a degree of control for the local population, it does not address broader questions about the model of intensive resource-extraction that is being promoted. These demands on the part of tribal leaders have been largely polemical, and the state has ignored these protests so far. There has also been low-key naxalite activity in Bastar for several years, but it has not seriously impeded the state's plans. At present, the state's partnership with private industry steams ahead unchallenged.

Committed to transferring natural resources to private industry, the only future envisaged by the state for those who depend on these resources for sustenance is that they join the pool of cheap compliant labour. At another level, though, the state proclaims its commitment to making rural livelihoods more prosperous by promoting greater industrial demand for agricultural produce. However, the tribal farmers of Alirajpur whose land barely provides enough food to see them through the year are unlikely to be the beneficiaries of this process.

Dalit politics and tribal assertion

Another new current has also affected the political climate of the state. This is the rise of dalit[8] (backward class) mobilization as an influential factor in state-level politics. In a state where 23 per cent of the population consists of Scheduled Tribes (compared to 7 per cent for India as a whole) and another 14 per cent belongs to the Scheduled Castes, and where the tribal population is heavily concentrated in particular districts, dalit support is crucial for electoral success. The success, if limited, of backward caste-based political parties in Uttar Pradesh (Bahujan Samaj Party) and Bihar (Janata Dal) has reinforced the belief of Madhya Pradesh's tribal politicians in the electoral potential of mobilizing on the basis of dalit identity. The confidence that dalit support can make or break the state government has led a tribal Member of Parliament, Jhabua's Dilip Singh Bhuria, to demand that Madhya Pradesh have a tribal chief minister. Since 53 out of the ruling party's 174 Members of the Legislative Assembly in 1997 were tribals, this demand had political potential.

Simultaneous to the above is the emergence of an inter-state movement of tribal assertion. In the entire Bhil belt stretching from Maharashtra through Madhya Pradesh, Gujarat to Rajasthan, there have been broadly two kinds of mobilizations attempting to remake *adivasi* identity. The first is illustrated by the Sangath's politics which focuses on rights to natural resources. This politics links the issue of *adivasi* identity to their material exploitation and argues that political control over resources is essential for gaining self-respect and dignity vis-à-vis non-tribal oppressors (tribal leaders like Vahru Sonavane, formerly with the Shramik Sangathana, exemplify this stream). The second movement seeks to remake *adivasi* identity by emulating upper-caste Hindu practices in order to shed the 'backward, savage' image of *adivasis*. Action consists of joining various religious sects such as the Ramanand *panth*, Kabir *panth*, Gayatri *parivar*, and so on. While the former movement asserts itself by espousing a revival of tribal culture, the latter seeks to erase its *adivasi* past in order to claim higher status in the Hindu hierarchy. The latter movement of religious revival has stayed aloof from any clear political stand, concentrating on self-improvement under the slogan '*tum badloge, yug badlega*' (if you change, the age will change).

While Sangath politics shares with state-level dalit politics an explicit orientation to capturing power, it differs from dalit politics in its understanding of what power should be used for. The failure of the sustainable development model espoused by activists is linked to the stumbling progress of the political project initiated by the Sangath which focuses on local control. The state not only refuses to relinquish its power over natural and financial resources, but in fact has launched a programme of industrialization that makes local control even more difficult. While people have fought as Sangath members to gain access to the forest and have gained a temporary reprieve through their efforts, their economic and political disadvantages vis-à-vis the state prevent them from realizing more fully the model of sustainable development. Under these circumstances, tribal leaders in the Sangath feel increasingly drawn towards dalit politics which promises to bring power into their hands. They are willing to shelve the model of sustainable development for the time being, arguing that capturing power would automatically enable them to use it for desired ends. This argument fails to persuade the non-tribal Sangath activists who point out that state-level tribal leaders such as Dilip Singh Bhuria or Arvind Netam are simply interested in controlling the lucrative flow of resources from the state for their own benefit. Though these leaders claim to represent exploited *adivasis*, they are not committed to challenging the model of development that impoverishes *adivasis*. Only a tribal leadership committed to the cause of sustainable development would genuinely transform the lives of poor *adivasis*. It remains to be seen how the recently passed Panchayats (Extension to the Scheduled Areas) Act of 1996, which has been hailed as giving 'radical governance powers to the tribal community and recogniz(ing) its traditional community rights over natural resources' (Mukul 1997) will be implemented. Supporters of the Act have repeatedly stressed that further political mobilization by *adivasis* will be crucial for realizing the potential of the Act (see also Krishnan, and Kothari et al. in this book).

Conclusion

The group whose fate—both present and future—is most directly linked to the fate of the forest and the land is the tribal community. Yet the exigencies of survival seem to subvert the best efforts of tribal groups and the Sangath to ensure ecological stability. At the same time, there is an emerging section of *adivasis* who have broken their links with the land and have adopted the lifestyles and aspirations of middle-class urban Indians in white-collar jobs. Acknowledging these two processes is an important corrective to our understanding of the relationship between tribal culture and nature. There is a great deal of literature today that celebrates the conservationist values of 'indigenous knowledge'. Some of this literature, especially that which constitutes 'ethnobotany', also tends to have an extremely instrumentalist view of tribal culture and seeks to selectively appropriate some aspects of tribal knowledge for pursuing the agenda of the state or the scientific establishment. While it is all very well to give due recognition to cultural practices and beliefs that have been systematically ignored or rejected as 'primitive' and 'unscientific', it is important to situate this knowledge within the larger context of a changing tribal reality. That is, before defending 'indigenous knowledge' one must speak up in defence of the tribal right to self-determination and cultural autonomy. This is by no means an easy move because, in a differentiated tribal polity, squeezed by the state and the market, tribal groups may demand futures that are antithetical to the goal of biodiversity conservation.

Portraying tribal communities as natural repositories of ecological wisdom may be politically correct, but does a disservice to tribal people who are struggling to secure social justice through control over the natural resource base upon which they depend. The imagery of 'ecological stewardship' can be a powerful, confidence-giving cultural statement and an important input into forging a distinct tribal identity. But this is a political use of symbols and should not be deployed to obscure present-day contradictions in tribal life. While tribal access to and control over natural resources are crucial conditions for biodiversity conservation, they are not enough. As long as the larger context compels greater extraction and commercialization of resources, and as long as *adivasis* remain vulnerable politically and economically, the pressure to destroy for short-term gain will prevail. In addition, it is crucial to recover or establish anew the value of conservation as an explicit ideology and not as something which occurs 'naturally' or which can be taken for granted.

NOTES

1. For a description of welfare programmes in Alirajpur, see Baviskar (1994).
2. After prolonged agitation by the Sangath, the Forest Department conducted a survey of encroachments in Sondwa block in 1988. The survey recorded that almost every cultivator had supplemented his legal holdings with several small plots of *nevad*. In Anjanvara, a village of 33 households, 14 cases of over 192 hectares of forest land were recorded, most of which had been cultivated at least since 1970.
3. These payments may be fines with proper paperwork and receipts, but they are more likely to just be under-the-table transactions. Villagers have little way of knowing if their practices are indeed actionable under the law or not.
4. 'Naxalites' refer to members of a movement which had its origin in armed struggle led by educated youth against oppressive classes in Naxalbari, West Bengal (eastern India); today the strongholds of this movement are in Andhra Pradesh and the central forest belt. The state views this movement as being 'terrorist' due to its often violent tactics; it also conveniently labels other militant movements across the country as 'naxalite', providing it the justification of using anti-terrorist provisions against them.
5. Most of this work is voluntary, based on a labour-sharing custom called *laah* where a household's labour-intensive tasks are shared by every household in the village which sends one person to help in the work, in return for a feast. All households, in turn, are entitled to reciprocal help. In 1990–91, the Sangath also paid people by using funds from the Jawahar Rozgar Yojna (a government employment programme) in *panchayats* where its members had been elected.
6. From childhood, all villagers learn to identify plants and their different uses. In one village, Anjanvara, a group of 10–15-year-olds identified more than 70 trees whose produce they harvested (Baviskar 1995). This knowledge, acquired while taking livestock to graze or by accompanying adults on collection trips, related to fuel, fodder, fruit and vegetable-bearing species, gum-yielding trees, medicinal plants and a host of other valuable produce. The ability to gather this wealth depends on being able to differentiate between species, knowing their seasonal cycles of flowering and fruiting, knowing which micro-habitat is most likely to suit a particular plant, and how to successfully extract maximum yield. Such local, context-specific knowledge would be essential for designing appropriate forest management practices, yet so far there has been no acknowledgement by the state of its relevance. See also Gadgil, Kothari et al., and Ramakrishnan in this book.

7. In the early 1990s, India moved into a phase of structural adjustments, partly under IMF influence. This included a thrust towards export-oriented growth, liberalization of the industrial environment, and privatization.
8. Though the term dalit (literally, 'oppressed') is usually used to denote only the Scheduled Castes, I shall employ it in the sense argued for by tribal activists in Madhya Pradesh, i.e., to refer to all the oppressed—Scheduled Castes, Scheduled Tribes as well as sections among the Other Backward Classes.

REFERENCES

Baviskar, A. 1994. Fate of the Forest: Conservation and Tribal Rights. *Economic and Political Weekly* 29 (38): 2493–501.

———. 1995. *In the Belly of the River: Tribal Conflicts over Development in the Narmada Valley*. Oxford University Press, New Delhi.

Buch, M.N. 1991. *The Forests of Madhya Pradesh*. Madhya Pradesh Madhyam, Bhopal.

Chakravarti, Sudeep. 1995. Madhya Pradesh: Trying to Keep Up. *India Today*. Special Survey: October 31.

Chambers, R., N.C. Saxena and **Tushar Shah.** 1989. *To the Hands of the Poor: Water and Trees*. Intermediate Technology Publications, London.

DRDA. 1993. *Jhabua: Handbook of Statistics*. District Rural Development Agency, Jhabua, Madhya Pradesh.

GOI. 1978. *Provisions in the Constitution of India for Scheduled Tribes*. Ministry of Home Affairs, Government of India, New Delhi.

———. 1981. *Village and Town-wise Primary Census Abstract*. Jhabua District Census Handbook. Part XIII-B. Madhya Pradesh. Census of India. Series-11.

Mukul. 1997. Tribal Areas: Transition to Self-Governance. *Economic and Political Weekly* 32 (18).

Pangare, G. and **V. Pangare.** 1992. *From Poverty to Plenty: The Story of Ralegan Siddhi*. INTACH, New Delhi.

Shah, G. 1991. Tribal Identity and Class Differentiation: The Chaudhri Tribe, in Dipankar Gupta (ed.). *Social Stratification*. Oxford University Press, New Delhi.

13

Conservation through community enterprise*

Seema Bhatt

Introduction

Biodiversity has always been of great relevance to communities who depend on it for subsistence; and of late its significance has been recognized globally, for scientific and economic reasons. For communities who live in and around biodiversity-rich areas, who are directly dependent on the resources therein and for whom these areas have a cultural or religious significance, there have been traditional ways of conserving it. There is also an enlightened community which lives far away from such areas, but realizing the significance of biodiversity richness has promoted other ways of conserving it. National parks, wildlife sanctuaries, and other types of protected areas (PAs) are at the forefront of modern efforts to conserve biological diversity. Initiated with all good intentions, many of these areas are today, however, in crisis. Conventional approaches to PA management are not working.

Why is this? Traditional societies in many instances have broken down, and subsequently traditional natural resource management systems have also become dysfunctional. Also, some communities have grown beyond the carrying capacity of the area and over-utilized certain resources. But perhaps one of the most important reasons for the failure of the conventional approach to conservation has been that people who have been living in and around these areas have

* Most information about the BCN-funded projects has been taken from the project proposals, some from technical reports and site visits. I am grateful to the BCN grantees for the same. The opinions expressed herein are those of the author and do not necessarily reflect the views of the United States Agency for International Development.

been overlooked quite completely in both the planning and management phases of conservation projects. In the design of such projects, the responsibility of managing resources lies not with the local communities who live closest to them, but elsewhere. The cost of conservation has fallen on relatively few people who may have otherwise benefited by exploiting these resources. Also, ironically, communities which live closest to areas richest in biodiversity are economically perhaps the poorest (McNeely 1988). Thus, over time, conventional management of PAs has resulted in resource conflicts, causing more harm to biodiversity than before, and many a traditional conservationist has turned poacher and exploiter.

If our remaining biological wealth is to be conserved, then the concepts of both conservation and management need to be re-examined and innovative alternatives to conserving biological wealth thought of. Several things need to be acknowledged in order to do so. First, that local communities need to be taken into consideration while planning for the biodiversity-rich regions. Second, there is no one solution or panacea for saving biodiversity in different parts of the world. Different approaches need to be tried and what may work well in one region may fail completely in another. Management plans have to be site-specific. Third, conservation does not mean non-use, but sustainable utilization of resources.

New directions

Since it has been acknowledged the world over that the conventional approach to conservation involving 'fences and fines' may not always work, several different approaches are being tried. One such is an attempt to motivate local communities to conserve instead of an approach which alienates them and often leads to a situation of antipathy and distrust.

Motivation of local communities to conserve requires incentives. Incentives could be of several kinds. There could be direct incentives like cash, in terms of fees, rewards, compensations, grants, subsidies, loans, etc. There could also be direct incentives in kind, like food, improved livestock varieties, and even limited access to resources which were once denied to the communities. There could be indirect inducements such as tax incentives and community development activities. Other incentives could be land tenure and even employment (McNeely 1988; see also Kothari et al., and Raju in this book).

An innovative programme that attempts to use this approach is the Integrated Conservation Development Project (ICDP). ICDPs were designed to integrate conservation with the social and economic needs of local communities (Wells et al. 1992). ICDPs vary in scale and scope. There are smaller projects in biosphere reserves, multiple-use areas and in buffer zones of national parks. There are other ICDPs implemented on a regional scale which include development projects which have links with PAs in the vicinity. ICDPs often look at more ecologically sound development alternatives, particularly when existing practices are exploitative in nature. Development activities are provided in ICDPs as incentives for sustainable management of resources (Brown and Wyckoff–Baird 1992). A project of this nature has been implemented in the Annapurna Conservation Area of Nepal. This area attracts tourists from all over the world. More than 30,000 trekkers visit this area every year (the permanent population in the area is around 40,000) (Wells et al. 1992). The project has attempted to focus particularly on increasing local benefits through tourism, and conserving the fragile ecosystem at the same time. Several activities such as introduction of fuelwood-saving technology, basic health services to the local community and training of local entrepreneurs in tourism have been carried out (Brown and Wyckoff–Baird 1992; see also Shrestha in this book).

Conservation through community enterprise

Biodiversity-based enterprises managed by local communities may also provide the right incentive for these communities to conserve. Local communities have always realized the use of local biological resources, both for subsistence and for revenue. Products that are sold commercially include bamboo, resin, lac, *tendu* (*Diospyros melanoxylon*) leaves for *beedies* (indigenous cigarettes), honey, *tasar* silk cocoons, honey, fruits, and other non-timber forest produce (NTFP). According to one estimate, NTFP income accounts for 55 per cent of the total employment in the forestry sector (Saigal et al. 1996).

However, local communities also do not get the full incomes they should from NTFP. They often get only collection charges even for products that have a very high market value. There are also products for which appropriate prices have not been set in the market. Sometimes marketing channels do not even exist. Another revenue-generating by-product of a biodiversity-rich area is eco-tourism. This kind of tourism relies on an area of natural beauty, and revenue generated as a result of this activity could be channelled back to the local communities. Tourism has been conducted the world over in naturally rich areas for quite some time. But often it has led to antagonism with local communities because they have gained nothing from this enterprise, or in fact ended up on the losing side. There has been a general feeling that the areas are protected for 'other' tourists at the cost of the local communities' welfare.

Revenue generation through conservation projects does happen occasionally, but generally the focus of these projects is not to raise the income of local communities. A case in point is WWF-India's Community Biodiversity Conservation Movement (CBCM). The central focus of the CBCM has been on practical field projects that demonstrate community-based conservation of biodiversity (WWF-India 1995). Of these projects, there have been a few where conservation has successfully taken place and the community also benefits economically through the projects. One such was implemented by the Swa-Sahyog Sanstha in Rajasthan. This organization, under the project, established a seed bank to store seeds of about 205 local and traditional crop varieties which may have gone extinct with the advent of modern agriculture. The seed bank has been set up with the active participation of the local farmers. The organization also possesses about 8 acres of land, leased for 10 years by a local farmer on which traditional seed varieties are produced. Local people have started procuring these seeds now, and the project is earning a marginal income which is being reinvested into the running of the seed bank (WWF-India 1995).

Another example of biodiversity-based enterprises being created as a 'spin-off' from a major community conservation programme is that of Joint Forest Management (JFM) (Agarwal and Saigal 1996; see also Sarin et al., and Raju, both in this book). Over the last few years, JFM has progressed beyond simply meeting subsistence needs of local communities, to contributing to income generation and improvement of livelihoods. The Nayagram Block, located in the south-western part of Midnapore, West Bengal, presents a good case study. This area is abundant in NTFP like *sal* (*Shorea robusta*) leaves, *sabai/babui* grass, *kendu* leaves and *mahua* (*Madhuca indica*) seeds. Sixty per cent of the inhabitants depend on *sabai* grass, as grass or as ropes. Seventy per cent of the population collects *sal* leaves and sells them as dried-paired leaves. However, there is no value-addition in either of the products and the income generated is meagre (Campbell et al. 1995). Now, the JFM project of the Indian Institute of Technology, Kharagpur is looking at value-addition and marketing channels particularly for *sabai* grass (for furniture and utility items), *sal* leaves (for moulded plate-making), oils and medicinal plants. In addition to this, the project also hopes to promote the cultivation and processing of mushrooms.

The Biodiversity Conservation Network (BCN)

The promotion of biodiversity-based enterprises raises a lot of ecological and sociological issues. How does one sustainably manage to prevent ecological degradation, but also to maximize output? How can the dependence of the local community on the resources of the area be met, while attempting to promote conservation through community enterprise? How best can traditional harvesting patterns be used? How can we assess indigenous knowledge? Also, how can simple harvesting and monitoring techniques be passed on to the local communities? Perhaps the most important question is: do the local communities see a link between the enterprise and conservation of biodiversity?

The Biodiversity Conservation Network (BCN) is a programme which is attempting to look closely at some of these issues. The BCN, which operates in Asia and the Pacific, is a component of the United States–Asia Environmental Partnership (US–AEP), funded by the Asia Bureau of the US Agency for International Development (USAID). BCN is administered by the Biodiversity Support Programme (BSP), a consortium of the World Wildlife Fund, The Nature Conservancy and World Resources Institute.

The BCN specifically, is a $20 million, six-and-a-half-year programme that was initiated with two fundamental goals: (i) to pursue enterprise-based strategies to conserve biological diversity at a number of sites across the Asia and Pacific regions; and (ii) to evaluate the effectiveness of these enterprise-oriented approaches for community-based biodiversity conservation.

The programme was created to test a specific, narrowly defined hypothesis that if conservation is to occur, the enterprise must have a direct link to biodiversity, generate benefits, and have a community of stakeholders who can act to counter threats to the biodiversity of the region. The enterprise must generate benefits for the community, both in the short term and also after the BCN funding ends. These benefits could be monetary (such as cash or shares in an enterprise), social (such as tenure rights), or environmental (such as watershed protection).

Approximately 63 per cent of the BCN grants support NTFP enterprises, 23 per cent fund terrestrial eco-tourism endeavours, another 8 per cent fund marine eco-tourism initiatives, and the remaining 6 per cent assist a variety of enterprises which include chemical prospecting, sustainable logging, biodiversity prospecting for flora and fauna that have pharmaceutical or industrial properties, the sale of insects collected or 'ranched' in forests and mineral water bottling from springs in critical watershed areas.

Because of the complexity of its mandate, BCN brings together several partners for each project. Partners range from non-government organizations (NGOs) and government agencies to universities in Asia, the Pacific and the United States, which form active partnerships with local and indigenous communities.

Crucial to BCN's mission is an intense monitoring programme which tries to address issues raised as a result of this specifically defined approach. Monitoring is expected to begin at the projects' inception, and, on an average, 30 per cent of grant funds at each site are allocated to it. The monitoring focuses on the biological, sociological and enterprise elements of the project. Biologically, monitoring looks at aspects like the sustainability of the enterprise, and indicators that biodiversity is being conserved at the project site. An important element being monitored from the sociological angle is if benefits are being distributed equitably among the community. The enterprise is itself monitored to see if it is reaching anticipated production levels and if there is a positive cash flow. Each project designs an elaborate monitoring and evaluation scheme that covers all these aspects with a definite work plan of how each will be carried out.

The focus of all BCN-funded projects is conservation through community enterprise. The following is a brief description of the Indian BCN-funded projects. A closer look at these projects brings up some fundamental questions about community participation which this paper will try and highlight.

NTFP in the Biligiri Rangan Hills, Karnataka

Background

Nestled in the Western Ghats of Karnataka in southern India, lies the Biligiri Rangaswamy (BR) Temple Wildlife Sanctuary. The BR Hills being at the confluence of the Western and Eastern Ghats, possess high floral and faunal diversity. This area harbours approximately 900 species of flowering plants and several faunal species such as elephant, gaur, sambar and leopard. The Sanctuary also houses approximately 4,000 Soligas, an indigenous tribe of the region. Traditionally engaged in shifting agriculture and hunting, the Soligas also collect a wide range of NTFP. Both shifting cultivation and hunting have been completely banned with the declaration of the BR Hills as a wildlife sanctuary in 1974. Limited exploitation of NTFP is still permitted in the Sanctuary under the aegis of the Large-scale Adivasi (tribal) Multipurpose Societies (LAMPS). These tribal co-operatives were created to ensure the full return of NTFP collection to the tribals (see also Lele et al. in this book).

The enterprise

The BR Hills region is rich in NTFP. However, the marketing of only very few NTFP is occurring at present, and that only through the LAMPS cooperative. The storage facilities provided by LAMPS are inadequate and spoilage losses are common. Also, there is no value addition at the point of origin. The situation is thus that extraction of the NTFP achieves very low economic return for the local people. Even with the few NTFP being extracted, preliminary research indicates that NTFP species are not regenerating adequately (Murali et al. 1994). However it is not clear whether this is due to overharvesting or other disturbances.

This project focuses on the sustainable extraction and local processing of three different forest products: (i) *amla* or gooseberry (*Emblica officinalis*), (ii) wild honey and (iii) some ayurvedic preparations from select medicinal plants. The biological sustainability of harvesting each of these products will be examined. By processing and improved marketing of *amla* (into pickles and jam), honey and some ayurvedic products, the project hopes to raise the income levels of the Soligas and provide them an economic incentive to conserve the region's biodiversity.

The proponents

The project works with the Vivekananda Girijana Kalayana Kendra (VGKK), a local NGO which has been working with the Soligas since 1981, primarily focusing on their primary healthcare. VGKK has since expanded its activities to education, vocational training and several social, political and cultural issues crucial to community development. VGKK's credibility and presence in the area makes it easy for the organization to look after the enterprise component of the project as also the socio-economic monitoring. The University of Massachusetts/Boston (UMB) and the Tata Energy Research Institute (TERI) coordinate the biological monitoring of the project.

Community involvement

VGKK has been actively involved in several welfare activities for the Soligas and has established a good rapport with the tribals. It has provided the much required link between the other project proponents and activities. As part of its activities VGKK has mobilized the setting up of the Soliga Abhivruddhi Sanghas (Soliga Development Organizations). The Sanghas now exist in most Soliga hamlets across the BR Hills and seek to revitalize Soliga traditions that have eroded over time.

The Soligas' ownership of land is limited to a small amount allocated to them for agriculture. They also have usufruct rights to collect selected NTFP for subsistence and commercial use, but the NTFP can be sold only to LAMPS.

The participation of the community in this project is at present limited to collection of selected NTFP and processing of the same. The enterprise is being managed by VGKK, but it is hoped that this component will eventually be taken over by the Soliga Abhivruddhi Sanghas' apex body. Community participation in socio-economic and biological monitoring is being initiated. However, before this can happen, an intense period of training on aspects of the enterprise, i.e., book-keeping, accounts and biological and socio-economic monitoring methods is visualized.

Tasar silk and honey in the Garhwal Himalaya

Background

This project is based in the Garhwal Himalaya in the state of Uttar Pradesh. It targets three catchments, Akash Kamini, Nagnath Pokhari and Mandal of the Okhimath, Pokhari and Dashauli blocks, respectively, all part of what is known as the Himalayan 'oak belt' which extends from Himachal Pradesh in the western range through Garhwal and Kumaon onto the north-western range in Nepal in the east. The project location is dominated by primary oak forests adjoining agricultural and agro-forestry lands at lower altitudes, and alpine pastures at higher altitudes. The area supports a number of threatened and rare high altitude faunal species such as the Black bear (*Selenarctos thibetanus*), Blue sheep (*Pseudois nayaur*), etc.. In fact, part of the high altitude project area lies in the Kedarnath Musk Deer Sanctuary, established in 1990 to protect the habitat of this endangered species.

This region is also significant because of the presence of a number of village-based institutions, i.e., Van Panchayats (village forest management committees), Mahila Mangal Dals (village women's organizations) and Yuvak Mangal Dals (village youth organizations). This suggests a high probability of people's participation in community resource management.

The enterprise

The main biodiversity-based enterprise activities in this project are oak *tasar* sericulture and village-based bee-keeping/honey production. The oak *tasar* sericulture involves village-based rearing and reeling enterprises. The silkworms are being reared on oak leaves sustainably harvested from surrounding oak forests. A district-wide company called Chamoli Tasar Udyog Private Limited will produce silkworm seed for sale to village rearers, purchase cocoons from the rearers and market the reeled silk. For honey production, a district-wide company to be called Dev Bhumi Madhu Udyog Private Limited will be launched to provide technical assistance and extension support, buy honey from bee-keepers and refine as well as package it. Both the companies are expected to be constituted and run by the local communities.

The proponents

The technical assistance on the enterprise side of the project is being provided by Appropriate Technology International and EDA Rural Systems, both organizations which have a long history in establishing micro-enterprises. Technical support for biological monitoring is being provided by the Kumaon University, which has a very strong Department of Botany, specializing in Himalayan ecology.

Community involvement

As mentioned earlier, this area has a history of people's institutions and community natural resource management with the presence of Van Panchayats, Mahila Mangal Dals and Yuvak Mangal Dals. Van Panchayats have existed in this region for more than 50 years. It was the Class I Reserved Forests that were given to the Van Panchayats. Van Panchayats are responsible for internal management, grazing, collection of fuelwood, fodder, timber and protection of forests (TERI 1995; see also Maikhuri et al. in this book).

The project has a distinct advantage because of the presence of these institutions. In the silk enterprise, since the silkworms depend exclusively on oak leaves, the harvesting of these leaves becomes crucial. Being an area where the forests are managed by the Van Panchayats, the harvesting of these leaves is already being controlled by them. At one stage in the project, a particular village Van Panchayat refused to allow the harvesting of oak leaves from their forest, because the silk enterprise did not exist in their village. It took a lot of convincing by the project team for them to allow harvesting; eventually this worked quite well with the project because silk-related activities came to this village also. Local people are also being trained in bee-keeping and sericulture to eventually be able to take over the enterprises. Over time it is hoped that more members of the community will be involved in the monitoring activities, but only after the appropriate orientation and training, which is an important project activity.

Equity is an important consideration in this project. The silkworm-rearing and/or bee-keeping activities require space, and clearly only the more affluent people of the village can participate because of this. However, it is anticipated that once the enterprises are up and running there will be scope for many more community members to participate in a range of activities.

Eco-tourism in Sikkim

Background

The state of Sikkim lies in the eastern Himalayan region and is known for its beauty and its rich biological heritage. The Eastern

Himalaya is one of the biodiversity 'hotspots' of India. It is estimated that Sikkim and its surrounding areas house over 3,000 flowering species, and are particularly famous for rhododendrons and orchids. At present the most significant fact about this area is the recent easing of regulations for foreign tourism. The project aims to focus on two particular sites in Sikkim: The Yuksom Dzongri trekking trail which lies in the buffer zone of the Khangchendzonga National Park, and the Kecheopalri Lake. The Khangchendzonga National Park in Sikkim harbours several endangered and/or threatened species such as the Snow leopard (*Panthera uncia*), the Clouded leopard (*Neofelis nebulosa*), the Red panda (*Ailurus fulgens*), and the Satyr tragopan (*Tragopan satyra*).

Yuksom literally means 'the meeting place of three superior ones'. Historically, three important Lamas met here and the place is since then held sacred. Yuksom is also considered to be the 'altar where offerings are made to protective deities'. It is said that Padmasambhava, who is worshipped by Sikkimese Buddhists, blessed Yuksom and its surrounding areas by placing within it a large number of hidden treasures. The people of Sikkim believe that these treasures must be protected from disturbing influences.

The enterprise

Sikkim occupies an interesting position between Nepal and Bhutan, two popular tourist destinations, and a challenge for developing ecologically sustainable tourism, particularly since the state is just opening up to tourism activities in a big way. At present there is a lack of understanding of what eco-tourism means, what it entails and how to go about developing it. The project is attempting to assist the Government of Sikkim and the Travel Agents Association of Sikkim (TAAS) in developing and marketing eco-tourism activities at two project sites: (i) the Yuksom–Dzongri trekking trail, and (ii) the Kecheopalri Lake. Essentially, four kinds of activities are being carried out: (*a*) Examining threats linked to tourism (eco-degradation, littering, etc.) and promoting activities that will reduce these, e.g., ways to reduce fuelwood use at the two sites; (*b*) Developing marketing strategies to promote eco-tourism in the region; (*c*) Developing economically profitable enterprise activities for local communities, such as lodge and restaurant operations, guiding, portering, pack-animal operation and vegetable production; and (*d*) Addressing policy issues related to tourism, such as the regulation and control of entry and the government's monopoly in coach transport.

The proponents

The project is being implemented by The Mountain Institute (TMI) alongwith the G.B. Pant Institute of Himalayan Environment and Development (GBPIHED), the Travel Agents Association of Sikkim (TAAS) and a local NGO, the Green Circle. TMI, along with TAAS and the Green Circle will provide all the technical and practical inputs for the eco-tourism activities, while GBPIHED will provide the technical expertise for the biological monitoring.

Community involvement

The major stakeholders in this project are the local communities at Yuksom and Kecheopalri Lake and the tour operators. The tour operators after enough orientation towards ecologically friendly tourism would be major participants in the project. Local community involvement will take place as the activities mentioned earlier are initiated and sustained. A preliminary exercise of planning for eco-tourism related activities has been carried out in Yuksom.

The historical significance of Yuksom has been kept in mind during the implementation of the project. It is hoped that the local community will publicize this fact when dealing with tourists. A naturalist guide handbook brought out by the project, emphasizes this fact by giving a good description of the historical importance of Yuksom.

Issues and approaches

The BCN was conceived to prove a narrowly defined hypothesis. However, at the core of every BCN-funded project is community involvement in the conservation of biodiversity, and these projects raise issues common to any community-based project.

Understanding land tenure and monopoly issues

Projects being implemented under the BCN are all operating in different land tenure systems. This has a major impact on the extent to which communities can participate in project activities. For projects operating within a wildlife sanctuary, such as the one in Karnataka, communities have only restricted rights to harvest/collect NTFP. Even if the project succeeds in its objectives, is this enough incentive for communities to participate? Linked to land tenure is also the question of monopoly of marketing. In tribal areas, marketing of NTFP is carried out through cooperatives such as LAMPS. In non-tribal areas there are other agencies. In Garhwal, for example, there exists the Bheshaj Sangh, a medicinal plants cooperative union. Ironically, this particular cooperative has ceased to function in most places, but it still constrains communities from doing their own marketing. This is not to say that illegal harvesting and marketing of NTFP does not take place. In fact, if communities had more management control over these cooperatives, perhaps, they may have a greater incentive to curb illegal activities. The existing cooperatives ensure a return to the collectors. There is no value addition and since the collectors are paid only collection prices, they are severely underpaid compared to the value of the product in the market.

Projects dealing with local communities will thus have to look closely at land tenure and ensure returns to the communities on the basis of that. Also, monopolies would have to be examined before any marketing strategies are designed.

Understanding what is the 'community' and 'local participation'

Community participation is generally a term loosely defined and used rather generically. In reality no community is a homogeneous structure. In any activity there will be one dominant faction of the community that will participate and reap the benefits from it. What any project will thus have to ensure is equity in participation and receipt of benefits. The lower castes, the poor and the women, are not generally permitted to participate nor do they benefit from such initiatives.

It is most important to understand what 'local participation' means. Local participation has to be viewed as a process and not simply the sharing of social and economic benefits. Cernea (1985), describes it as, 'empowering people to mobilize their own capacities, be social actors rather than passive subjects, manage the resources, make decisions, and control the activities that affect their lives'. Local people can and usually do participate in the following ways: (i) Information gathering; (ii) Consultation; (iii) Decision-making; (iv) Initiating and continuing action; and (v) Monitoring and evaluation (Wells et al.

1992; Pimbert and Pretty in this book). Projects should ensure that at least one, if not all, of these activities take place.

Ensuring awareness

Very often when projects like these are initiated, only a part of the community or worse, only a few individuals, are aware of the project and its implications. For any project not initiated by the community, the implications of the project should be completely understood by them. Implementation of such projects without the community's knowledge or agreement could have disastrous consequences. A dialogue with the community has to be initiated even before a project starts and should be continued throughout the duration of the project. Results from all the activities carried out (particularly monitoring) should be shared with the community on a regular basis, perhaps through village meetings, and members' opinions should be sought.

Establishing a rapport with the community

Community involvement in projects such as those discussed, depends on how the project is being implemented, and what are the links of implementing agencies with the communities. In the Karnataka project, for example, the involvement of the NGO, VGKK, is beneficial because close links have already been established with the community. In the Garhwal project, links with the community are also rather strong because the project team consists of local people who are recognized and accepted by the community. However, what works at one place may not necessarily work some place else, and each project has to be site-specific. What is important to remember is that whatever the means, a rapport with the community has to be established before any activity is implemented. Ideally, the community should be consulted from the planning stage itself. It is also important that whatever the composition of the project team, over a period of time, a conscious effort be made to involve more and more community members.

Assessing the capacity of the community

It is important to first assess the capacity of the communities to be involved in enterprise and conservation activities. If communities are to take over a biodiversity-based enterprise, then community members have to be trained in various aspects of the enterprise, e.g., book-keeping, accounting, etc. In an eco-tourism-based enterprise, community members would have to be trained as guides, porters and cooks first, in order to participate in the desired activities. There also could be some activities where community members are more than proficient and should be encouraged to share their skill and knowledge with others.

As projects proceed, there should be provisions for the capacity-building of community members, particularly for those involved in the projects. Training programmes for different target groups and at different levels need to be planned.

Integrating traditional systems of conservation

Most communities living in and around biodiversity-rich areas have traditional systems of conservation. Over the years, some of these systems have naturally eroded or have been modified because of other imposed systems of conservation and management. Incorporating traditional systems of conservation would prove beneficial for any project. The presence of Van Panchayats in Garhwal is extremely favourable for the BCN-funded project because communities are already managing some of the forests in the region. However, the introduction of JFM here, for example, could have unforeseen/negative repercussions if the existing institutional framework is changed. This, however, remains to be seen. Baseline surveys about the area and about the local community should be available before the inception of any project. The surveys should be able to ascertain the existence of traditional systems of conservation and management, if any.

Indigenous knowledge could be an extremely useful tool to involve local communities in conservation programmes. But the steps towards this are first acknowledging that indigenous knowledge exists and is useful, understanding it and finally incorporating it in conservation management systems. A case in point is the BR Hills project. A researcher from the University of Wisconsin, Madison and the social worker on the project have attempted to look at systems of knowledge that have guided traditional Soliga land use practices. The study found that the Soligas classify four distinct vegetation types and grasslands. Along with this the Soligas have an extensive knowledge of the fauna and flora found therein. Soligas collect four types of honey from four different species of bees. From the project's point of view, it is interesting that they can discern between the taste of honey.

Promoting interaction with state departments and other institutions

No project can function in isolation, especially in India where a majority of significant biodiversity areas fall under the jurisdiction of the State Forest Department or the Revenue Department. A relationship with the state departments will have to be established at the very beginning, which would help to obtain support and technical help. There should also be a continuing dialogue with the state departments, to inform them of what is going on and to seek their inputs.

In addition, a lot of creative and innovative work is being carried out on several related aspects, the world over. Exchange of information and collaboration should enrich any project. There is always scope for broadening horizons, but no time to reinvent the wheel. Sharing of information thus becomes extremely important. The JFM network, for example, is looking at some of the same issues that BCN is, particularly relating to sustainable harvesting of NTFP, and there is now exchange of information between the two networks.

Ensuring sustainable use

Value-addition could lead to more sustainable use, but it could also lead to greater incentives to exploit the resources for quick economic gains. Greater economic returns from forest products may also attract people not living in the area or adjacent communities who were earlier not interested in the resource. Such a circumstance would certainly lead to over-exploitation. In such cases it is hoped that since the local community manages the resources as well as the enterprise, it would also devise methods to prevent such over-use. Economic returns could, on the other hand, provide a positive incentive to adjacent communities to also take up value-addition and the subsequent monitoring of resources. However, it is important to reiterate that solutions to a situation such as this must be site-specific. It is possible that a resident NGO could help in arbitration in some cases.

Conclusion

The actual implementation of the use of economic incentives for the better conservation of biodiversity is far from simple or easy. The

Biodiversity Conservation Network is attempting to see how well this approach works. Whether these enterprises are sustainable, whether they are entirely managed by the community, and finally whether conservation is taking place, remains to be seen. However, whether success or failure ensues there will be a lot of lessons learnt, new relationships forged, and most importantly, a new approach to conservation will have been tried.

REFERENCES

Agarwal, C. and **S. Saigal.** 1996. Joint Forest Management in India: A Brief Review. Society for Promotion of Wastelands Development, New Delhi. Draft.

Brown, M. and **B. Wyckoff–Baird.** 1992. *Designing Integrated Conservation and Development Projects*. Biodiversity Support Program, Washington, D.C.

Campbell, J., R.N. Chattopadhyay and **C. Das**. 1995. Income Generation through Joint Forest Management in India with a Case Study of the Participatory Forest Management Project in Nayagram, West Bengal. Paper presented at the Regional Seminar on Income Generation through Community Forestry, 18–20 October, 1995, Bangkok, Thailand.

Cernea, M. 1985. *Putting People First: Sociological Variables in Rural Development.* Oxford University Press, New York.

McNeely, J.A. 1988. *Economics and Biological Diversity: Developing and Using Economic Incentives to Conserve Biological Resources.* IUCN-The World Conservation Union, Gland, Switzerland.

Murali, K., U.S. Agarwal, R. Uma Shanker, K. Ganeshaiah and **K.S. Bawa**. 1994. Impact of Non-Timber Forest Product Extraction on Forests of B.R. Hills, Mysore State. Unpublished paper.

Saigal, S., C. Agarwal and **J.Y. Campbell.** 1996. Sustaining Joint Forest Management: The Role of Non-Timber Forest Products. Society for Promotion of Wastelands Development, New Delhi. Unpublished paper.

TERI. 1995. Community Participation in Van Panchayats of Kumaon Region of Uttar Pradesh. Tata Energy Research Institute, New Delhi. Draft Report.

Wells, M., K. Brandon and **L. Hannah.** 1992. *People and Parks: Linking Protected Area Management with Local Communities*. The World Bank, World Wildlife Fund, and U.S. Agency for International Development, Washington, D.C.

WWF-India. 1995. *Community Biodiversity Conservation Movement: A Profile.* World Wide Fund for Nature—India, New Delhi.

14

Conserving a community resource: Medicinal plants

Darshan Shankar

Introduction

In this paper, it is argued that given the widespread and long-standing use of thousands of medicinal plants in India and the rich and diverse cultures that support their utilization, medicinal plants should be declared a 'national resource' and a national policy be formulated for their conservation and sustainable utilization.

A living cultural heritage

India has one of the oldest, richest and most diverse cultural traditions associated with the use of medicinal plants. The remarkable fact is that it is still a living tradition. This is borne out by the fact that there exist, today, around a million traditional, village-based carriers of the herbal medicine traditions in the form of traditional birth attendants, bone-setters, herbal healers and wandering monks (see Table 14.1). (These numbers have been extrapolated from microstudies on local health traditions in rural communities, carried out by the Foundation for Revitalization of Local Health Traditions [FRLHT], Bangalore.) Apart from these specialized carriers, there are millions of women who have traditional knowledge of herbal home remedies and of food and nutrition.

Complementing the village-based carriers, there are over 400,000 licenced, registered medical practitioners of the codified systems of Indian Medicine like Ayurveda, Siddha, Unani and the Tibetan system of medicine (GOI 1987). The codified systems have sophisticated theoretical foundations and there are hundreds of medical texts in the form of *nighantus* (lexicons) and texts on *bhaisaj kalpana* (pharmacy) that specifically deal with plants and plant products.

Table 14.1: Carriers of village-based health traditions

Traditional Carrier	*Subjects*	*Nos*
Housewives and elders	Home-remedies, food and nutrition	millions
Traditional birth attendants	Normal deliveries	7 lakh
Herbal healers	Common ailments	3 lakh
Bone-setters	Orthopaedics	60,000
Visha Vaidhyas (Snake, Scorpion, Dog)	Natural poisons	60,000
Specialists	Eyes, Skin, Respiratory, Dental, Arthritis, Liver, Mental, Diseases, GIT, Wounds, Fistula, Piles	1,000 in each area

Apart from the traditional use of plants for human health, India has a rich veterinary health tradition (there are medical texts that deal with the treatment of cows, horses, elephants and birds), as well as for plant health (*Vrksh-ayurveda* and *Krsi-sastra*). Though the use of plants for medicinal purposes is by far more common, animals and animal products too have been used, e.g., the *Charak Samhita* mentions 112 bird and animal species used in traditional medicines.

Not science, a '*shastra*'

In passing, it may be worth observing that the knowledge of the Indian people about plants and plant products is not based on the application of Western categories of knowledge and approaches to studying natural products, like Chemistry and Pharmacology. It is based on a sophisticated, indigenous knowledge category called '*dravya guna shastra*'. Unfortunately, due to lack of rigorous cross-cultural studies and in the absence of a well-accepted methodology for such cross-cultural study, there exists no reliable bridge to cross over from Chemistry and Pharmacology to '*dravya guna shastra*' or vice-versa.

Conservation status of Indian medicinal plants

Number of medicinal plants known to the people of India

Over 7,500 species of plants are estimated (GOI 1995) to be used by 4,635 ethnic communities (Singh 1995) for human and veterinary healthcare, across the various eco-systems from the trans-Himalayas to the southern tip of India and from the west coast to the far corners of the north-east. In the codified medical texts of Ayurveda, a recent study (by FRLHT) enumerates around 1,700 species of plants that are fully documented in terms of their biological properties and actions and over 10,000 herbal drug formulations that are recorded for a range of health conditions from the common cold to the raising of the body's general immunity.

There are, however, still big gaps in the work of completing an exhaustive inventory of the medicinal plants of India. While there exist several ethno-botanical studies of different scattered geographical pockets in the country which have documented the locally used plants for medicine, there is at the moment no exhaustive and reliable inventory available of all the medicinal plants of India. State-wise studies of the plants known to various ethnic communities in the various states of India have not yet been undertaken. While checklists do exist, there are no rigorously referenced inventories from primary sources of the plants used in Ayurveda,[1] Unani, Siddha or the Tibetan medical systems. (FRLHT has recently computerized plant names mentioned in 24 primary texts which cover the period 1500 B.C. to A.D. 1900.)

Analysis of medicinal plants: Distribution, habits and families

Macro-analysis (by FRHLT in 1995) of the distribution of medicinal plants shows that medicinal plants are distributed across diverse habitats. Around 70 per cent of India's medicinal plants are found in the tropical forests spread across the Western and Eastern Ghats, the Vindhyas, Chhota Nagpur plateau, Aravallis, the *terai* region in the foothills of Himalayas and the North-East. Less than 30 per cent of the medicinal plants are found in the temperate forests and higher altitudes. Micro-studies (Rani et al. 1995) show that a larger percentage of medicinal plants occur in the dry and moist deciduous forests as compared to the evergreen or temperate forests.

The same research also indicates that medicinal plants are equally distributed across various habits. One-third are trees, an equal proportion is of shrubs and the remaining one-third are herbs, grasses and climbers. A very small proportion of medicinal plants are lower plants like lichens, ferns, algae, etc. The majority of medicinal plants are higher flowering plants. Around 158 families of plants are represented in the Indian medicinal plants.

Consumption of medicinal plants

In recent years, the growing demand for herbal products has led to a quantum jump in volumes of plant material traded within and across countries. Conservative estimates put the economic value of medicinal plant-related trade in India to be of the order of Rs 1,000 crores/year (AIADMA personal communication) and world trade to be over US$ 60 billion. These figures are growing. In fact apprehensions are being expressed that with the trend pointing towards an inexorable monetization and commercialization of the medicinal plant economy, we could have a scenario where the rich alone would be able to afford herbal products while the poor would have to make do with cheap, mass-produced, synthetic, chemical drugs.

More immediately, however, the availability of medicinal plants is already under serious threat. Over 95 per cent of the medicinal plants used by the Indian industry today are collected from the wild. Less than 20 species of plants are under commercial cultivation, while over 400 species are used in production by industry. FRLHT research in 1996 shows that over 70 per cent of the plant collections involve destructive harvesting because of the use of parts like roots, bark, wood, stem and the whole plant (in the case of herbs). This poses a definite threat to the genetic stocks and to the diversity of medicinal plants. A threat-assessment exercise carried out by FRLHT in 1996, as per latest IUCN guidelines, has already listed 74 species of medicinal plants in southern India that are threatened. Estimates suggest that over half-a-million tonnes of dry raw material is indiscriminately, and mostly destructively, collected from the wild each year. Notionally, this is equivalent to over 165,000 hectares of forest being clear-felled each year.

Traditional practices relating to medicinal plants

One does not really have to look for discrete and specific instances of medicinal plant-conservation practices amongst traditional communities in India, because the ways of conservation are pervasive and an integral part of living culture.

Analysis of habits of plants used by communities shows that all the plant forms including lower plants are used. Micro-studies show that the largest single use of plant and animal diversity in any ecosystem is for the purposes of human and veterinary medicine. Thus 30–70 per cent of biodiversity in an ecosystem is used as medicinal resource. Tables 14.2 and 14.3, based on field studies in the Western Ghats, illustrate this point.

The intimate knowledge of local communities about their bioresources is clearly seen in the tremendous diversity in local names

Table 14.2: Some of the medicinal and other plants used by Mahadev Koli tribals (western India)

Purpose	*No. of Plants*
1. Medicinal uses	202
2. Veterinary uses	109
3. For fish poison	23
4. For pest control	51
5. For water purification	3
6. Wild edible plants	87
7. Fodder plants	65
8. Fuel plants	30
9. Hunting purposes	3
10. Cultural and religious purposes	38

Source: Kulkarni 1993.

Table 14.3: Medicinal plants used by the healers of Karjat tribal block of the Western Ghats

Habits of Plants	*Nos Used as Medicine*
1. Trees	168
2. Shrubs and Herbs	207
3. Climbers and Creepers	105
4. Grasses	13
5. Epiphytes and Parasites	16
Total	509

Source: Palekar 1992.

and uses of the same plant and animal species as one moves across what may be referred to as 'ethno-bio-geographic regions'. The diversity of names and uses shows how every community has made an independent assessment of its local resources (see Table 14.4).

The etymology of local names and the content of local knowledge also reveals the understanding communities have of properties, morphology, phenology, reproductive biology and habitats of plants. Given in Table 14.5 are illustrative examples of the meanings of plant names.

There are local traditions regarding the 'collection times' of plants that also illustrate the conservation ethic. Collection times are related to seasons, e.g., tender leaves are to be collected in monsoon, barks of trees and sap are only to be collected in autumn, tubers in early

Table 14.4: Diversity in uses of some plants in South India

Examples of ethno-medicinal plants with 10 and more uses reported across ethnic communities in South India.

Sl.No	*Plant Name*	*Reported No. of Uses*
1.	*Centella asiatica*	33
2.	*Pergularia daemia*	23
3.	*Aristolochia indica*	22
4.	*Ichnocarpus frutescene*	22
5.	*Alstonia scholaris*	19
6.	*Holarrhena antidysentrica*	18
7.	*Trachyspermum ammi*	16
8.	*Hygrophila auriculiculata*	15
9.	*Trianthema portulacastrum*	15
10.	*Semecarpus anacardium*	15
11.	*Hemidesmus indicus*	15
12.	*Catharanthus roseus*	14
13.	*Apama siliquosa*	13
14.	*Anacardium occidentale*	12
15.	*Costus speciosus*	12
16.	*Justicia gendarussa*	11
17.	*Pergularia extensa*	10

Source: Data generated by the Research Department of the Foundation for Revitalization of Local Health Traditions.

Table 14.5: Meanings of traditional plant names

Sl.No	*Botanical Name*	*Local Name*	*Locality*	*Meaning*
1.	*Tridax procumbers*	Jakam Jod	Karjat, Maharashtra	Healer of wounds
2.	*Zarnia diphylla*	Harun Khure	Konkan region, Maharashtra	Ears like those of a 'deer'
3.	*Mucuna pruriens*	Khaj Khujli	Thane District, Maharashtra (Warli tribe)	Causes severe itching
4.	*Holarrhena antidysentrica*	Girimallika	(Sanskrit literature)	Habitat on hill sides
5.	*Catharanthus roseus*	Sadaphuli, Sadabahar	Maharashtra, U.P.	Always in flower

Source: Data generated by the Research Department of the Foundation for Revitalization of Local Health Traditions.

winter, roots in late winter, and so on. They are related to the stage of the plant life cycle, e.g., leaves of certain plants should be collected before the flowering of the plant (e.g., *Adathoda vasica, Occinum sanctum*).

There are practices which involve non-destructive collection, e.g., only the north-facing root of a plant should be collected, or in tribal areas it is preferred to use only fresh plants, so that a plant is collected only when needed and in the limited quantity that is actually needed. There are guidelines related to the time of the day when a plant should be collected, e.g., plants are usually not collected after sunset or at mid-day.

In certain cases plants are collected only at certain times of the year when particular constellations occur, e.g., the *Pushya Nakshatra* or on a full-moon or a new-moon day. All these regulations which are part of 'culture' can be seen as indicators of a conservation ethos. More obvious forms of conservation like sacred groves and gardens for the 'snake gods' (*Sarpa Kavu*) are only highlights of a more generalized conservation ethos.

Today, for various reasons, many of these traditional conservation practices related to medicinal plants are being eroded.

Community-oriented programme for conservation of medicinal plants in South India

A community-oriented conservation strategy for medicinal plant resources has been implemented in South India since 1993. This project is the most comprehensive post-independence effort in India to conserve plant diversity. Fifty-five conservation sites (Medicinal Plant Conservation Areas [MPCAs] or Medicinal Plant Conservation Parks [MPCPs]) have been established across the three states of Karnataka, Tamil Nadu and Kerala. The conservation strategy consists of six elements. All the six elements of the strategy involve community participation, as outlined in Table 14.6.

The politics–science nexus and indigenous cultures

The impact of the politics–science nexus on indigenous cultures can be explained through the example of bio-prospecting. Bio-prospecting is understood to mean seeking access to the biological resources of a country, region, or locality with commercial intent (which may

Table 14.6: Conservation strategy for medicinal plants in South India

Sl.No.	*Elements of Conservation Strategy*	*Role of Local Communities*
1.	Establishment of a network of *in situ* reserves (MPCAs) across all the vegetation types of South India where an attempt is being made to conserve viable breeding population of medicinal plants.	Local community management committees work alongside the Forest Department to manage the MPCA.
2.	Establishment of a network of ethno-botanical parks (MPCPs) to create living collections of all the medicinal plants known to various ethnic communities of South India.	Community-oriented NGOs manage the MPCPs. High community participation is involved.
3.	Establishment of a chain of nurseries to multiply and supply three basic medicinal plant packages namely: a. Plants for Primary Health Care b. A package of economic plants that are needed by industry c. Threatened plants for possible re-introduction into their natural habitats	The choice of species in nurseries is advised by folk healers and community leaders.
4.	Establishment of a network of JFM projects (MPDAs) to propagate medicinal plants in degraded forest areas and harvest them sustainably.	The key partners in Joint Forest Management are village households.
5.	A focused research programme to guide conservation action.	Barefoot taxonomists are involved.
6.	Development of communication materials like films, illustrated books, posters, folk media to create public awareness and involve local communities in conservation action and sharing of benefits from MPCAs, MPCPs, and MPDAs.	The focus of the communication programme is on creating public awareness and participation.

be preceded by a research programme), sometimes with (and often without) material and information transfer agreements and subsequently sharing of patent rights and royalties, if a commercial product is developed.

There are controversies of an ethical nature surrounding bio-prospecting initiatives, and these generally relate to no acknowledgement of source of materials and information, and total absence of benefit-sharing agreements. This situation is labelled as bio-piracy. Controversies also arise in cases when exchange agreements exist but are perceived to be unfair and one-sided, and it is held that weaker parties in the deal are not adequately compensated for their knowledge and resources.

There are also typical complications that arise in bio-resource rich countries with 'living' cultural traditions in the context of rewarding all the carriers of indigenous knowledge and owners of the bio-resources. These complications relate to practical problems in identifying the 'rightful' owners of indigenous knowledge. The example of a recent benefit-sharing arrangement between the Tropical Botanical Garden and Research Institute (TBGRI) and the Kani tribe of Kerala is illustrative. A part of the sales of the drug developed out of a herb *arogya paccha*, whose medicinal properties were pointed out by the Kanis to TBGRI, are going to the tribe. In this case, who should the benefits go to: the individuals who first alerted TBGRI to the plant, the hamlet of which these individuals are a part, or the whole tribe? The plant was known to the entire Kani community. Furthermore there have also been 'unverified' claims that the knowledge, use and natural distribution of the *arogya paccha* plant is not restricted to the Kanis in southern Kerala, but is also known to tribals in some parts of Karnataka. The TBGRI is at least trying to be fair to the tribe, and has helped in establishing a trust run by the Kanis, into which the money will go (Anuradha 1997; Pushpangadan 1996).

Even if in this case the knowledge is in fact restricted to the Kanis of south Kerala, it is more normal to expect indigenous knowledge of a particular bio-resource to be widely distributed. This can be seen in the case of the plant *Phyllanthus niruri* (*Amarus fraternia*), whose use for treatment of *Kamla* (infective hepatitis) is known throughout South India and in the Andaman and Nicobar Islands. The Fox Chase Research Centre in the USA has filed a controversial patent claim on a hepatitis drug developed from this plant. The practical problem in such cases is, how does one share benefits with a widely distributed owner constituency? One of the solutions offered to this problem is that the benefits should be placed in a community biodiversity fund, from where they could be distributed to communities. This proposal raises further issues, viz., who will control the biodiversity fund and how will it be allocated equitably?

The advocates of bio-prospecting make two claims. First, they project 'substantial' economic benefits that prospecting agreements can provide to the communities who are the owners of bio-cultural resources and by the same token they therefore suggest that commercializing such resources holds the modern day motivation for 'conservation' because it can provide economic incentive to the communities to conserve their natural resources.

Both these claims appear to be fragile on closer examination. In the example of the Kani tribals, the per capita benefit to a community of a few hundred households from a licence fee of Rs 500,000, would amount to a less than Rs 5,000 income per household. If the distribution of the compensation is to be divided over several dozens of villages as in the case of the knowledge of *Phyllanthus niruri*—which represents a more typical example—the per capita benefit would be of the order of perhaps Rs 50 or Rs 100. In other words, in per capita terms, the benefits to members of a community from the sale of indigenous knowledge to a bio-prospector could amount to 'peanuts'.

In the conservation context, it also seems to be clear that market cultures do not really promote conservation substantially because the market uses only a fraction of the biological resources known to communities. In the case of medicinal plants, out of 30,000 species estimated to be used by indigenous cultures worldwide, around 150 species are used in the global market. In India, out of 7,500 species known to tribals, less than 500 are used in the market place with real benefits going to traders rather than primary collectors. Out of over 300 medicinal species known to the Kanis, only one found its way to the market. Thus the claim of market incentives promoting conservation appears to be hollow.

This is not to argue that one should not attempt to commercialize indigenous knowledge. The argument is that commercialization is not enough. Ways to ensure really wholesome and substantial community benefits from utilization of indigenous knowledge and native resources seem to lie in pursuing other approaches also, rather than depending on the fruits of bio-prospecting alone.

A central issue that is totally missed in the midst of these ethical and benefit-sharing debates is that of the serious cultural erosion that bio-prospecting entails. Bio-prospecting actually involves one culture prospecting on the intellectual and biological resources of another culture. Specifically it involves 'science' and its carriers (a Western knowledge system) prospecting on some indigenous knowledge system of non-Western origin. One culture is a prospector and the other the prospected.

One culture is assumed to be the creator of new and superior knowledge and products, and the other, the donor of raw material and imperfect, crude and unrefined knowledge which is only good enough to provide 'leads'. One culture is considered advanced and the other viewed as essentially backward. It also happens to be the case that economic and political powers lie within the prospecting cultures, whereas the donor culture belongs to a society which is today politically and economically weak.

Given the domination of the prospecting culture, intellectual property is also only defined in terms of the parameters of one cultural tradition. The parameters, categories and concepts of diverse ethnic knowledge systems cannot be applied to claim IPRs under the rules of the currently expanding global market.

When bio-prospecting is viewed thus in terms of a cross-cultural transaction, the question that arises is, can any financial 'compensation' and reward be adequate to the donor ethnic culture for permitting itself to be prospected upon and thus demeaning its own integrity, identity and value of its own culture? The current debates ignore this destructive consequence of bio-prospecting. Whereas mutually respectful exchanges across cultures are to be welcomed, the political, sociological and epistemological foundations for such cross-cultural dialogue have not yet been established and bio-prospecting represents the typical one-sided transaction.

It must be recognized that cultural diversity is as essential for civilizational evolution as genetic diversity is for biological evolution. Therefore, for the long-term survival of human societies, it may be worth sacrificing the more short-term and short-lived gains of economic prosperity that the present monocultural model of global economy promises to 20–30 per cent of the world's population.

Medicinal plants are a 'bio-cultural' resource. While the need to conserve the plants themselves is more easily appreciated by policy-makers, the importance of the indigenous cultures associated with the plants are not substantially understood. The general perspective of 'scientists' is that while the traditional knowledge about the use and properties of plants is good enough to serve scientists as raw material or a lead for more sophisticated scientific work, viz., to discover bio-active molecules or to evolve more refined drugs, it is not good enough to be promoted in its essential traditional form. This perspective, with perhaps a two-centuries-old political genesis in Europe, in fact has no scientific basis. It is based on neither a deep understanding nor an objective evaluation of the traditional knowledge systems.

Box 14.1 gives some specific illustrations of the political discrimination faced by traditional systems of medicine in India and also in global forums like WHO.

Box 14.1

Examples of Political and Economic Intervention Regarding Traditional Medicine

Colonial Period in India

- Banning of indigenous small-pox inoculation by the British government in 1805
- Closing down of ayurvedic teaching institutions in 1835
- No legal registration of native doctors
- Total cut off of government aid to Indigenous Systems of Medicine (ISM)
- No serious evaluation of capabilities of ISM

Post-independence (in India)

- ISM get less than 4 per cent of the National Health Budget
- Sub-critical investments in research, teaching and public health services

World Health Organization

- Invests less than 5 per cent of its budget on traditional medicine
- Policies on traditional medicine are usually framed by Western medicine professionals
- Ethno-studies on plants and drugs are based on Western medicine hypotheses and are only aimed at pirating from the materia-medicas of ISM, and not revitalizing or strengthening the foundations of ISM

Source: Data generated by the Research Department of the Foundation for Revitalization of Local Health Traditions.

It is beyond the scope of this paper to explore the social motives for the promotion of scientific culture at the cost of the denigration of indigenous non-Western knowledge systems, but it is clear that the underlying motives have been economic and political.

The point however must be made that with respect to medicinal plants and their culturally evolved traditional products, it is important to carefully examine the local cultures and discover their genius. The traditional knowledge systems may already yield fully matured products for the contemporary use of the world, which need no upgradation by scientists and bio-prospectors.

Protection of traditional knowledge

So far investments on the conservation of medicinal plants of India have been rather skewed. No one has shown any interest, for instance, in investing public or private funds for ensuring long-term availability of the large number (over 7,500 species) of eco-system-specific medicinal plants that have been traditionally used by numerous village communities throughout the length and breadth of India for their basic health needs. This is because self-reliance of 300–400 million rural poor in primary healthcare is on no one's political or economic agenda.

Perhaps not so surprisingly, there has in fact been considerable public investment but marginal private investment, on medicinal plants needed by the modern pharmaceutical and perfumery industry. The Council of Scientific and Industrial Research (CSIR) and Indian Council of Agricultural Research (ICAR), Government of India, have invested public funds to serve the needs of industry by working on the agro-technology of about 40 odd species for the last three decades or so.

In recent years, there has been a new worldwide, commercially fuelled research interest in the traditional knowledge of medicinal plants. Controversial patents have been filed on products derived from Indian plants like *Azadirachta indica* (*Neem*), *Curcuma longa* (turmeric), *Phyllanthus amarus* (*Bhumi Amla*) and *Piper longum* (*Pipalli*). In each of these patent claims researchers have superficially modified traditional knowledge of these culturally well-known plants and claimed the modification to be their original innovation. They have thus attempted to privatize public domain knowledge without seeking informed consent and without negotiating compensation or sharing of benefits with the traditional custodians of the knowledge of medicinal plants. Whereas there are a lot of energetic exercises underway worldwide to work out model agreements and draft legislation that can prevent bio-cultural piracy, the long-term developmental interests of traditional communities with reference to revitalization and utilization of their indigenous knowledge for their own current and future needs is ignored.

While it is essential to take practically implementable steps to check bio-cultural piracy, it is rather uncertain what per capita benefit members of the traditional communities are going to ultimately derive from compensation paid for the commercial utilization of a small fraction of their vast local knowledge. In the long-term interests of local communities, it seems to be far more important to put in place a programme for the revitalization of local health cultures and conservation of bio-resources which could certainly ensure self-reliance in healthcare to millions.

Suggested goals of national policy on medicinal plants

Given the length of the unbroken tradition in the use of medicinal plants across over two millennia and given the continuing social and growing economic importance of this bio-cultural resource, it is necessary for the Indian people to view medicinal plants as a national community-based resource. A national policy on medicinal plants conservation, and immediate implementation of effective programmes in line with the national policy, are absolutely essential if we are to save elements of this valuable resource from extinction and be in position to use it sustainably over the next decade and into the next century.

The goals of a national Medicinal Plants Conservation Policy should include the following:

a. To ensure long term *in situ* conservation of viable breeding populations of medicinal plant diversity in a range of natural habitats corresponding to their natural distribution.
b. To regulate, improve and where necessary ban medicinal plant collections from the wild.
c. To carry out appropriate legislation and/ or modify existing ones to regulate internal and external trade in medicinal plants and to ensure adequate staff training of the agencies involved in regulation.
d. To encourage and support a range of *ex situ* conservation measures as supplementary and complementary measures to the above.

e. To encourage medicinal plant requirements of industry to be met sustainably through policies favouring large-scale cultivation of medicinal plants.
f. To ensure access of 'region-specific' medicinal plants to village communities for their primary healthcare needs and address issues relating to participation and empowerment of these communities in conserving and managing forest and non-forest resources.
g. Considering the low value of many medicinal plant species in raw material form, to promote community-level enterprises for value addition to medicinal plants through simple, on-site techniques like drying, cleaning, crushing, powdering, grading, packaging, etc. This will also increase the stake of village communities in conservation.
h. To ensure community benefits arising from commercial utilization of indigenous knowledge of plant resources relating to the medical traditions of the country and to safeguard them from unfair patent claims, in the context of the international understanding reached on this issue in the Convention on Biological Diversity at the Earth Summit in 1992.
i. To develop a cadre of taxonomists, para-taxonomists and forest scientists to support and sustain rapid, prioritized baseline surveys with respect to medicinal plants conservation status and to monitor changes on a long-term basis.
j. To identify, review and clearly mandate appropriate institutions (governmental and non-governmental), to implement the national policy; and to strengthen/build adequate institutional structures and capacity.
k. To encourage and facilitate time-bound projects through governmental and non-governmental institutions, to further the goals of the policy.
l. To continually monitor, evaluate and analyze the impact of this Medicinal Plants Conservation Policy.
m. To create widespread public awareness on the need for medicinal plants conservation.

REFERENCES

AIADMA 1996. Personal communication with the All India Ayurvedic Drug Manufacturers Association.

Anuradha, R.V. 1997. Sharing with the Kanis: A Case Study from Kerala, India. Kalpavriksh, New Delhi. Submitted to the Secretariat, Convention on Biological Diversity, as a case study on benefit-sharing.

GOI. 1987. *Handbook of Health Statistics 1987.* Department of Indian Systems of Medicine, Ministry of Health, Government of India, New Delhi.

———. 1995. *All India Ethno-Biology Survey.* 1995. Ministry of Environment and Forests, Government of India.

Kulkarni, D.K. 1993. Medicinal Plants Used by Mahadev Kolis. Ph.D. Thesis, Agharkar Institute, Poona.

Palekar, R.P. 1992. Medical Traditions of the Thakur Tribals of Karjat Tribal Block. Unpublished paper, Academy of Development Science, Kashele, Karjat, Maharashtra.

Pushpangadan, P. 1996. Tropical Botanical Garden Research Institute: People Oriented Sustainable Development Programme. Paper presented at the UNEP/GEF Indigenous People's Consultation Meeting, Geneva, 29–31 May 1996.

Rani, M., D.K. Ved and **B.R. Ramesh.** 1995. Ascribing Priority for Mapping Medicinal Plants Distribution in South India. Research Paper for FRLHT, Bangalore.

Singh, K.S. (ed.). 1995. *People of India.* Anthropological Survey of India, Calcutta.

15

Institutional structures for community-based conservation

G. Raju

Introduction

Indian forest management is gradually shifting towards participatory or joint management, involving a partnership between state agencies and local communities. The now famous Joint Forest Management (JFM) has been in existence as an official policy endorsed by guidelines issued by the Ministry of Environment and Forest (MOEF) in 1990, though informally it has existed for much longer.

There is growing evidence in the country today that people's institutions (PIs) are involved in protection of forests and managing them well enough for researchers to call the phenomenon an environmental movement. The notable feature of this movement is that it had started well before any official statement on JFM became an accepted policy. An estimate puts the figure at over 10,000 PIs protecting and managing forests over 1.6 million hectares in the states of West Bengal, Orissa, Bihar, Madhya Pradesh, Gujarat and Rajasthan (Poffenberger 1995; see also Das and Christopher, and Sarin et al. in this book).

Many of these PIs started protecting forests as a response to the alarming deforestation and resultant difficulty in access to forest produce (Raju et al. 1993, Poffenberger 1995). These efforts were often self-initiated, and at times catalyzed by local leaders, youth clubs, NGOs and even Forest Department (FD) personnel.

The salient features of this trend are: (*a*) Strong sense of ownership of the forest and a commitment to protection; (*b*) Evolution of local institutional structures; (*c*) Formation of self-made regulations for use and management of forests; (*d*) Increase in biodiversity in PI protected forests; and (*e*) Symbolizing a form of people's protest against the past and present policy of commercialization of forests and its anti-people character.

Characteristics of people's institutions

PIs have a life of their own. In order to understand their evolution patterns, how some PIs do better than others, complexity of decision-making, and existing conflict resolution mechanisms, an interactive workshop of about 11 dynamic PIs was called in Gujarat in 1994. This workshop, 'Strengthening People's Institutions Network', included the supporting NGOs. The object of the workshop was also to explore the possibilities of networking among the participating NGOs and the PIs. The following is a summary of findings, which may have relevance for PIs in other parts of India too.

Evolution

PIs have come into existence having felt the need for collective action (*a*) to regenerate forest resources to meet consumption needs, or (*b*) to gain access and develop such resources for economic empowerment to emerge from existing social oppression. While some of the PIs are self-initiated, some are organized by NGOs, while others are inspired by similar work in their neighbourhood. The key lessons from their evolution are that the strengthening of PIs in the initial years involves a great deal of effort, in which dealing with minor to major conflicts contributes a great deal. External support, especially from NGOs during the evolutionary period helps facilitate the institution-building process. As PIs gain confidence and build their capabilities, they diversify their activities and become more self-reliant. With the strengthening of PIs, support from government departments becomes more forthcoming.

Structure

Basically, PIs have a two-tier structure:

(a) The general body of its members which is all powerful. Membership is open to all residents of the village (which is the area of operation) with one or more representatives from each household. Women are being inducted in PIs where concerns regarding the participation (or lack of it) are being discussed. Some PIs have worked out a compensatory contribution for late joiners as members.

(b) The executive committee (EC), a representative body elected by the general body for functional purposes. The EC usually has representation from the hamlets/castes/clans sub-groups within the PIs. Some PIs have an NGO or a government representative on it without voting rights. The EC meets at least once a month. Some PIs have kept these meetings open to all members. One vote per member, a quorum for meetings and decision-making are norms found across PIs.

By and large, whether registered or not, PIs are governed by the principles of cooperation founded on democratic norms. Registered PIs as in Gujarat follow the cooperative model. The organizational structure and practices of most PIs reflect the reverence for collective wisdom of its members as well as their democratic action and management of common resource such as forests.

Activities

(a) Protection of forest lands with the object of meeting needs is the primary activity. This initial activity usually inculcates a sense of ownership, collective responsibility in members and they take turns to patrol the forest in groups under the *vara* (rotational) system or contribute towards the salary of watchmen either in kind or cash, or spontaneously assist in a crisis. PIs further levy fines to deter offenders and take them to task. Firm action in enforcing protection establishes the PI's intent, building its confidence and image in the area, further boosting its efforts.

(b) Regulating forest use for consumption needs usually begins simultaneously with protection activity. Each PI evolves its own norms for the purpose, such as:

- Fuelwood holidays; fixed days of fuelwood collection; sometimes a small fee for collection; collection of only dried twigs and branches; distribution of cut produce.
- Specification of the number of collectors per household, or pooling of the produce to be shared equally.
- Fixed time for collection of fodder, usually after seeding of grasses; a permit to graze only after that; grazing of cattle allowed only when the forest has grown sufficiently.
- Free access to non-timber forest produce (NTFP), sometimes with seasonal restrictions.

- Access to small timber provided on a permit basis, sometimes with a small fee which is credited to the village development fund. No sale of small timber permitted.
- Compulsory participation in forest protection, usually carried out in groups on a roster basis.
- Keeping the forest closed to non-members, according to norms evolved by themselves.

The norms evolved by the PIs reflect a concern for inter-household equity on items of common interest. Besides, they demonstrate the PI's capacity to manage forest use. PI norms are evolved and modified collectively over a period of time. The inherent flexibility in evolution of norms and its effectiveness in implementation in the interest of its members reinforces collective action and commitment.

(c) Forestry activities like nursery, plantation and soil-moisture conservation are being taken up by the PIs that have forged linkages with the FD. While a few PIs have contributed voluntary labour (boosting their cooperative effort) some have had access to funds and thus larger control over implementation. However, in most PIs, their role in such activities is limited by the FD's control over both funds and implementation. Together with forestry activities, other related and integrated development activities serve to provide employment, increase resource productivity and address member's needs, thus sustaining member interest.

(d) Few PIs have entered the area of procurement, processing and marketing of forest produce. Nationalization of some forest produce and difficulties in obtaining permits to sell produce are some of the government-imposed restrictions that have hindered development of such enterprise. This leaves very little room for PIs to earn a better margin and has limited their efforts at value addition and direct participation in the market.

Decision-making

The general body of the PIs is vested with the ultimate power in decision-making. The general body elects the EC and authorizes it to take certain decisions and actions. The bye-laws of registered PIs specify duties and responsibilities at each level. However, PIs adopt a fluid decision-making process with the general body taking important decisions and the EC implementing them. With the increase in the number of activities the decision-making becomes more complex. Transparency is retained in the entire process.

Financial management

PIs, both registered and unregistered have bank accounts with joint signatures operated by two EC members selected for the same. In some cases NGOs also include a signing member for operating accounts. Being a vulnerable area of management, institutional procedures have been developed to keep members informed and to check any malpractice. These include keeping records of transactions, discussions in committee meetings and in the annual general body meeting. Some PIs also have a system of internal audit, this role sometimes being performed by appointed members or an NGO. The number and range of funding sources tapped by PIs depend on the level of activities and the contacts established with the outside agencies. Besides access to such funds, some PIs also mobilize funds from savings of members and sale of produce. They then go on to exercise their autonomy by utilizing funds to provide credit to members and by investing a part of their finances.

Benefits and benefit-sharing

Benefit-sharing in PIs is largely determined by the local conditions and needs of the people, the objectives of the PIs and the perceived importance and priority of the benefit. PIs have a variety of experiences in utilization of benefits arrived at from their activities. The nature of benefit could be: economic (e), material (m) and/or social (s). These benefits could accrue to an individual (i), the weaker sections (w), and/or the collective (c) as a whole. A combination of these two sets of factors yield different types of benefit-sharing arrangements as indicated in the following list:

ei: Wages through employment generated during resource development paid to individual members
mi: Increased availability of fuelwood to individual members for domestic use
si: Courage gained by members to speak at forums
we: Exemption/concession for poorer members from making cash contribution
mw: Allowing landless an extra share of fodder from the forests
sw: Enrolling and making space for the weaker group to participate
ec: Building a fund for the PI or collective
mc: Forest regeneration as a result of regulated use by all members in the collective

sc: Increased unity and a positive image of the PI in the eyes of the world outside.

While individual economic (ei) and individual material (mi) benefits are the starting points for collective action in PIs, they also serve to sustain members' interest over time. Special consideration for women and the weaker section in terms of concessions in contribution or encouragement to participate has been a feature in some PIs. With the increase in activity levels, PIs have been feeling the need to generate a collective fund or create common assets. PIs from backward areas and oppressed groups have been giving emphasis to creating collective wealth to serve members with credit and to generating employment opportunities, thus also building money power and a positive image of the PI in the area.

Federations of PIs

It is also seen that with the coming up of several PIs in a cluster, disputes at times occur between the PIs over boundaries, access to resources that are being protected, and people trespassing into another forest. As a result, the need for a common forum to discuss the issues, and for conflict resolution arises. Under such circumstances PIs in some areas have felt the need to establish linkages with each other and federate, as seen in Orissa and Gujarat. In Orissa, such processes, both formal and informal, have a three-tier structure, viz., village-level PIs, cluster- or block-level committees, and an overall federation. Their objectives are: (*a*) To spread the idea of community protection to newer villages, build unity and cooperation in cluster committees and PIs; (*b*) Conflict resolution within and between PIs if unresolved at local levels; (*c*) Liaisoning and lobbying with the government and critiquing the government order for necessary changes; and (*d*) Running a newsletter service to provide information and links with its constituency.

Some examples of federations will illustrate the above. The Bon o Jeevan Bandhu Parishad, based in Bolangir, Orissa, is a federation of 419 villages. It was PIs that initiated networks to form cluster committees (CCs). The CCs have a varying number of villages in their ambit and their members include all members of PIs which may be registered or unregistered. The PIs have their own by-laws, rules and regulations and retain their own identity. These CCs have formed a federation called the Maha Sangh. The Maha Sangh has open membership and has no control over PIs or CCs. This has been a conscious decision to respect the independent status of PIs and CCs. The federation has an annual meeting of its members to vote its 30 executive committee (EC) members by secret ballot. The EC has representatives of women, and Scheduled Castes and Tribes. It meets every two months to carry out its functions.

A similar federation exists in Nayagarh district of Orissa, consisting of 347 villages covering over 85,000 hectares of forests. In Mayurbanj, Orissa, a 40-village federation exists with headquarters at the village Budhikhamari.

The Lok Van Kalyan Parishad is an informal federation of about 35 tree growers' cooperatives in Bhiloda and neighbouring talukas of Sabarkantha district, Gujarat. The Parishad's constituency consists of these 35 PIs and the total membership is about 4,289 of which the female membership is 517. Roughly 8,000 hectares of forest land is covered. Each PI is represented on the Parishad by one man and one woman. These representatives have chosen a coordinator to look into routine activities and represent the Parishad as and when required.

The need for the Parishad arose from the experience of two PIs in villages Malekpur and Abhapur which are described briefly below. Malekpur's constant bickering with the neighbouring villages on the issue of forest boundary and trespassing by cattle and people brought out the need to spread the idea of JFM in these villages too, so that the communities there would not only stop trespassing Malekpur's forest but also start protecting their own forest. To this end, the PI in Malekpur organized a meeting of leaders of neighbouring villages with the support of the FD and an NGO, the Vikram Sarabhai Centre for Development Interaction (VIKSAT).

In the case of Abhapur, when VIKSAT took up cudgels on behalf of some restive members who were unhappy with the proceedings in the PI, the EC members took a strong stand against VIKSAT. The experience pointed out the need for a peer group to take up issues related to the functioning of PIs, their accountability to members, etc. Around the same time some documentation of a Parishad type of organization was being done in the state of Orissa (Raju et al. 1993), and the idea of a Parishad for this area's PIs took shape.

The Parishad has a set-up striving for democratic principles in forest management. It helps establish a platform for communication with the FD on an equal basis. Since its establishment in 1993, it has been dealing with a range of issues (see Figure 15.1). For example, the Parishad reviewed the agreement to be signed with the FD and offered the following recommendations: (*a*) the agreement does not mention tree growers' cooperatives as one of the PIs' models for participation in JFM; hence this is to be specifically stated; (*b*) PIs should be provided *all* of the poles that result out of thinning and cleaning apart from fuelwood, as against 25 per cent provided in the agreement; (*c*) the distribution of forest produce to the members should be left to the PIs and not to the FD as provided in the agreement. The impact of the Parishad is seen in the expansion of the JFM programme to newer villages, better communication with the FD, information dissemination, creation of a conducive atmosphere for conflict resolution, and strengthening of PIs in overall terms. The FD has recognized the role of the Parishad and is using it as a forum for dialogue with the PIs (Raju 1996).

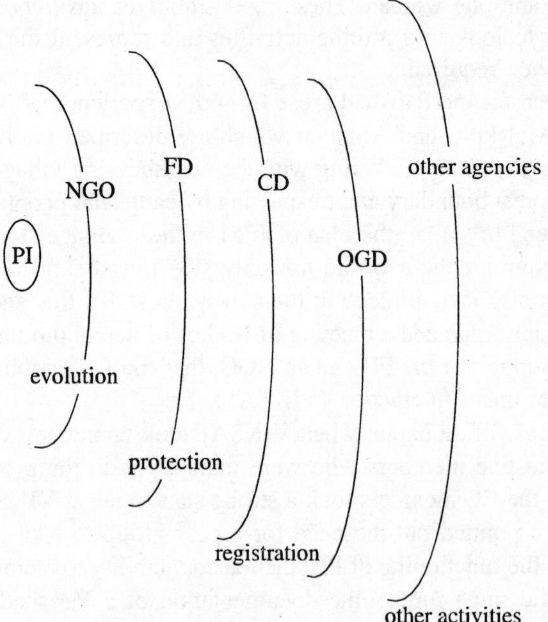

Figure 15.1: Ripples of People's Institutions

In the state of Bihar, the Bisrampur sub-beat in the Singhbum district has seen the emergence of 42 village communities protecting about 1,600 hectares of forest lands for the last 20 years. The 16 PIs formed in these villages have also formed a central committee (Singh 1994).

From available experience, it can be seen that federations provide an important platform for peoples' voices. Respect for independence and individuality in member organizations and adoption of democratic processes are noteworthy features. Federations also play an important role in balancing power equations between the FD and PIs.

Box 15.1

Saksham: A Forum of PIs

In Gujarat, as a result of an interactive learning opportunity (detailed below), representatives of PIs and their federations felt the need to explore collective action on areas of common concern. This culminated in the formation of Saksham, a broad-based collective forum of PIs to take up issues at the state level while also enabling the PIs to gain strength and confidence from it. Its role has been envisaged as follows:

- Take up state-level issues.
- Provide information on policy and schemes to all PIs.
- Lobby for pro-people policies.
- Organize training and visits for the capacity-building of PIs.

Saksham's membership consists of two representatives (one man and one woman) from each of the eight clusters of PIs representing about 300 PIs in different parts of the state. NGOs would also be represented. Thus, Saksham would have a 22-member working group. An NGO is performing the role of the secretariat. Currently, Saksham runs a newsletter to share experiences. It has also prepared a memorandum to the state government, urging it to:

- Induct five representatives of Saksham in the state-level working group for JFM.
- Change JFM agreements to enable an equal partnership of, and fair benefit-sharing between the PIs.
- Allow good (standing) forest (not just degraded forests) to be taken under JFM.
- Hand over implementation responsibility to PIs once an action plan is agreed upon.
- Return a share of the fines collected to PIs.
- Declare degraded forests as village forests and provide legal backing to JFM.
- Simplify procedures for speedier registration of PIs as cooperatives.
- Provide permits to PIs, rather than outside traders, for carrying out value-addition and sale of forest produce.

Conflict resolution

Conflict resolution has been one of the most pervasive and sensitive aspects of PIs (see Box 15.2). During the course of their existence, PIs experience conflicts over a variety of issues: rights related to membership, benefits, power, decision-making and facilities, boundary demarcation, and enforcing of sanctions, especially at the intra-village and inter-village levels. The exposure of a PI to conflicts of different kinds and its ability to deal with them are related to the degree of interaction with outside agencies. The increasing levels of such interactions and of experience, skill and maturity brings about corresponding changes in the nature of conflicts they experience.

Most PIs opt for a compromise or *panch* (mediation by a body) approach through leaders/neutral mediators of credibility. Only when this approach fails do PIs take the issue to court.

Box 15.2

Conflict Resolution: Some Examples

The Lok Van Kalyan Parishad (see the section on federations of PIs) of Sabarkantha district, Gujarat, has been involved in conflict resolution exercises for some years:

- Vansli and Kamthadia are two hamlet villages falling in a single resource village boundary. Vansli communities were interested in a separate PI, which was not possible administratively. The Parishad with the FD conducted meetings in these villages and resolved their differences to form one single PI. The FD has also provided an *adhikar patra* (permit) for 332 hectares of forest under JFM.
- Itawadi, Chitrodi and Virpur are three villages having one contiguous forest patch of over 1300 hectares. As per records, the forest area is shown in Itawadi. Mediation is on to find a way of allocating forest area to all the three villages under JFM.
- Patiakuva and Budhrasan PIs have a running dispute over their forest boundary. Although the issue is yet to be resolved, the Parishad's intervention has helped in a continuous dialogue and prevented aggravation of conflict.
- The Parishad confronted the Assistant District Registrar (Cooperation) and strongly reasoned out the hardships on account of delay in the registration process of PIs. Although the ADR's reaction initially was one of anger, subsequently, instructions have been given to the Taluka Officer for speedy processing.

Interaction with other agencies

As the PIs come into existence and develop over time they interact with other agencies for specific purposes. The following steps outline their increasing transactions: (*a*) with NGOs for organizing, initial guidance, and in some cases financial assistance; (*b*) with the banks to open accounts and other financial dealings; (*c*) with the village *panchayats* for necessary documents to apply for registration, availing of government schemes, etc.; (*d*) with the cooperative department for registration; (*e*) with the police for assistance in conflict resolution and security; and (*f*) with the traders/contractors for market links, even though the PIs would like to replace them ultimately.

During their initial stages PIs deal more with NGOs/government departments, facilitating their formation. Then, as they stabilize, they start dealing with other agencies as seen earlier. Further, with growth and maturity they interact with other agencies not explored earlier, like existing federations. Thus a series of ripples are generated (see Figure 15.1) corresponding to the formation of the PI, its increasing spheres of activities leading to increasing transactions with other agencies. Successful experience in the series of interactions enable PIs to build their capacities and reduce dependence on NGOs/government departments. They increasingly take over the implementation of activities and programmes and move towards self-reliance.

During the initial stages, the PIs face difficulties in getting a favourable response from the other agencies due to their socio-economic standing and low credibility in the eyes of the outside world. Initial successes and subsequent perseverance enable PIs to elicit positive responses from the same agencies who now see their trustworthiness. The process of federating and aligning forces with other PIs strengthens them further. PIs are able to expand their sphere of activities easily having reached this stage. They also become pro-active and influence policy. Thus, as Figure 15.1 depicts, PIs go on to expand their sphere of influence.

Equity concerns

Within PIs there exists a diversity in the extent of dependence on forests by socio-economic categories across class, caste and gender. The most forest-dependent users are socially disadvantaged groups including tribals and artisans. These groups have been rendered voiceless and invisible by years of oppression and prevailing power structures. Field studies by the author have indicated that these forest-dependent users continue to be marginalized even in the emerging PIs (see also Sarin et al. in this book).

PIs, both self-initiated and facilitated (by FDs/NGOs) have structures that are democratic. Yet the powers that be generate processes that usurp or misuse these democratic structures. Much concern has been raised by researchers, NGOs, FDs and other agencies regarding this inequity in PIs. One concern has been the lack of representation of some of these groups on the general body and the EC of PIs. Their limited presence at meetings, limited participation even when present and their lack of participation in decisive pre-meeting processes is an additional cause for concern. However, it needs to be acknowledged here that PIs themselves have raised this concern when discussing their internal problems (e.g., at the Saksham meeting).

It is commendable that PIs have evolved their own norms to protect and use forest resources with a clear concern for inter-household equity regarding items of common interest (fuelwood, fodder, timber for domestic use). Yet in some cases the differing needs of the most forest-dependant groups tend to suffer neglect in PIs.

Some PIs are giving heed to the special needs of their weaker sections, and ensuring such recognition in their collective forums (see Box 15.3). Considering the fact that PIs are embedded in a

Box 15.3

Addressing Equity Issues the Bhiloda Way

The PI in Malekpur (part of the area in Sabarkantha district of Gujarat, where the Lok Van Kalyan Parishad operates) started the silvicultural operation called 'cutback' in the forest in a collective way a couple of years ago. The fuelwood so collected by members of the PI is distributed equally. This process lasts for a few days but generates sufficient fuelwood to meet the demand for a few months. It is worth noting that men, women and children take part in this activity leading to the lessening of burden for women. The experience was shared in the Parishad resulting in many other PIs taking up similar activities. In Malekpur and Abhapur each member took away 400 kilograms of fuelwood from the cutback operation in 1995.

PIs in Malekpur, Abhapur, Indrapura, Siladri and Makroda also undertook fodder-management activities in 1995. Accordingly, the PIs protect the forest till the beginning of winter by which time the seeding is completed. On a predetermined date the members get together and harvest the grasses. In this way each member takes away about 400 to 600 kilograms of grasses from a few days of collective action. This activity again provides much-needed relief to women.

Nearly 400 biogas plants installed in the villages have reduced the burden of fuelwood collection for many women in these families. The Parishad plays a linkage role between PIs and the biogas agency.

society that remains largely inequitous, they have made significant strides towards equity. Positive discrimination at the PI level coupled with people-to-people processes at collective forums could give further necessary impetus in this direction.

Official policies towards PIs

The official policies and programmes that call for participation of communities often tend to view participation in a narrow sense, restricting it to only protection. The management continues to remain with the government agency and sops are offered to communities in lieu of their protection effort. Although the discourse on indigenous knowledge has forced the agencies to take notice, this recognition is being done as a matter of privilege that is accorded to the communities. So every project proposal includes participatory rural appraisal (PRA) as a matter of thumb rule as if merely this would ensure participation.

Official JFM policies

An examination of the JFM orders issued by three states (West Bengal, Gujarat, Orissa) reveals the following:

a. Role of PIs is sought to be restricted for protection only, with management functions still remaining with the FD.

b. No role of PIs is envisaged in good (standing) forests; their participation is restricted to degraded forests.

c. With the exception of Gujarat, no state provides a separate legal entity status to PIs.

d. Sharing arrangements and quantum have been unilaterally decided by the FD.

e. No authority has been devolved to PIs for protection of forests.

f. The implementation authority is vested with the FD, with PIs participating as labourers.

g. The preparation of JFM action plans is sought to be done in a participatory manner, but the *modus operandi* is left entirely to the FD.

h. The constitutions of PIs and their ECs are predetermined by the FD with the exception of Gujarat. FD has even taken on the secretary's role of the PIs in the case of West Bengal and Orissa. The Karnataka state order even goes to the extent of conferring the power of dismissal of primary members from PIs to the Range Forest Officer.

i. In case of disputes or controversies, the word of the FD is treated as final.

Watershed Development Programme

The Ministry of Rural Development of the Government of India has been more forthcoming in the aspect of community participation. The guidelines for watershed development suggest the formation of user groups and self-help groups in consultation with the *gram sabha* (village council). Further, a Watershed Committee (WC) would be formed from the representatives of the self-help and the user groups with a nominee from the *gram panchayat.* A bank account is proposed to be opened in the name of the WC to be operated by the chairperson and the secretary. A watershed fund is also to be created for post-programme maintenance. The watershed plan is to be made by the user groups in consultation with the WC and the local team of the implementing agency, be it the NGO or any other body. The implementation is to be done by the local WC together with self-help and the user groups. Thus, the guideline calls for a decentralized approach with the devolution of decision-making powers to the community institution. As the scheme has become operational only from April 1996 it is too early to comment on the outcome.

Nepal user group forestry

India's neighbouring country Nepal has gone far ahead in legitimizing the functioning of the forestry user groups. User groups have been provided with explicit roles and responsibilities and benefit-sharing arising out of forest protection and management. The Forest Act, 1993, states (Section 25): 'The District Forest Officer may hand over any part of a national forest to a users' group in the form of a community forest in the prescribed manner entitling it to develop, conserve, use and manage such forest and sell and distribute the forest produce by independently fixing their prices, according to an operational plan. While handing over a community forest, the District Officer shall issue a certificate thereof.' Further, in Section 26, the user groups are provided the scope of amending this operational plan and merely informing the District Officer. The user groups have also been bestowed with the authority to inflict punishment on any offender and recover the loss from the offender. Section 43 states that the user groups shall be an autonomous and corporate body with perpetual succession. Section 45 provides for a separate fund for the user group and provides a prescription for the use of the funds.

This positive law has led to a spurt in the evolution of users' groups and to the formation of a national-level federation called the Federation of Community Forestry Users in Nepal (FECOFUN). It is estimated that nearly 4,500 users' groups are protecting and managing 200,000 hectares of forest (see Shrestha, in this book).

Institutional structures for community participation

The experience of PIs in JFM suggests that the PIs should be treated as autonomous bodies. Further, the form of the organizational structures in PIs should not be imposed and is best left to themselves. The role of PIs should not be restricted only to protection but be expanded to cover the management aspect as well.

The operational plan is best prepared by the PIs themselves with guidance from the FD or the NGOs. The PIs should be entrusted with the responsibility of implementation and be provided with funds. PIs can develop their own fund-management plans and some guidelines may be provided such as the need for joint accounts, submitting annual accounts to the general body and to the funding department. The membership of PIs needs to be kept open to all adult residents of the community; the executive membership may also represent all the sections of the community and have a gender balance. PIs should be provided with the authority to punish offenders and have full access to forest products and fix their own price for the sale of produce. As in Nepal, the law should provide for explicit handing over of village forests to the PIs.

The 1988 Forest Policy is considered by many as progressive in the sense that for the first time people's needs are given a priority. But this needs a paradigm shift in legal provisions, which has not yet happened. It also requires a massive re-orientation exercise amongst 100,000 foresters across the country. Then the communities living in and around the forest must be approached with this conviction of making them partners in the conservation programme. Considering the fact that the communities have a great dependence on the forest, how does one deal with the situation? There is no absolute answer to this question; it depends on the particular situation. This is where a truly participatory planning process would help. A survey of resources, extent of dependence, who depends on what, considerations for flora and fauna, sustainability of resource access and use, estimating future use, do's and don'ts, etc., will have to be done through a joint exercise. The JFM experience has shown that the present arrangement and practices have aggravated the condition of the poor, making them losers. Extra effort is required in carrying out the above exercise. The operational plan that follows from this exercise should remain flexible and be reviewed from time to time. The sort of benefits that can accrue to the communities can then be ascertained from this plan.

JFM and eco-development programmes also now provide for some benefit-sharing in non-core conservation areas. In Biligiri Rangaswamy Hills Sanctuary some experiments are underway to harvest NTFP in a sustainable way (Lele et al. in this book). The community in Rajaji National Park (the Gujjars) are proposing a detailed management plan, and access to grass resources has recently been granted to surrounding communities of rope-makers (Vania 1997). Another important aspect of the operational plan is the zonation of the conservation area. In general, the aspect of benefit-sharing within PAs is also now gaining ground (Kothari et al. 1996; Sarkar et al. 1995). For a large conservation area it may be desirable to evolve some federal body of PIs as in the JFM experience.

Finally a joint management council consisting of representatives of the FD, PIs and their federation, NGOs and researchers needs to be formed. This council must be empowered under the law with a clearly defined mandate and responsibilities. Such a council needs to meet at least once a quarter when the progress is reviewed and action taken as required. Financial authority must rest with the council to implement the operational plan.

Linkages with other issues of CBC

Who are the stakeholders?

In conservation programmes there are various stakeholders with varying degrees of dependence on the resources that need conservation. Among the rural stakeholders it is the community which is proximate to the resources which constitutes the primary stakeholders. This has been the organizing principle that the PIs have used. This leaves the question open as to the interests of other stakeholders such as nomads and distant users (Kothari et al. in this book). In the

JFM context, there is also the issue of gender and equity, the grounds on which the PIs are seen to be exploitative. These issues are solvable. There is nothing in the PI structure that prohibits the involvement of those who are presently excluded (VIKSAT 1995), provided there is enough capacity-building, negotiation and legal empowerment.

Local common knowledge

There is a lot of concern about the erosion of local knowledge. Gadgil (in this book) attributes this to the loss of community control over resources. Gupta (1997) blames it on insensitive state systems which have taken control over resources. Others identify the forces of urbanization, indiscriminate logging, and uncontrolled harvesting of minor forest produce. Research done in the JFM area points to a diversity of species where local community has taken a measure of control (Malhotra et al. 1992). These studies testify that if local community is allowed the management of resources the biodiversity increases. At another level is the question of power over decision-making. It is our belief that PIs and their federation, if given adequate control over the resources, can revive at least some part of the local knowledge. The experience of Van Panchayat (Ballabh and Singh 1988) justifies this belief. Gupta (1997) brings in a new dimension by his argument that all knowledge is not necessarily held in the communal domain, but individuals do exist who are bearers of such knowledge. Also, there are numerous examples of innovations by the community or an individual which cannot be called traditional. In the following section on rights and benefits we explore how this dimension can be reconciled with community-based conservation programmes.

Creating a stake/sharing the benefits

Various articles in this book deal with this issue. Gadgil suggests value addition to biodiversity by building capacity of local communities. He further adds that financial inputs should be organized as a national biodiversity conservation fund which should be rationally allocated to local communities. Shrestha (in this book) points to the creation of autonomous committees in Nepal in the PAs that are now collecting tourism revenues for allocation to community development. The institutional structure of PIs and their federation can act as a referral point for operationalizing the above suggestions. Gupta (1997) suggests the need for a tailored approach to different contexts in which conservation takes place and raises the challenge of devising an incentive system which fulfills four conditions of sustainability that he has identified. The financial-management capacity of the PIs has been discussed earlier in this paper, and it suggests that the ability exists to handle the funds that may be generated by implementing the ideas expressed by the various authors. Again referring to the Nepal experience, the national federation of the user groups in forestry has registered itself as an autonomous entity and has received funds from an international donor. Had a similar federation been promoted in West Bengal, an award from WWF International for participatory forestry in the state could have been received by the federation rather than the Forest Minister.

What exactly is the precise role of the PIs and their federation in developing and implementing a satisfactory mechanism for benefit-sharing is something that defies clear answer at the moment. Value addition at the PI level and marketing at the federation level is something that is much practised in India in various sectors. The famous Anand pattern in the dairy sector is one such exemplary model that could be tailored for some resources that have a marketable character. Venture fund idea mooted by Gupta (1997), as also networking and information exchange, could be operationalized by the federation and the PI network.

Conflict resolution

Kothari et al. (in this book) suggest that conflict resolution needs to be built within the institutional structure. The federations themselves act as the forum for conflict resolution for intra-PI and inter-PI conflicts. Such federations can be considered as the modern-day counterpart of the traditional *thain* or the *gram sabha* (Singh in this book). But for conflicts between the FD and the PIs no such forum exists. The word of the FD is treated as final. This is an unfair way of dealing with conflicts as there is no accountability of the Department. There is a need for a forum within the institutional structure that is broad-based and represented by all the stakeholders and maintains a balance of power between the constituents. At present the formal law courts seem to offer the only option to resolve conflicts or to take up grievances.

Capacity-building

Various strategies have been suggested: consultative meetings and workshops with the stakeholders, participatory training programmes for the staff and communities (Rathore 1997). Gupta (1997) suggests networking of innovators and conservators. Creating awareness about the biodiversity, educating the communities about the law (Datta in this book), and restructuring the relationships within conservation organizations (Pimbert in this book) are other suggestions. Our own experience suggests that the people-to-people process that the federation facilitates, encourages the sharing of experience, hand holding, dissemination of information, guidance regarding procedures, discussion on policies, conflict resolution and confidence building. These are the processes that built the Parishad and Saksham referred to earlier.

The question of capacity-building in the staff of conservation agencies, and the re-orientation required for the new approach is a mammoth task that defies any simple strategy. The FD alone has a strength of over 100,000 staff at various levels.

Linking with *panchayati raj* institutions

The *panchayat* represents the official democratic processes, whereas the PIs that we discussed represent people-oriented processes. In the state of Orissa where the maximum PIs exist, the involvement of *panchayats* is at a minimum. The PIs have been organized around smaller units than the *panchayats*, so much so that within a *panchayat* there may be even a dozen PIs. Moreover, the PIs resent the imposition of *panchayat* representatives that the Orissa order on JFM suggests. As against this the West Bengal experience demonstrates a positive role that the *panchayats* could play. The *panchayat*, by its deep grassroots movement understands the pulse of the people well and in the JFM context provides the much-needed balancing of power between the FD and PIs.

Unfortunately, the recent constitutional amendments empowering *panchayats* with much greater local functioning, do not provide these bodies with *locus standi* vis-à-vis the government-owned forest areas as defined in the forest and wildlife protection laws. In such a situation, how the linkages will be established is something only time and experience can tell.

REFERENCES

Ballabh, Viswa and **Katar Singh.** 1988. Van (Forest) Panchayats in Uttar Pradesh Hills: A Critical Analysis. Research Paper. Institute of Rural Management, Anand.

Gupta, A. 1997. Getting Creative Individuals and Communities their Due. Paper presented at the South and Central Asian Regional Workshop on Community-based Conservation, 9–11 February, Indian Institute of Public Administration, New Delhi.

Kothari, A., N. Singh and **S. Suri** (eds). 1996. *People and Protected Areas: Towards Participatory Conservation in India.* Sage Publications, New Delhi.

Malhotra, K.C., N. Satish Chandra, T.S. Vasulu, L. Majumdar, S. Basu, M. Adhikari and **G. Yadav.** 1992. Role of Non-timber Forest Produce in a Village Economy: A Household Survey in Jamboni Range, Midnapore District, West Bengal. IBRAD Working Paper, Calcutta.

Poffenberger, M. 1995. India's Forest Keepers. *Wastelands News*, August–October. Society for Promotion of Wastelands Development, New Delhi.

Raju, G. 1996. People's Institution to Parishad. VIKSAT, Ahmedabad. Unpublished paper.

Raju, G., R. Vaghela and **M.S. Raju.** 1993. Development of People's Institutions for Management of Forests. VIKSAT, Ahmedabad. Unpublished paper.

Rathore, B.M.S. 1997. New Partnerships for Conservation. Paper presented at the South and Central Asian Regional Workshop on Community-based Conservation, 9–11 February, Indian Institute of Public Administration, New Delhi.

Sarkar, S., N. Singh, S. Suri and **A. Kothari.** 1995. *Joint Management of Protected Areas in India: Report of a Workshop.* Indian Institute of Public Administration, New Delhi.

Singh, R.P. 1994. From Degradation to Sustainability: Kudada shows the Way, in *People's Institutions in Forest Management: A Training Report.* VIKSAT and Centre for Environment Education, Ahmedabad.

Vania, F. 1997. Rajaji National Park: Prospects for Joint Management, in A. Kothari, F. Vania, P. Das, K. Christopher and S. Jha (eds). *Building Bridges for Conservation.* Indian Institute of Public Administration, New Delhi.

VIKSAT. 1995. *Report of the National Conference of People's Institutions in Forestry.* Ahmedabad.

16

Gender and equity concerns in joint forest management[1]

Madhu Sarin with
Lipika Ray, Manju S. Raju, Mitali Chatterjee,
Narayan Banerjee and Shyamala Hiremath

Introduction

Recognizing the symbiotic relationship between the *adivasi*[2] and other poor people living within and near forests, the 1988 Forest Policy of India states:

- The villagers' customary rights and concessions should be fully protected;
- Their domestic requirements of fuelwood, fodder, non-timber forest produce (NTFP) and construction timber should be the *first charge* on forest produce;
- Their income and employment should be enhanced by improving and increasing the production of NTFP; and
- A 'massive people's movement *with the involvement of women*', should be created for achieving the policy objectives (GOI, 1988a).

This represents a reversal of the 1952 Forest Policy priorities of meeting the industrial and commercial demand for forest produce, and maximizing state revenue, and a shift from totally state-controlled forest management to decentralized, participatory and local *need*-based co-management.

To translate the new policy objectives into practice, the Ministry of Environment and Forests, Government of India, issued a circular (No. 6.21/89-FP) on 1 June 1990 to the Forest Secretaries of all states and union territories providing guidelines for the 'Involvement of Village Communities and Voluntary Agencies in the Regeneration of Degraded Forests'. In accordance with these guidelines, to date, 17 state governments have issued orders specifying their respective basis of working in partnership with local communities. This approach has come to be called 'joint forest management' (JFM).

State-owned forests are 23 per cent of India's land area and the country's largest land-based common pool resource. Vast sections of the Scheduled Tribe population (68 million), and people of other disadvantaged communities living in or near forest areas, continue to depend on forests for livelihood and subsistence needs.

However, environmental degradation and deforestation have severely curtailed the traditional occupations of *adivasi* (tribal) communities based on gathering, hunting and agriculture without providing them alternative livelihoods. This has been compounded by the *adivasis* losing much of their control over natural resources through the nationalization of forests, transfer of tribal lands to non-tribals, and the implementation of large 'development' projects in forest areas. Although *adivasis* are only 8 per cent of the total population, they are estimated to have comprised 40 per cent or more of those displaced by 'development' projects (Fernandes 1993).

There has also been a resurgence of grassroots community initiatives for regenerating degraded forests to deal with hardships caused by resource scarcities. Thousands of such self-initiated forest protection groups (SIFPGs) are protecting several hundred thousand hectares of state-owned forests (Mohanty 1996; Sarin 1994a; 1995c; Poffenberger 1995; Vasundhara 1996a). Many of these groups have already gained, or are attempting to gain, formal recognition under JFM.

Several indicators suggest a positive impact of both JFM and self-initiated forest protection on the condition of forests. Studies in Gujarat, Haryana, Madhya Pradesh, Karnataka, Jammu & Kashmir, Orissa, Bihar, Andhra Pradesh and West Bengal have recorded increases in forest productivity, often with increased flows of NTFP to the communities (Vaghela and Bhalani undated; Bahuguna 1992; Malhotra et al. 1992; Ravindranath et al. 1996; Singh 1996; Poffenberger et al. 1996).

However, while degraded forests are improving, field observations and studies are beginning to raise disturbing questions about the gender and equity impacts of the present framework and practice of JFM. Unfortunately, very few studies have looked at who, within communities and households, has gained and who has lost by class, caste, ethnicity and gender.

The communities participating in JFM are not homogeneous entities but consist of diverse groups differentiated by caste, class, tribe, religion and/or ethnicity and within and between each of these groups, by gender and age. It is normally the poorest and most marginalized constituent groups within the communities who are acutely dependent on forest resources for survival livelihoods, whereas the relatively better off and more powerful ones often have limited or no forest dependence. Yet it is the latter who have the greatest visibility and voice.

These differences are permeated by gender differences in access to and control over natural resources, perpetuated through social institutions such as *purdah* (social and physical segregation of women) and women's exclusion from the public sphere, which place multiple constraints on women's equal participation in JFM.

Second, initiating JFM or SIFP on state-owned forest lands often entails bringing effectively open-access forests under common pool resource (CPR) management, *changing prevailing norms of access* to local forests by transferring control to local institutions (LIs). Who is included, and who is excluded from participating in 'community' decision-making and articulation of CPR management priorities will tend to determine who gains and who loses.

Further, as LIs enter into formal partnerships with state agencies, their rules also represent a new regime of property rights to common pool forest resources. These overlap with the often complex existing regimes of customary as well as legal usufruct rights to forest resources. Thus, the extent to which women's equitable access and entitlements to the produce of public forests being brought under JFM is being institutionally ensured needs to be examined.

Gender and development in the national context

The principles of gender equality are enshrined in the Constitution of India. *Towards Equality*, the report of the Committee on the Status of Women in India (CSWI) (GOI 1974), was a landmark in changing government thinking on women's role in development. Based on gender-specific empirical evidence, the CSWI highlighted that rather than improving women's status, the dynamics of 'development' had created new disparities such as the declining sex ratio, lower life expectancy, higher infant and maternal mortality, declining work participation, illiteracy and rising migration (GOI 1988b). The CSWI report brought about the recognition that *women as a group are adversely affected by the processes of economic transformation.*

The report of the National Commission on Self Employed Women and Women in the Informal Sector, set up by the Government of India, found 'that one third of all households were solely supported by women and in another third, 50 per cent of the earnings were contributed by women' (GOI 1995). Such economic productivity of women is particularly critical for the 60 million Indian households below the poverty line. The poorer the family, the more it depends for its survival on the earnings of its female members (World Bank 1991).

The government policies for making development gender sensitive now increasingly focus on inculcating self-confidence among women; generating awareness about their rights; training them for economic activity and employment; and providing joint or individual titles to assets like land (GOI 1988b). The most well-known initiative is the reservation of one-third elected seats at all levels, including positions of chairpersons, in all *panchayati raj* bodies through the 73rd Constitutional Amendment. Through this measure, an estimated one million women could emerge as leaders at the grassroots levels in rural areas alone with 75,000 of them being chairpersons (GOI 1995).

Gender relations in *adivasi* and non-*adivasi* culture

Adivasi women enjoy a better status within their own communities than women in mainstream Indian society. Their critical role as gatherers of forest foods and subsistence goods make them economically valued members of the community. Till today, *adivasi* women's income from the sale or processing of NTFP that they collect from common lands is respected as *their* income which *they* control although this is under threat of rapid erosion. In comparison, despite the hard work they invest in cultivating private lands, all income from private lands is strictly considered the male landowners' income (Kelkar and Nathan 1991).

Even for *adivasi* women, however, there have been two crucial areas of inequality by tradition—property rights and political participation. Among the Jharkhand tribes of eastern India, for instance, property, particularly land, passes through the male lineage and under customary law, women do not have inheritance rights to land. And in most parts of India, the traditional village assembly (*panchayat*) is virtually an all-male institution (Kelkar and Nathan 1991). In line with this tradition, the majority of SIFPGs, many of which are gaining formal recognition under JFM, are exclusively male institutions.

Women have been, and continue to be, major gatherers and users of a much more diverse range of forest products than the men. Gender roles often vary widely between different communities, sometimes within the same hamlet or village. They are also undergoing rapid transformation with socio-economic change. Male-dominated migration, combined with men becoming more integrated in the market economy is not only increasing women's role in the subsistence sector, but also making many of them the primary bread earners of their families.

In the socio-cultural hierarchy of castes, tribes and occupational groups, there are also strong associations of superiority and inferiority, as well as specialization, in the collection and use of different forest products. Thus, in north Haryana, women of the Gujjar community consider it beneath their dignity to process a local fibrous grass (*Eulaliopsis binata*) into rope as that is the traditional vocation of the lower-status Banjara community.

Thus, women are by no means a homogeneous category. Different groups of women within the same community may have widely different, possibly conflicting forest management priorities, often for the same plant species.

The JFM framework: Where do local needs fit?

In essence, the state JFM orders assure participating villagers free access to specified NTFP and a 25 to 50 per cent (net or gross) share of poles/timber (in cash or in kind) on 'final felling'. In return, the villagers are expected to protect the forests after forming an organization conforming to the membership and structure specified by the Forest Departments (FDs).[3] Most states prescribe a forest officer as the member-secretary of the LI, responsible for conducting all the LI's proceedings.

Although JFM is often considered synonymous with 'participatory' management, the most critical decision, namely its *primary management objective*, has implicitly been pre-defined and made non-negotiable by the government in almost all the state JFM orders. The major 'benefit' being offered to the LIs is income from timber on 'final' felling. Village women and men's 'participation' has been confined to helping FDs protect degraded forests to regenerate timber for subsequent felling by the FDs.

'Block felling' of poles or timber from regenerating degraded forests implies waiting for a minimum of 10 to 20 years before the LI as a whole, or its individual members, become entitled to a share of the 'major' benefit.[4] However, to get this 'benefit' after several years, all those *currently* dependent on the forest area for multiple products and uses, such as firewood through lopping bushes, branches or stems; tree leaf fodder through lopping; supporting livestock through grazing, etc., not only have to forsake current consumption in the short term but, in many cases, are expected to do so permanently.

Curtailed access to firewood and its differential impact

One of the *immediate* consequences of forest closure for protection *without* prior diagnosis of existing forest use patterns, is that instead of receiving priority consideration, such users are the first to be perceived as villians. The incentive of a share of revenue from timber tends to make the more powerful and 'larger farmers, with little dependence on the forest, and who previously were largely uninterested in forest products, become new stakeholders to gain rights within the forest' (Femconsult 1995). They do not have to incur significant opportunity costs of forsaking current consumption while waiting for timber to mature. The most common rule for firewood collection imposed by FDs, as well as by the male leaders of most LIs is that only 'dead, dry and fallen twigs and branches' may be collected. This overlooks the fact that degraded forests do not have much of these. The expectation that more firewood will become available once forests have regenerated, also does not necessarily hold true unless the forests are specifically *managed* for increasing firewood availability.

Even where firewood is not so scarce and the collection rules more liberal, sale of firewood for income is almost always forbidden. It has been estimated that two to three million people are engaged in headloading, making it the biggest source of employment in the energy sector in India (CSE 1985). The majority of headloaders are tribal and other poor women. Suddenly stopped or restrained from this act, headloaders are compelled to either 'steal' firewood from the closed areas or to go to more distant areas for collection (AKRSP 1995; Sarin 1994a; 1994b; 1996a; Sarin and SARTHI 1994; CES 1995). This also makes the sustainability of such community forest management precarious, particularly during lean agricultural work seasons, when 'mass loots' of regenerating forests even under JFM or community protection take place (Mukherjee 1995; Shramjeevi Unnayan 1994).

In a rare assessment of overall impact of community forest closure by 45 villages, an NGO estimated that about 19,000, mostly poor *adivasi* women of Churchu, Mandu and Sadar blocks of Hazaribagh district in Bihar, had been acutely affected. While being compelled to switch to using leaves, lantana and dung as cooking fuel, in lieu of headloading, the majority had had to start brewing alcohol for sale, working in brick kilns on exploitative terms or as unskilled manual labourers for digging earth, for survival income. Many women continue resorting to 'thieving' from the closed forests, getting fined and humiliated when caught (JSPH 1994).

Women resort to vicious abuses and accusations of attempted molestation, sometimes filing police cases against male watchers or even physically attacking them on being prevented from collecting firewood; communities like the Lodhas have taken up armed fights with traditional weapons to retain access to their source of livelihood (Singh 1996; Poffenberger et al. 1996; Sarin and SARTHI, 1994; Sarin 1994a; 1994b; 1996a; Singhal 1995a; 1995b).

In the well-known pilot project at Arabari in West Bengal, approximately Rs 100,000 per year was spent for many years (with such expenditure continuing till today) on generating wage employment for the poorer villagers to compensate them for forsaking income earned through headloading (A.K. Banerjee 1996). The amount was based on the estimated income the villagers were earning from the forest prior to closure (Chatterji 1995). However, none of the present large donor-funded JFM projects have planned for such compensation.

Access to NTFP: Illusion and reality

In lieu of the opportunity costs for existing users while waiting for timber to regenerate, most states permit LI members free access to a number of NTFP. These primarily include fodder grasses; dry and fallen twigs and branches; leaf litter and leaves; and where available, mushrooms, edible tubers, flowers, fruits and medicinal herbs. However, the more valuable NTFP are either excluded from free access (e.g., cashew nuts, bamboo and fibrous grasses), or are nationalized.

Monopoly collection rights for such NTFP are often auctioned to private contractors (e.g., *tendu* [*Dispyros melanoxylon*] leaves in Gujarat and most NTFP in Orissa and Karnataka) or leased to other agencies (e.g., LAMPS—Large-scale Adivasi Multipurpose Cooperative Societies—in tribal areas). In Orissa, only one private company has been given collection rights for 29 items of NTFP for 10 years with no check on the prices it pays to the *adivasi* collectors (Saxena 1995).

The nationalized and other high-value NTFP thus neither come under the purview of 100 per cent usufructs nor under revenue sharing as a JFM benefit (Agarwal and Saigal 1996). Collectors of such NTFP continue to receive only wages for their labour, often at abysmally low rates. The market price for the NTFP, or the profits from value-addition through processing them, go to contractors, traders, industry, LAMPS or state agencies.[5]

Access to relatively low-value NTFP under JFM (to which villagers already have legal rights in most cases) is *assumed* to be adequate for taking care of local 'needs' as mandated by the 1988 Forest Policy. Where begun (as in West Bengal, Rajasthan and Karnataka), 'microplanning' has consisted of FD staff preparing aggregate estimates of the villagers' forest product requirements without co-relating them in time and space with who needs how much of what, when and for what purpose (Suess 1995; Rathore and Campbell 1995; Femconsult 1995). Even in such calculations, only the villagers' 'bona fide domestic needs' are normally considered, ignoring the processing and/or selling of NTFP as a primary source of livelihood.

Further, a common assumption pervading the practice of JFM is that forest regeneration through protection will also result in increasing production of all NTFP used by local women and men which will provide them a regular flow of benefits. However, due to the primary focus on timber, the production of 'less valuable' bushes, shrubs, grasses, creepers, climbers and herbs, becomes incidental. There is often a decline in the availability of some of the more important NTFP for disadvantaged women and sub-groups.

For instance, large numbers of women FPC members in West Bengal are complaining that their incomes from *sal* and *tendu* leaves have declined under JFM due to reduced yields (Sarin 1995b). In a *sal*-coppice forest under JFM in West Bengal, it was calculated that while the net worth of employment generated by the FD increased from zero to Rs 4,67,000, the income of 30 per cent of the households selling firewood as headloaders declined from Rs 13,65,000 to zero during the 30-year period (Femconsult 1995).

Impact of grazing bans

In addition to firewood extraction, unregulated grazing is considered to be a major cause of forest degradation. The GOI 1 June 1990 circular prescribes enforcing grazing bans as an LI responsibility in JFM forests.

In many ecological regions, particularly the arid and semi-arid areas where livestock rearing is a major economic activity, 'forest' areas have traditionally been sustainably managed for pasture and grass/fodder production. The single largest group disenfranchised by grazing bans is that of nomadic pastoralists. Due to their nomadic lifestyle, the pastoralists have remained almost invisible in the JFM process.

Even for the settled villagers in such regions, stopping grazing altogether without developing fodder alternatives invariably effects the landless and small/marginal farming households more adversely than the larger landowners with irrigation facilities.

Bans on goat grazing, considered the 'worst enemy of forests' by the forestry establishment, create pressure to sell them off. While the better-off households are able to substitute them with stall-fed cattle, the poorest households often loose access to even small amounts of milk for tea and children's nutrition besides getting deprived of saleable assets during emergencies (Raju 1997; Sarin 1996b).

In the well-known pilot project in Sukhomajri in Haryana, it was realized fairly early that the Gujjar grazier community could not be expected to start 'social fencing' by voluntarily stopping goat and cattle grazing in the forest unless first provided with a more secure and improved source of livelihood. Instead of recurring investment in generating wage employment as in Arabari, social fencing in Sukhomajri took birth only after irrigation from a rainwater harvesting dam quadrupled agricultural yields, and *all* families were assured access to equal shares of water.

Recent research studies indicate that where the villagers have framed the rules themselves, they have seldom resorted to outright grazing bans. Instead, they adopt diverse management systems, including rotational or seasonal grazing, without hampering vigorous regeneration of their protected forests (CES 1996).

Do shares of timber revenue satisfy local 'needs'?

The very basis of present 'benefit-sharing' under JFM violates the mandate of the 1988 Forest Policy to treat the villagers' needs, including for construction timber, as 'the *first charge* on forest produce'. The local institution's present 25 to 50 per cent share of the harvested timber will not necessarily be adequate for meeting the participating villagers' own timber requirements. For example, it was estimated that the well-regenerated forest of the SIFPG of Kaimati in Sarangi range of Orissa, cannot 'meet more than 5 to 10 per cent of the hamlet's fuelwood and timber needs' (Poffenberger et al. 1996). According to Orissa's JFM order, the FD will take away 50 per cent of even this produce as its share if the villagers decide to participate in JFM. In some states (e.g., West Bengal), such 'benefit-sharing' is applicable even to firewood available from coppicing species.

The problem with sharing the revenue from sale of timber is that there may be considerable unsatisfied need for poles or timber among the FPC members while the poles regenerated by them are sold off by the Forest Development Corporation. Not surprisingly, many FPCs in West Bengal are demanding that their share of the *sal* poles be given to them in kind, either standing in the forest or after harvesting (N.K. Banerjee, 1996).

Haryana's draft JFM rules specify that Hill Resource Management Societies (HRMSs) must satisfy their own members' consumption and/or livelihood needs before selling any produce outside the community. Non-enforcement of this provision, however, is threatening poorer women and men's traditional, often legal, access to fodder grasses from forests. This has happened due to powerful non-forest users appropriating control of the LI's decision-making. Encouraged by the Haryana FD to maximize the HRMSs' cash income from the sale of *bhabbar* grass as fibre, dominant leaders of Dhamala and Sukhomajri villages have stopped access of the landless and other poor to forests for fodder grasses when they need them most (Sarin 1996a; 1996b).

Grassroots priorities and management alternatives

Contradictions in the assumptions of the official JFM framework become more evident when seen against the motivations of SIFPGs to start forest protection.

Some of the community initiatives go well beyond simple protection to sophisticated local forest management systems designed to accommodate diverse needs of the community's different user groups on principles of *equity*, and not mechanical prescription of *equal shares* to all member households, irrespective of their wealth status. In Gadabanikilo village in Orissa, instead of closing their entire forest area, the community leadership has for the last 45 years set aside *different* areas for firewood collection, grazing and NTFP by closing different patches to different extents (Vasundhara 1996b). Imposition of the FD's standard rotational block fellings for timber sharing on such existing systems could only destroy the delicate balance between satisfaction of diverse needs and sustainable resource management.

Instead of shares of timber from block fellings done by the FD, SIFPGs want the *authority* to undertake selective felling of timber and poles for house repair or construction and agricultural implements, as and when required. Although selective timber/pole harvesting is already being done by many self-initiated and JFM groups, this constitutes an 'illegal' act making the villagers vulnerable to harassment, fines and bribery. Due to such disagreements with the official JFM framework, and the fear of losing management control over the forests regenerated by them, many SIFPGs in different states have refused to participate in JFM altogether (CES 1995: 17).

JFM partnerships need to be revised on the principle that no forest produce shall go outside the participating villages until the requirements for it within them has been satisfied. Any sharing between the LIs and FDs should apply to only that part of the produce which is surplus to village needs, if any surplus is available. Second, the interpretation of local needs should not be restricted to only 'bona fide domestic needs' but should explicitly include the need for secure incomes and livelihoods based on processing and selling NTFP.

Gender and social relations in local institutions

The LI in JFM is expected to articulate and represent the interests of *all* user sub-groups of a forest area in the partnership agreement with the FD (Sarin 1993). Most state JFM orders specify formal membership norms and organizational structures for the LIs eligible for participating in JFM. Formal LIs, such as registered or cooperative societies, as well as most forest protection groups being specifically created for JFM, have a three-tier pyramidal structure: a *general body of members*, which elects or selects a *managing or executive committee* (of between seven to 15 members), and from amongst the management committee, one to three or more office bearers who are delegated the *authority* to act *on behalf* of the institution as a whole.

Both the leadership structures and general body membership of the SIFPGs, have much greater diversity. Some of the SIFPGs in Orissa, for example, are youth groups which may or may not have broad-based membership of the concerned village or hamlet. The leadership of the SIFPGs similarly is different from 'natural' leaders, groups of elders, village councils or groups of elite men (Kant et al. 1991; Sharma 1995; Sarin 1995a).

Discriminatory membership

The traditional village assembly (*panchayat*) in most parts of India has virtually been an all-male institution. This is true of the majority of the SIFPGs also.

While women as a category are excluded from groups by tradition, inclusion and exclusion from them by caste, class or ethnicity varies with more complex, context-specific dynamics of wealth, power and cultural norms. Relatively homogeneous *adivasi* or caste groups may be highly democratic and inclusive at least for the men. In more heterogeneous communities, however, the most forest-dependent sub-groups may be labelled as 'destroyers of forests' or 'criminals' by the dominant groups, resulting in their systematic exclusion.

Mimicking the cultural norm, most state JFM orders prescribe eligibility of only one 'representative' per household as a general body member. This single rule automatically denies the majority of women, and many marginalized men, the right to participate in JFM on their own behalf. This is because the one representative is almost always a man, socially and culturally perceived to be the 'head' of the household.

In states with the LI membership rule of only 'one representative' per household, the vast majority of women forest users have remained more or less forgotten and invisible in the JFM process (Suess 1995; Ray 1996a; Correa 1995). A survey of 72 FPCs in the Midnapur West Forest Division found that out of 8,158 members only 241 (less than 2 per cent) were women, most of whom were widows (Narain et al. 1994). The only exceptions have been either where the local forest officer has taken personal interest in involving women or an NGO or women's organization has facilitated women's participation.

Several studies indicate that male household 'heads' neither necessarily consult their household women nor 'represent' the women's priorities. They are often remarkably unaware about the negative impact of sudden and total forest closure on women's gendered responsibility for procuring firewood (Sarin 1994a; N.K. Banerjee, 1996; Correa 1995).

In response to criticism of women's exclusion from the FPCs,[6] some states (Himachal Pradesh, Andhra Pradesh, Tamil Nadu, Orissa, Madhya Pradesh), now provide for membership of one man and one woman per household to avoid total exclusion of women from the LIs. However, even this does not ensure membership access to the poorest and neediest users. The problem lies in the continued use of the household as the qualifying unit of membership, on the implicit assumption that most households consist of standard nuclear families. However, nuclear households comprise only 43.7 per cent of the total households in India according to the 1981 Census.

Membership of LIs should be open to all adult women and men, irrespective of their status within the household. Such a change has been envisaged in the draft JFM rules proposed for Haryana. However, changing institutional rules does not necessarily lead to changes in practice. Even where membership of one man and one woman per household has now been provided for, this has often only resulted in women's names being added to membership lists.

Some FPCs in West Midnapore Division receive 25 per cent share of *sal* multiple shoot (MS) cuttings each year as firewood. Because husband and wife have 'joint' FPC membership in West Bengal, although it is primarily women who are responsible for firewood, it is mostly men who come to collect the household's share of the MS cuttings. *However, about 80 per cent of the men sell off their firewood shares* (Sarin 1995b). Realizing the potential contradictions, a woman forest officer posted in West Midnapore Division has tried to ensure the presence of the women when shares of firewood are distributed.

Women's access to managing committees and becoming office bearers

Most states arbitrarily provide for a minimum of one to three women members in the managing committee (MC) of the LIs. Bihar's JFM order is remarkable for specifying a *maximum* of five women MC members. Tamil Nadu and Himachal Pradesh's orders are the only ones requiring that 50 per cent of the village representatives on the MC are women although in both cases, due to several outsiders being nominated as MC members (such as the forester) who would normally be men, women's actual representation will be less than 50 per cent even in their case.

Only Himachal's order requires a representative of the village *mahila mandal* (women's association) to be a member of the LI's MC to link it with a local women's forum. Tamil Nadu's order requires a woman representative from each hamlet in the LI's area to be a member of the MC. None of the orders either require that the women representatives be selected by more forest-dependent women within the community or that they should themselves be dependent on forests to be able to articulate the problems and priorities of women forest users.

As far as the LI's office bearers are concerned, practically all the orders are totally silent about any woman's presence at that level.

Further, MC membership does not necessarily result in women having a say in decision-making. As it is, women have to muster exceptional courage to speak up in a traditionally male domain. However, *often they are not even invited to the MC meetings*. In many cases, they remain unaware that they have been made MC members as the men select the women members on their own simply to complete the formality (Sarin 1993; 1996b).

State JFM orders need to be amended to provide that:

- At least one-third (half where women are the predominant forest users) of the LI's MC members, as well as the office bearers, are women.
- The women MC members should be selected by women forest users in separate meetings.
- All MC members and office bearers should themselves be forest users, to prevent powerful non-users from usurping control of LIs.
- Women's proportionate presence must be made mandatory to complete the quorum of all valid LI meetings.

Without clear guidelines for involving actual forest users in JFM, powerful non-forest users are tending to find disproportionate representation in the MCs and leadership posts of LIs. This is not surprising as FD field staff find it easier to interact with people they already know in the villages. In West Bengal, where a large number of SIFPGs already existed before the state JFM order of 1989, when the FD started registering FPCs, 'the timber merchants or the existing leaders of self-initiated groups automatically became the leaders of FPCs' (N.K. Banerjee 1996, Saxena et al. 1997).

Institutional norms and inequitable access to property rights

Where the state JFM orders provide for distributing shares of income from 'final harvests' to individual members of the LIs, this involves an explicit privatization of property rights to timber. Such provisions for individual entitlements, combined with the existing formal and informal membership norms of LIs, will lead to serious gender and class differentiated changes in access to forest resources not only in the short term but on a long-term legalized basis.

This is highlighted by the Arabari experience in West Bengal, where JFM has been in operation for many years.[7]

The Executive Committee of Arabari's FPC and the West Bengal FD decided that every household (defined as a family unit sharing one kitchen) should receive an equal share of the 25 per cent cash profit from annual timber harvesting. 'It was also decided that each household (HH) head would be nominated as a beneficiary who would be the official recipient of the money within that particular family and that it would be upto each beneficiary to determine further distribution of the money within his/her family' (Chatterji 1995).

Since 1980, the number of total households has increased by 31.3 per cent to about 900. However, no mechanism has been evolved to accommodate the new households. Further, the process of inheritance and actual transfer of beneficiary status from a deceased beneficiary to the descendant is often fraught with legal constraints, especially if there is more than one son within the household. In such cases, sons vie with each other for control, causing conflict within the family.

However, 'most villagers in Sankrui (one of the 11 villages in Arabari's FPC), especially the women, do not have a general idea of the total profits made by the FD or the beneficiaries, or of the actual process of profit distribution among the 601 beneficiary households'. Only 11.6 per cent of the sample respondents claimed having been present in discussions regarding the project's management issues.

Further, at least 81.3 per cent of the households owning 1.5 *bighas* or less of land sell fuelwood and NTFP regularly from the FPC's forest. Lower caste and *adivasi* groups comprise 46.2 per cent of these collectors, and a majority are women.

Thus, 25 years after Arabari gave birth to the concept of JFM in West Bengal, women of its disadvantaged *adivasi* and lower-caste households are still cutting *sal* and other species for survival income, while the new structure of legal property rights to shares of revenue from timber has been almost exclusively transferred to male 'heads of households'. The income shares have neither given the poor women forest users a stake in regenerating timber nor reduced their vulnerability to humiliation due to the continuing compulsion to cut firewood 'illicitly'.

Such questions of class, gender and generational equity underlying the granting of individual property rights to shares of timber revenue from public forests demand far more thorough examination than they have received. The states where JFM orders have such a provision need to revise them and leave distribution-related decisions to the LIs. The FDs' role should be to facilitate transparency and respect for principles of equity in the LIs' decision-making.

This question of *who* should get the household's share was posed to combined groups of women and men members of three FPCs in south Bankura Division. Once the women understood the question, in all the three villages, without any hesitation, all the women felt that the shares should be divided into two equal parts and given separately to husband and wife (Sarin 1994a). Whispering that they wanted their half of the share of income, the women had said it was they who had to take care of household sustenance.

Strategies for making JFM gender and equity sensitive

Moving from forest protection to participatory planning and management

The starting point for 'participatory' forest management needs to be identification of the users of the concerned forest and understanding the diversity of their needs. Although use of participatory rural appraisal (PRA) techniques has become commonplace in JFM, it has become a ritualized preparation of village and/or forest maps, doing transects or time lines with 'some' villagers. These do not ensure participation of the most critically dependent users and can leave them as invisible and voiceless as before.

An adaptation of a four-step PRA methodology, developed by IDS, an NGO based in Karnataka, to ensure participation of actual forest users in the entire process, begins with identification of the diversity of users and their needs, and goes on to planning and evolving equitable management options (Hiremath 1996).

Being state agencies, FDs are in the advantageous position of introducing progressive changes in traditional attitudes, practices and organizational norms by insisting that priority is given to gender and equity concerns in JFM. This approach was adopted for Haryana's JFM programme from 1989. To ensure that all women and men had a right to participate in JFM, all adults were made eligible for the Hill Resource Management Society (HRMS) membership and acceptance of women's independent eligibility by village men was made a pre-condition for HRMSs being able to participate in JFM.

To ensure women's physical presence during the process of day-to-day interaction with the HRMSs, FD field staff were told to insist on the principle that the maximum number of village men and women must be present for all JFM-related discussions. This in itself conveyed a strong message both to the FD's male staff and the village men who often attempted to bypass this resolve by saying that the women were busy with cooking, livestock or childcare, often questioning the need (or use) of their presence. This insistence not only raised male awareness about the state's commitment to gender equality but also involved men in inviting the women to the meetings.

The women often disproved the men's initial apprehensions by turning up in large numbers. Even if they feel inhibited to speak in public gatherings in the beginning, their sheer presence provides them access to information about JFM and their equal rights to participate. All states would do well to emulate the above approach and make the presence of 30 to 50 per cent women mandatory for completing the quorum for valid LI general body as well as MC meetings.

Another strategy used is to encourage respected male leaders to promote respect for women's equal rights and active participation in LI's decision-making.

Role of facilitators in giving women a voice

Getting the women accepted as members and to come to village meetings, however, is only the first step. Almost invariably, seating arrangements for the meetings mirror the organizers' perceptions of the power hierarchy—tables and chairs for the FD officers and support team, *charpais* (cots) for the men and the ground for the women and children!

To initiate a process of questioning this hierarchy, Haryana's JFM support team often gently but firmly proposed that *everyone* (including FD officers) should sit on *charpais* (or on the ground if there weren't enough *charpais* readily available). Most village men and male Haryana Forest Department (HFD) staff often reacted by saying that women would not sit on *charpais* in front of elder household men. However, if additional *charpais* were brought out, at least some, often many, women would sit on them (Sarin 1996b).

To avoid women being ignored at meetings, the strategy of always talking to them *first*, on the grounds that their views are less known, although equally important, can be adopted. Creating such space and actually *listening* to what women have to say, can prove both empowering for the women and focus attention on a previously unarticulated set of priorities.

Other strategies for increasing gender sensitivity in the day-to-day interaction with LIs include: (*a*) Ensuring that the facilitating agency interacts with the LIs through staff teams with equal male and female members; and (*b*) Designing documentation formats which record the number of women and men present in village meetings and the different concerns/views expressed by them.

Facilitating the adoption of more equitable management options

The vast majority of FD staff, and even many NGOs, continue to perceive women's 'participation' in JFM as an instrument for conserving forests rather than for understanding their variable needs and priorities to evolve appropriate management options. Such an understanding can be obtained using simple tools of gender analysis such as: (*a*) Identifying the different uses of the forest area and *who*—women, men, the elderly or children—*within* households are responsible for each activity. This needs to be done by having discussions with village women and men, both individually and in small homogeneous groups; (*b*) Exploring what impact the protection rules framed only by the men have had on women's gendered responsibilities such as firewood and fodder procurement through informal discussions with different local women.

For two years after the Malekpur tree growers' cooperative society (TGCS) in the Sabarkantha district of Gujarat initiated forest protection, the village women had to walk to the still unprotected forest of Vagheswari village several kilometres away to collect firewood. Their problem remained 'invisible' to village men, the Gujarat FD and even the supporting NGO, till the NGO started facilitating separate meetings of village women. Subsequently, the TGCS initiated annual multiple shoot cutting of teak in different forest patches, which now yields several quintals of firewood for each household every year.

Not only did this alleviate women's problem of firewood scarcity but it also led to a progressive change in the gendered division of responsibility for procuring firewood for the household. Earlier, regular fetching of headloads from distant forests was only 'women's' work. Now, when the multiple shoot cutting is done, men not only do the cutting but also help the women carry the large quantities of the firewood to their homes (Sarin 1994a).

Strategies for empowering women to participate

Public village meetings are one of the most commonly used mechanisms for facilitating participatory consultation and decision-making for JFM. However, experienced grassroots facilitators have observed that decisions are seldom taken at public village meetings. Those villagers who are aware of the meeting's objectives, and have high stakes in the outcome, often come to the meeting *after* having had caucus group discussions and pre-meetings in which key decisions are negotiated between the relatively powerful male interest groups in advance. The primary function of public meetings for such interest groups is often simply to get public ratification of the decisions they have already taken. Women of all classes and men of disadvantaged groups remain excluded from such behind-the-scenes bargaining and negotiations.

External facilitators can facilitate development of more equitable, alternative management options based on disaggregation of local needs by gender and socio-economic status. They also need to be aware of the power dynamics between different interest groups and have a strong commitment to protecting the interests of the weakest.

Many NGOs are adopting the strategy of holding separate meetings with women. For its forestry plantations, the Aga Khan Rural

Support Programme (AKRSP) has introduced participatory ranking of species with separate groups of women and men. The two sets of rankings are then discussed in a combined meeting with AKRSP staff facilitating negotiation between the women and the men to finalize the species for planting. This strategy is being accompanied by making women members of the Gram Vikas Mandals (Village Development Associations). Separate Mahila Vikas Mandals (Women's Development Associations) are also being formed to build up women's self-confidence and skills.

Most FDs and some NGOs have promoted a few all-women JFM groups as a strategy for overcoming cultural constraints. Women members of such groups tend to be much better informed (GOWB 1996), although they have continued being oriented towards forest protection within the JFM framework rather than evolving management options more responsive to the women's needs and priorities.

An interesting grassroots initiative by a small *adivasi* women's NGO, Pragatisheel Mahila Sabha (PMS), has emerged in the Dumka district of Bihar to form all-women Village Forest Management and Protection Samitis (VFMPSs). Due to their men easily getting bribed with alcohol by vested interests to break group rules, PMS has decided not to permit men to become members. This represents a reversal of typical roles. Members of the women *samitis* are exclusively from the poorest castes and tribes with maximum dependence on forest products as higher-caste women are unwilling to work in the forests. The women have framed their own forest use and management rules and do daily patrolling duties in rotation. Within two years of initiating forest protection, they have started getting a variety of forest products (tubers, wild vegetables, mushrooms, fruits and seeds besides *mahua* [*Madhuca indica*] flowers and fruit) which provide the women regular income and subsistence benefits for six months of the year. Women of other villages are not permitted entry and there are strict punishment rules for any violators.

The difference in the rules framed by such groups and those framed by predominantly male groups is interesting. The women's *samitis* have not imposed an outright ban on firewood harvesting. Neither is firewood collection seen as being inconsistent with sustainable forest management. On the contrary, in contrast to the dominant male view about women being reckless forest destroyers, the women's *samitis* 'do not want men to become members of their committees because they are the ones who destroy the forests so mercilessly as they do not know the difference between timber and firewood' (Murmu 1995).

Promoting alliances

Promoting federations of women's groups to lobby for gender-sensitive changes in policy and practice is an effective strategy for increasing women's voice. Nari Bikas Sangh (NBS), a federation of grassroots *mahila samitis* in Bankura district of West Bengal, has, with the support of the Centre for Women's Development Studies (CWDS), been lobbying for one-third representation of women on FPC executive committees, in line with the reservation for women in *panchayati raj* institutions.

Another NBS/CWDS strategy has been to define management domains based on gender roles. NBS's membership consists of women NTFP collectors who have been lobbying for better prices and marketing support for over two decades. NBS is arguing that while men should manage *sal* poles, women should manage NTFP of greater interest to them. NBS and CWDS are lobbying for women being granted exclusive rights to the collection, processing and marketing of *tendu* leaves. It has been fighting for the women's right to raise *tasar* cocoons on *sal* trees which forest officers fear will harm tree growth. Articulate and self-confident after years of struggle and organized action, the women are able to confidently assert that their methods of raising cocoons on *sal* do not damage the trees (Campbell 1996).

Another strategy being developed to increase women's voice is to promote alliances between women of JFM groups and other local women in leadership positions. The recently elected women *panchayat* representatives fall in this category. Many villages now have women ward *panches* or *sarpanches* through whom women involved in JFM can develop better links with other *panchayat* activities. Alliances can also be built with *anganwadi* workers, DWCRA groups, and women organizers trained under state or central government programmes such as Women's Development Programme in Rajasthan and Mahila Samakhya in six or seven states (Gujarat, Karnataka, UP, Bihar, etc.) where these programmes overlap with JFM areas.

Forest department strategies and institutional constraints

Out of all the different actors involved in JFM, it is the FDs who face the biggest and most difficult challenge. Affecting a role reversal from policing to partnership can neither be easy nor rapid, given their inappropriate institutional structures and technical expertise. Despite that, at least in some states, most notably West Bengal, remarkable progress has been made. As mentioned earlier, responding to criticism of women's exclusion, many states have revised their JFM orders to open LI membership to women.

It is in designing effective strategies for translating the rules into practice that there has been a big gap. FDs can use their organizational culture of strict discipline to advantage if clear directions for ensuring presence and participation of the poorest forest users are issued from the top accompanied with step-by-step procedures for implementation.

Feeling that the task of promoting women's participation required the undivided attention of an officer at the state level, a post of a 'Women's Coordinator' was created in the West Bengal Principle Chief Conservator of Forests' (PCCF) office in 1994 and a woman officer posted there with responsibilities for: (*a*) motivating women to become active FPC members and orienting male FPC members to support their participation; (*b*) motivating the department's field staff to promote women's participation; (*c*) promoting formation of women's sub-groups to ensure that the benefits accruing from income-generating activities percolate to women FPC members; and (*d*) collecting information and data about women-related activities from different divisions for dissemination to all concerned (Ray 1996b).

By all accounts, concerted efforts of the Women's Coordinator had a marked impact on increasing women's presence in FPC meetings and under her guidance and monitoring, departmental field staff also started taking greater interest.

However, the strategy is a double-edged weapon, placing the concerned person (invariably a woman) in a difficult role. Other staff tend to expect such an officer to deal with *everything* to do with women or 'gender', thereby shrugging off their own responsibility in the matter.

Senior leaderships of FDs can promote the participation of women and other marginalized forest users by assigning clear responsibility to specific officers at different levels accompanied by clear procedures for implementation and monitoring performance.

The importance of increasing women staff within the forest departments

As long as FDs remain almost exclusively male institutions, their effectiveness in convincing village women and men of their commitment to gender equality will remain questionable. Yet the number of women officers in FDs remains extremely small. The Indian Forest Service was opened to women only in 1979; by the 1995 batch, there were 72 women IFS officers out of a total strength of 2,576 (less than 3 per cent).

At other levels in the departmental hierarchy, especially forest guard and forester levels, many states still do not permit women to apply for FD jobs. In West Bengal, the rule for recruiting Range Officers and Foresters was amended recently to permit women to apply for the posts. Orissa and Karnataka have reserved one-third of all fresh recruitment by FDs for women.

NOTES

1. This is an abridged version of a working paper by the same title written by a core team of the gender and equity group within an informal network coordinated by the national support group for JFM at the Society for Promotion of Wastelands Development (SPWD), New Delhi. SPWD's permission for using the paper is gratefully acknowledged. For tables of state-wise data on women's access to JFM institutions and benefits, please see the original paper.
2. The term *adivasi*, i.e., indigenous inhabitant, is commonly used for people belonging to Scheduled Tribes, particularly those inhabiting the forested regions of central and eastern India. The terms *adivasi* and tribal have been used interchangeably in the paper.
3. Most FDs reserve the right to unilaterally cancel a JFM agreement (and in most cases, to even dissolve the LI itself) if the LI is considered to be violating any condition in the agreement. In such a situation, the LI has no right to any compensation for its investments of labour, time or capital during the validity of the agreement. If the FD fails to honour its commitments, the villagers have no reciprocal rights for penal action against the FD.
4. New plantations on totally barren lands are also being done under JFM but on a relatively smaller scale. In their case, the waiting period could be even longer. The discussion here is focused on the issues related to the more common regeneration of forests from existing rootstock through community protection.
5. For a more detailed discussion on the multiple contradictions in the continued nationalization of NTFP and JFM, see Saigal et al. 1996.
6. Nari Bikas Sangh, a grassroots peasant women's organization based in Bankura district of West Bengal, was the first to raise its voice due to its members' high economic dependence on NTFP; in March 1990, NBS appealed to the State Minister of Forests, the DFO (Bankura) and the Chairman, Ranibandh Panchayat Samiti, that 50 per cent of FPC members should be women (N.K. Banerjee 1996).
7. Most of the information about the Arabari case has been drawn from Chatterji (1995).

REFERENCES

Agarwal, C. and **S. Saigal.** 1996. *Joint Forest Management in India: A Brief Review.* Society for Promotion of Wastelands Development, New Delhi.

AKRSP. 1995. Soliya Harvesting—Gender Perspective. Mimeo. Aga Khan Rural Support Programme, Ahmedabad.

Bahuguna, V.K. 1992. *Collective Resource Management: An Experience in Harda Forest Division*. Regional Centre for Wastelands Development, Indian Institute of Forest Management, Bhopal.

Banerjee, A.K. 1996. Some Observations on Community Forestry. *Wasteland News* XI (3) Feb–April.

Banerjee, N.K. 1996. Draft note on history and struggles of Nari Bikas Sangh. Centre for Women's Development Studies, New Delhi.

Campbell, J. 1996. Personal communication with Jeff Campbell, Ford Foundation.

CES. 1995. Community Forestry: An Ecological, Economic and Institutional Assessment in Western Ghats Villages. Report for submission to the Ford Foundation. Mimeo. Centre for Ecological Sciences, Indian Institute of Science, Bangalore.

———. 1996. Ecological and Economic Studies on Community Forestry and JFM in India: Summary of Preliminary Findings, Phase I. Centre for Ecological Sciences, Indian Institute of Science, Bangalore.

Chatterji, A.P. 1995. The Socio-economic Project at Arabari, West Bengal; Participatory Enquiry Toward an Understanding of Socio-cultural and Subsistence Issues. Mimeo. SPWD, New Delhi.

Correa, M. 1995. Gender and Joint Forest Planning and Management, A Research Study in Uttara Kannada District, Karnataka. Mimeo. Indian Development Service, Dharwad.

CSE. 1985. *The State of India's Environment 1984–85: The Second Citizen's Report*. Centre for Science and Environment, New Delhi.

Femconsult. 1995. Study of the Incentives for Joint Forest Management, Main Report. Mimeo. The Netherlands.

Fernandes, Walter. 1993. The Price of Development. *Seminar*, December 1993, New Delhi.

Government of India. 1974. *Towards Equality*. Report of the Committee on the Status of Women in India, New Delhi.

———. 1988a. *National Forest Policy Resolution*. New Delhi.

———. 1988b. *Draft National Perspective Plan for Women 1988–2000 A.D.* New Delhi.

———. 1995. *Country Report, The Fourth World Conference on Women, Beijing, 1995*. Department of Women and Child Development, Ministry of Human Resource Development, New Delhi.

Government of West Bengal. 1996. *Role of Women in the Protection and Development of Forests*. Forest Department, Calcutta.

Hiremath, S. 1996. The Four Step PRA process used by IDS for a Government of Karnataka Integrated Watershed Development Project. Unpublished paper, Dharwad.

JSPH. 1994. *A Brief Report of Jan Sewa Parishad Hazaribagh's Environment Based Activities*, Churchu, June 1994.

Kant, S., Neera Singh, M. Singh and **K. Kundan.** 1991. *Community Based Forest Management Systems: Case Studies from Orissa*, IIFM, SIDA and ISO/Swedforest, New Delhi.

Kelkar, Govind and **Dev Nathan.** 1991. *Gender and Tribe: Women, Land and Forests in Jharkhand*. Kali for Women, New Delhi.

Malhotra, K.C., D. Deb, M. Dutta, T.S. Vasulu, G. Yadav and **M. Adhikari.** 1992. *Role of Non-Timber Forest Produce in Village Economy*. IBRAD, Calcutta.

Mohanty, S.C. 1996. A Brief Note on Implementation of JFM in Orissa. *Wasteland News*, XI (3) Feb–April. SPWD, New Delhi.

Mukherjee, N. 1995. Forest Management and Survival Needs; Community Experience in West Bengal. *Economic and Political Weekly* XXX (49): 3130–32. December 9.

Murmu, E. 1995. *Report of the Meeting Dated 19.4.95 in Village Taldangal*, PMS.

Narain, Satya, A.K. Mandal and **Viji Srinivasan.** 1994. *A Case Study of JFM in Santhal Parganas*. Adithi, Patna.

Poffenberger, M. 1995. India's Forest Keepers. *Wasteland News* XI (1) Aug–Oct. New Delhi.

Poffenberger, M., P. Bhattacharya, A. Khare, A. Rai, S.B. Roy and **N. Singh.** 1996. *Grassroots Forest Protection: Eastern Indian Experiences*, Asia Forest Research Network Report No. 7, Berkeley.

Raju, M.S. 1997. *Seeking Niches in Forest Canopy: An Enquiry into Women's Participation*. Study supported by Ford Foundation, New Delhi.

Rathore, B.M.S. and **J.Y. Campbell.** 1995. Evolving Forest Management Systems: Innovating with Planning & Silviculture. *Wasteland News* XI (1) Aug–Oct.

Ravindranath, N.H., M. Gadgil and **J. Campbell.** 1996. Ecological Stabilization and Community Needs: Managing India's Forests by Objective, in M. Poffenberger and B. McGean (eds), *Village Voices, Forest Choices: Joint Forest Management in India*, Oxford University Press, Delhi.

Ray, Lipika. 1996a. Analysis of Data Gathered During Visits to 25 Randomly Selected FPCs as Women's Co-ordinator, West Bengal Forest Department, Calcutta.

———. 1996b. A Brief Note on the Responsibilities of the Women's Coordinator, West Bengal Forest Department, Calcutta.

Saigal, S., C. Agarwal and **J.Y. Campbell.** 1996. *Sustaining Joint Forest Management: The Role of NTFPs*. Society for Promotion of Wastelands Development, New Delhi.

Sarin, M. 1993. *From Conflict to Collaboration: Local Institutions in Joint Forest Management*. JFM Working Paper No. 14, SPWD and The Ford Foundation, New Delhi.

———. 1994a. Regenerating India's Forests: Reconciling Gender Equity with JFM. Paper presented at the International Workshop on India's Forest Management and Ecological Revival organized by the University of Florida and TERI, New Delhi, 10–12 February, 1994.

———. 1994b. Leaving the Women in the Woods. *Down to Earth*, 30 September 1994, New Delhi.

———. 1995a. Delving Beneath the Surface: Latent Gender Based Conflicts in Community Forestry Institutions. Paper written for the FTPP, FAO, Rome.

———. 1995b. *Gender and Equity in JFM, Discussion and Emerging Issues*. Proceedings of the Gender & Equity Sub-group meeting, 27–28 November, 1995, Society for Promotion of Wastelands Development, New Delhi.

———. 1995c. Joint Forest Management in India: Achievements and Unaddressed Challenges. *Unasylva* 46: 30–36.

———. 1996a. Actions of the Voiceless: The Challenge of Addressing Subterranean Conflicts Related to Marginalised Groups and Women in Community Forestry. Theme paper for the FAO E-Conference on Addressing Natural Resource Conflicts Through Community Forestry, Jan–April.

———. 1996b. *Joint Forest Management: The Haryana Experience*. Environment & Development Series, Centre for Environment Education, Ahmedabad.

Sarin, M. and **SARTHI.** 1994. The View from the Ground, Community Perspectives on Joint Forest Management in Gujarat, India. Paper presented at the IIED symposium In Local Hands: Community Based Sustainable Management, Sussex, 4–8 July.

Saxena, N.C. 1995. Forest Policy and Rural Poor in Orissa. *Wastelands News* XI (2) Nov–Jan.

Saxena, N.C., M. Sarin, R.V. Singh and **T. Shah.** 1997. Western Ghats Forestry Project: Independent Study of Implementation Experience in Kannara Circle. Mimeo. Department for International Development, British High Commission, New Delhi.

Sharma, C. 1995. Community Initiatives in Forest Management: Issues of Class and Gender; A Case Study of Panchmahals District, Gujarat. M.Phil thesis, Centre for Development Studies, Trivandrum.

Shramjeevi Unnayan. 1994. *Sanyukt Van Prabandhan Mein Prashasnic Kamiyan evam Apekshayen*. Paper presented at a workshop with Bihar Forest Department, Ranchi.

Singh, R.P. 1996. Degradation to Sustainability—Kudada Shows the Way. In *New Voices in Indian Forestry*, Society for Promotion of Wastelands Development, New Delhi.

Singhal, R. 1995a. *Behavioural Factors in Institutional Effectiveness*, JFM Study Series, Indian Institute of Forest Management, Bhopal.

———. 1995b. *Gender Issues in Joint Forest Management: A Force Field Analysis.* Indian Institute of Forest Management, Bhopal.

Suess, W. 1995. *How 'Joint' is Forest Management in the Actual JFM Implementation? Observations on Practice, Problems and Prospects of JFM from Selected Cases in Udaipur District.* Seva Mandir, Udaipur.

Vaghela, R.N. and **D. Bhalani.** Undated. *Management Options for Regenerating Forests, Case Studies of Abhapur and Malekpur.* VIKSAT, Ahmedabad.

Vasundhara. 1996a. Community Forest Management in Transition: Role of the Forest Department and Need for Organizational Change. Mimeo. Bhubaneswar.

———. 1996b. Ecological, Institutional and Economic Assessment of Community Forest Management in Village Gadabanikilo. Mimeo. Bhubaneswar.

World Bank. 1991. *Gender and Poverty in India.* Washington, D.C.

17

Legal and policy issues in community-based conservation

B.J. Krishnan

Introduction

Indigenous societies and other local communities have depended on their local environments for survival for a long period of time and therefore have developed a stake in conserving the same. In the process they have accumulated a detailed empirical and qualitative knowledge base 'handed down through generations by cultural transmission, about the relationship of living beings, including humans, with one another and with their environment' (Berkes 1993). Their world-view, treating humans as a part of the natural world, and their belief systems, stressing respect for the rest of the natural world, are of great current value for evolving sustainable relations with the natural resource base. However, this world-view has been continuously challenged since the colonial forces of Europe set out on their journey of conquest.

The local communities of the 'Independent Village Republics of India' of the 19th century had a defined environment and natural resource base to protect, care for, improve and sustainably use (Krishnan 1996). The common natural resources including forests, rivers, village commons and grazing lands were regulated through diverse decentralized community control systems. Tribal communities that lived inside the forest did hunt and cut trees for the fulfillment of their minimum needs, but effective traditional, social and cultural norms regulated such activities and ensured adequate protection and regeneration of natural resources. Tribals and other forest communities led a life of frugality and simplicity and took from the forest what was absolutely necessary for their subsistence. This explained the strange nexus of high diversity and high poverty.

The British presence from the late 18th century onwards started making a difference to land and forest use in India. Guided by commercial interests, the British viewed forests as crown lands, limiting private property rights to continuously cultivated lands. Often, such forests were under community management and their annexation by the government alienated the people from their erstwhile common resources, leading to their over-use by the same people (Saxena 1996). The Imperial Forest Department was established in 1864 and the Indian Forest Act was enacted in 1865. This Act stated that declaration of an area as government forest should not automatically abridge or affect any existing rights or practices of individuals and communities, and also provided that the reservation could be contested within three months. In actual practice, however, the illiterate communities were hardly able to do so (Saxena 1996). And thus, by the turn of the present century, some 20 million hectares of land were brought under the category of Reserved Forests (Stebing 1922–27). These were exclusively for the use of the Forest Department (FD) and the people of the surrounding villages had no rights other than the ones explicitly permitted by the state. The area of Reserved Forest has now increased to 46 million hectares out of a total of 67 million controlled by the FD.

The Wild Life (Protection) Act, 1972, provided for the constitution of sanctuaries and national parks for the protection of wild animals. The FD had by 1996 constituted 80 national parks and 441 sanctuaries covering 4.3 per cent of the total land area of the country and representing 20 per cent of the land in the custody of the FD. The basic difference between reserved forests, sanctuaries and national parks lies in the legal stringency in relation to rights of people in these forest zones, with the last category being the most exclusionary.

Forests in India: Survival resource

Biodiversity in India is widely spread over a variety of ecosystems: marine and coastal, wetlands, deserts, etc. However, this paper essentially looks at forests and grasslands while examining the legal and policy context of community-based conservation (CBC).

Forests in India are concentrated in the North-East, the Himalayas and Shivalik ranges, the central belt, strips along the Western Ghats and other hill areas and in patches of coastal mangroves. Though from the narrow legal point of view the presence of humans inside the forest, especially reserved forests, sanctuaries and national parks, can be treated as trespass, a punishable offence, in actuality, forest dwellers continue to live inside these areas. More than 50 per cent of forest land is located in the central plateau. This is the poorest region in India with low agricultural productivity and poor soils. This region also has a heavy concentration of tribal populations. India's forests, generally speaking, have not been uninhabited wilderness (Saxena 1996). Today, there are about 100 million forest dwellers in the country living in and around forest lands and another 275 million for whom forests continue to be an important source of livelihood. There are some 5,000 villages with a population of about 2.5 lakhs in the core areas of protected areas (PAs) alone (CSE 1996). As of 1992, India had an estimated population of 63 million indigenous people representing 7 per cent of its national population (Durning 1993).

The subsistence rural economy is mainly based on agriculture and livestock. Very few tribal communities thrive purely on hunting and food gathering. However, agriculture and livestock economy are also deeply linked with forests. Furthermore, the polity, culture and religion of many local communities have evolved in close interaction with forests. Not surprisingly, one finds numerous instances of tribal uprisings in the history of India which emerged to protect rights of access to and control over forests (PRIA 1993).

The process of alienation from forests for local communities began during colonial rule. The industrial and commercial interests were instrumental causative factors of such alienation. After independence the same 'interests' were considered as national interests and continued to be pursued. During the 1970s and thereafter, the government vigorously pursued the interest of wildlife conservation as national priority, and used special legal apparatus of creating sanctuaries and national parks for the purpose. This has contributed significantly to the further alienation of the people from their natural environment.

Forestry laws in general and the Wild Life (Protection) Act, 1972, in particular, ignore the historically evolved symbiotic relations between the forest and forest dwellers and drastically curtail their traditional rights to use forest resources. In this perspective of declining access to and control over natural resources by the traditional communities, it becomes imminent to critically examine the issue of CBC from the legal point of view.

Customary and community laws—An overview

Indigenous and rural communities, notwithstanding the attempts of the colonial systems and enactments to dominate and assimilate them and centralize the control and management of their community resources in the last 150 years, continue to use their immediate natural habitat for their basic biomass needs. This practice of long usage had given rise to certain claims and customs. A custom is such a usage as has obtained the force of law.

A custom is a particular rule of conduct, which is observed by the community concerned spontaneously for a sufficiently long period of time, without the sanction of any express provision of law giving rise to a right. Both national and international courts play an important role in the determination and application of custom.

Valid customs give rise to customary rights. And as a rule those community-customary rights are the collective rights of the people of the community; they are not individual rights. A customary right may be enjoyed by anyone who inhabits a particular locality or who belongs to the particular class entitled to the benefit of the custom. In short, customary rights are in the nature of public rights annexed to the place and/or the community in general.

The basic features of customs are elaborated below.

A custom must be ancient, immemorial: In order that a custom be considered legally valid, all that should be necessary to prove is that the usage has been acted upon in practice for such a long period and with such invariability as to show that it has, by common consent, been submitted to as the established governing rule of a particular locality (AIR 1952). There is no fixed period during which the continuance of the local practice has to be proved. In a Madras case the right to catch fish in a tidal river at a certain place by putting stake nets across a certain river was claimed on the footing of a custom, and was held proved as customary right of the locality on proof of 30 years' use (Narsaya Vs. Sami 1889).

A custom must be reasonable: A custom derives its validity from its being reasonable in its inception and also present exercise (AIR 1931). A customary right, namely, the right to take earth for making pots, claimed by the Kumhar community of a village, was upheld in the Nagpur case of Bhiku Vs. Sheoram (AIR 1928).

A custom must be certain and invariable: Unless the custom is certain in its extent and mode of operation, the courts would not recognize it as valid. It should be definite and invariable.

Customs are however, different from: (i) Public Rights, which are in favour of the general public at large, whereas a customary right is in favour of a limited section of the public like the inhabitants of a village or the members of a caste or community; (ii) Right of Easement, or private rights belonging to a particular person while customary rights are public in nature annexed to the place in general. Customary rights are specifically excluded from the purview of the Indian Easement Act, 1882.

Customary right, by its very definition, cannot be a creature of a written instrument (AIR 1958). Neither are the principles of customary law codified nor are customs listed out separately by the legislature in India. Customary laws are the creation of Indian courts. Facts relevant for the proof of customary law are detailed under Section 13 of the Indian Evidence Act, 1872. Article 13 of the Constitution of India treats customary law along with the other branches of civil law. A custom or usage, if proved, would be law in force within this Article (AIR 1980). Also, customary law is 'law' according to the Indian Constitution which can be taken judicial notice of by Indian courts under Section 57 of the Indian Evidence Act, 1872.

However, a customary right should stand the test of constitutional rights and should not be inconsistent with or in derogation of the Fundamental Rights (AIR 1965). Infringement of Fundamental Rights by a pre-Constitution custom or usage renders the custom inoperative (AIR 1951).

Community- and customary-level laws reflect the ethos and traditions of the local people. The local communities observed these binding regulations voluntarily; there were no formal law enforcing agencies to oversee their observance. These customary principles helped the local communities to be self-reliant and self-sufficient, to an extent. However with the advent of colonial rule and the introduction of formal legislative laws, customary regulation gradually receded to the background.

It should be said to the credit of the colonial British rulers, that their legal mechanism did not altogether discard the customary–oral–legal tradition of the local communities. On the other hand, the English

Courts in India tried to formulate a more rational legal structure into which the customary regulations could be integrated. In the absence of any written law or guidance in this regard it was left to the colonial courts in India to develop customary laws as a new branch of civil law. However, most of the decisions rendered by the courts in this context were related either to religious ceremonies or hereditary offices. Though community conservation, control and use of community-based natural resources came under the purview of customary rights, the concerned communities seldom moved the courts to assert these rights even when they were threatened. This was due to the lack of awareness of the nature of their rights in some communities and also a lack of proper guidance. The other reason is more telling: the compact indigenous communities did not recognize the courts that were functioning outside their customary community jurisprudence and therefore the question of their seeking any legal relief through such courts did not arise. It is no wonder then that there were hardly any case laws relevant to CBC under the customary laws.

The colonial courts, working within the framework of Anglo-Saxon jurisprudence, also had their limitations in comprehending the situation in a holistic way. Indigenous communities had (and still have) a concept of custodial association with the community-controlled natural resources (Roy–Burman 1993). This is a combination of custodial responsibility towards and custodial right over endowments of nature. The ecological concern of the indigenous people was basically a manifestation of their ethical world-view. It would not be correct to project this concern only in terms of control of resources. Nature is not just an ensemble of resources; it is much more. The ethical–custodial relationship is in the nature of the custodial relationship among the members of a family.

In the custodial association of the traditional communities with nature, the right to utilize resources has its counterpart in the responsibility to replenish the same. This community right is different from the concept of common property resources. Common property resources frequently means open-access public resources where state power prevails over community norms, i.e., ownership still vests with the state. The element of 'ownership' is totally absent in the custodial association of community rights; the natural endowments belong to nature itself. The Eurocentric notion of ownership cannot assimilate and integrate these principles of community custody and collective management of resources.

But in the fast-changing modern world, local communities can neither remain homogeneous nor static. Social, political and economic forces have affected them; these forces include 'modern' science and technology, state control of natural resources through various legislations and administrative regulations, severe curbs on the access of the traditional communities to their biomass catchments, out-migration to urban areas (and occasional immigration from other communities). By and large, the consequent changes have had a negative impact on customary practices and law. However, for compact communities and indigenous people that live deep inside the forest and not within the easy reach of the outside civilization, the customary practices and laws continue to be as relevant as before.

CBC and Indian laws

Forest and wildlife laws

The colonial government established the Imperial Forest Department in 1864 and in the following year enacted the Indian Forest Act, which was consolidated in 1878. With these acts, forests, the most important common natural resources, were brought under government control. The Indian Forest Act, 1878, was subsequently amended and consolidated by the Indian Forest Act, 1927 (IFA, 1927). This Act provided for the constitution of reserved and protected forests, and the major chunk of the country's forests were brought under this classification.

An elaborate procedure was prescribed for the constitution of reserved forests under the Act: a simple notification by the state government, appointment of a Forest Settlement Officer, a summary enquiry into the claims or objection of the interested persons, and passing the final order declaring the constitution of a particular reserved forest. Under Sections 11 and 12 of the IFA, 1927, the Forest Settlement Officer had the powers for either admitting or rejecting claims in or over any lands including right of way, right of pasture, right to forest produce, the right to water-course, and other rights relating to the land. There was enough scope under these provisions to admit all claims of the local communities with respect to their customary rights in forests, now declared as 'reserved'. The least that the Forest Settlement Officer could have done was to recognize the right of residence of the forest dwellers in the forest. However, this was not done. The subsistence needs of these traditional communities were termed 'biotic pressure' on the forest; the intention of the colonial government was to restrict the rights of these people and not to legitimize or expand them. Further, the Forest Settlement Officers were drawn from the 'lower' rungs of the Revenue Administration of the state, who did not have the necessary legal acumen, breadth of vision or social insight to appreciate and recognize the local traditional practices and the customary laws.

There was another missed opportunity. Under Section 28 of the IFA, 1927, state governments had the option to assign any village community, the rights of the government over any land which had been constituted a reserved forest. The forests so assigned to the village community were called the village forests. The state governments were empowered to make rules for regulating the management, protection and improvement of such forests and also for providing timber, or other forest products or pasture. Had this provision been used with certain insight and imagination by the colonial administration, the foundation for CBC could have been well laid within the existing legal structure and administration. That did not happen.

The colonial rulers considered forests from the narrow viewpoint of timber harvest and revenue generation. They could neither comprehend nor appreciate the traditional systems of CBC and sustainable use of natural resources. What is surprising is that the independent and democratically elected Government of India continued in the same colonial framework.

The period 1972–76 was momentous for nature and wildlife preservation from the legal point of view. The Wild Life (Protection) Act (WLPA), a comprehensive legislation to protect wildlife, was enacted in 1972 by the Government of India. The WLPA, 1972 laid down the procedure for the establishment, maintenance and administration of areas as wildlife sanctuaries and national parks.

The procedure laid under the WLPA, 1972 for the constitution of wildlife sanctuaries (Sections 18 to 26) is almost identical with that prescribed for constitution of reserved forests under the Indian Forest Act, 1927. There is a formal declaration about the intention of the government for constitution of a sanctuary, followed by a proclamation by the Collector of the district inviting claims of rights from interested persons within two months of the proclamation, an enquiry into the claims of the people with respect to the land proposed to be converted into a sanctuary, and the passing of the final notification. If the Collector admits the claim, he may exclude certain areas in question from the limits of such a sanctuary, or proceed to acquire the said land under the Land Acquisition Act (1894), or allow for certain rights to continue. A similar procedure is prescribed for the constitution of national parks, with the exception that no rights can be allowed to continue after final notification.

The customary rights of forest dwellers and local communities could have been legally recognized by the enquiry officers at the time of settlement of rights while constituting sanctuaries. However, the officers who enquired into the customary claims of these communities lacked the necessary social vision and concern; through their bureaucratic blinkers, they could only see the narrow and limited interests of the government. Consequently, very few of the customary rights of the traditional communities, were generally recognized. On the contrary, entering into or residing inside the PAs without permission are punishable offences under the Act. This situation has led to increasing conflicts between local communities and forest authorities. There is no scope for CBC under the WLPA, since the interests of people and wildlife are perceived to be antagonistic to each other.

Constitutional provisions

The Constitution of India, as originally adopted, was blind to conservation. Though Gandhian thinkers like J.C. Kumarappa were thinking and talking in terms of 'Economy of Permanence' (Kumarappa 1945), there was hardly any awareness of protection of the environment. The only silver lining in this regard was Article 40 of the Constitution, dealing with local self-government, which ultimately led to (after nearly a quarter of a century) the 73rd and 74th Amendments to the Constitution, giving constitutional guarantees to the self-rule of the people at the village level.

In 1976, the 42nd Amendment Act incorporated two new Articles into the Indian Constitution. Article 48A provides for protection and improvement of environment and safeguarding of forest and wildlife as part of the Directive Principles of State Policy, 51-A on Fundamental Duties. Article 51-A (g) declares that it is the fundamental duty of every citizen to protect and improve the natural environment including forest and wildlife. This Article gives a broad constitutional orientation to CBC.

The Scheduled District Act, 1874, provides for delineation of Scheduled Areas, and certain tribal areas were declared as Scheduled Areas in 1921. The modified Act of 1935 formalized two categories of Scheduled Areas—the North-East province and Central India. The Act wholly excluded the North-East from the operation of the national laws and Central India was partially excluded. The Act provided for Tribal Advisory Councils. However, the Presidential Order of 1950 on Scheduled Areas left out the tribal areas of West Bengal, Karnataka, Kerala and Tamil Nadu.

In the absence of the legislative laws of the state, the Scheduled Areas were governed by their respective customary laws, social, religious and cultural practices and traditional management practices of communities. It was a simple case of CBC without the intervention of the state. The Fifth Schedule of the Constitution of India provides for establishment of a Tribal Advisory Council in each Scheduled Area. Under Article 244(1) of the Constitution of India, administration and control of Scheduled Areas and Scheduled Tribes in any state, other than the North-East, shall be governed by the Fifth Schedule. The Sixth Schedule provides for the formation of Autonomous Districts and Autonomous Regions in Tribal Areas. Article 244 (2) makes the Sixth Schedule applicable to the North-East region.

In 1976, the Parliament amended the Fifth Schedule to allow rescheduling of tribal majority areas with a view to bring more areas under community control. However, the expected rescheduling never took place; the states were not inclined to do so.

The most significant Constitutional amendment came in the 1990s. The Constitution (73rd Amendment) provides for people's participation in the preparation and implementation of development plans and strengthening democratic institutions (*panchayats*) at the grassroots. However, for various reasons, 80 million tribals remained outside the scope of this *panchayati* system. The Bhuria Committee, a parliamentary committee appointed by the central government to resolve this problem, in its report of 1995, advocated the constitution of a *gram sabha* (village council) for every tribal habitation consisting of all adult members of the village which will manage the natural resources like land, forest and waters of the locality. It also recommended that all tribal-majority villages of the country should be brought under the Scheduled Areas.

After considerable dilly-dallying, during which tribal and other citizen's groups launched major national campaigns, the Parliament enacted the Provisions of the Panchayats (Extension to the Scheduled

Areas) Act, 1996 (Act 24 of 1996). According to this Act, the state legislations that may be made shall be in consonance with the customary laws, social and religious practices and traditional management practices of communities. The Act provides for the constitution of a *gram sabha* for every village, to safeguard and preserve the customs of the people, to approve development programmes, and to identify beneficiaries of such programmes. *Panchayats* at the appropriate levels shall be endowed with ownership of minor forest products, the *gram sabha* or the *panchayat* at the appropriate level shall be consulted for granting prospecting licences or mining lease of minor minerals and their prior recommendation obtained for acquisition of land in the Scheduled Areas for development projects or for resettlement of project-affected members of the Scheduled Tribes. The *gram sabha* shall have the power to prevent alienation of tribal lands and to take appropriate action to restore any unlawfully alienated land of a Scheduled Tribe.

This Act breaks new ground in vesting legislative powers to community management of natural resources that were hitherto based on customary/traditional practices. If implemented with a sense of commitment and vision, the underlying principles of the legislation can be enlarged and extended to the non-Scheduled Areas where traditional/community resource management is still in vogue.

Policy implications of CBC

The Man and Biosphere (MAB) Programme of UNESCO represents a totally new initiative in the strategy of CBC, in advocating that conservation must be an open system in which undisturbed natural areas can be surrounded by areas of sympathetic and compatible use. The MAB envisages the constitution of different zones: Core Zone (minimum human interference), Buffer Zone or Manipulative Zone (traditional activities like hunting, grazing, fishing and timber extraction, with excellent scope for CBC), Restoration Zone (reclamation activities in degraded ecosystems), and Cultural Zones (stable land use areas). The MAB programme cculd have been a new beginning in involving local communities in the conservation of natural resources. But in India, it was not to be.

When the first biosphere reserve in India, the Nilgiri Biosphere Reserve was set up a decade ago, there was great euphoria among the conservation-conscious people of Nilgiris. The Save Nilgiris

Campaign (SNC), an active grassroot environmental voluntary organization, offered its wholehearted support to the forest bureaucracy to make the MAB programme a people's movement in the Nilgiris. Unfortunately, the FD, with its colonial mindset did not allow this. From the point of view of community conservation and people's participation in the protection of natural resource, the Nilgiri Biosphere Reserve Programme died before it could take any roots.

From the broader policy viewpoint of CBC and people's participation in the protection of natural resources, the National Forest Policy, 1988, was another milestone. It was for the first time that any forest policy, either before or after independence, had recognized that tribals and forest communities were living in the forest and that they were entitled to share the benefits of conservation. Some of the provisions are quoted below:

> The holders of customary rights and concessions in forest areas should be motivated to identify themselves with the protection and development of forests from which they derive benefits. The rights and concessions from forest should primarily be for the *bona fide* use of the communities living within and around forest areas, specially the tribals (Article 4.3.4.2).
>
> The life of tribals and other poor living within and near forests revolves around forests. The rights and concessions enjoyed by them should be fully protected. Their domestic requirements of fuelwood, fodder, minor forest produce and construction timber should be the first charge on forest produce (Article 4.3.4.3).
>
> Having regard to the symbiotic relationship between the tribal people and forests, a primary task of all agencies responsible for forest management, including the Forest Development Corporation, should be to associate the tribal people closely in the protection, regeneration and development of forest as well as to provide gainful employment to people living in and around the forest. While safeguarding the customary rights and interest of such people, forestry programmes should pay special attention to the following... (Article 4.6).
>
> Appropriate legislation should be undertaken, supported by adequate infrastructure, at the Centre and State levels in order to implement the policy effectively (Article 4.6).

However, major changes are necessary in the main existing laws relating to wildlife and forests, to integrate them into the above vision.

At the international level, during the Earth Summit, 1992, the Convention on Biological Diversity (CBD) was adopted. India has ratified the Convention and expressed its willingness to legally abide by its provisions. The scope of the CBD with reference to the various policy and legal aspects is too wide to attempt a single comprehensive umbrella legislation. There are many issues relating to conservation of natural resources under the CBD which are not covered either by the National Forest Policy, 1988, or by any other policy statement of the Government of India. It is therefore necessary and appropriate that the Government of India enlarges the scope and objectives of the present forest policy and adopts a more comprehensive National Conservation Policy statement with a view to harmonize its policy on biological conservation with that of the CBD.

It is often said that the administrative structure of the FD remains basically colonial in nature. The FD has under its direct custody and control 23 per cent of the total land area of the country. The centralized bureaucratic management of these public lands and the land use pattern needs fundamental reorientation. This becomes all the more imminent and imperative considering that a substantial part of India's population depends on these. These public forest lands serve a threefold purpose at present: Ecology (conservation imperatives), Equity (livelihood security), and Economics (wood-based commercial interests). Considering the inter-twined issues of growing human population and the correspondingly shrinking forested areas, the commercial pressure on public forest lands should be completely shifted to private lands. Commercial plantations should be raised as a kind of cash crop on private agricultural lands and the small farmers should be encouraged to develop tree farming with adequate market guarantees. Public forest lands should exclusively serve the twin purposes of ecological security and livelihood imperatives of the people. These in turn would correspond to Conservation Forests and Community Forests. Forests that fall under the present PAs may be classified as 'Conservation Forests' and those that are outside the PAs, including Reserved Forests may be classified as 'Community Forests'. 'Conservation Forests' would be critical for ecological security and may be protected with minimum external interference without displacing the compact tribal communities that live inside the PAs. Involvement of these forest dwellers in the conservation of the biodiversity-rich PAs is not only relevant but essential. Their needs are limited to subsistence and resource use is usually sustainable. They should thus have access to their biomass catchments.

On the other hand, marginalized rural communities who depend on forests, now degraded, for their subsistence and survival should be fully involved in the conservation of 'Community Forests'. The legal equivalent to the Community Forests in the Indian Forest Act is 'Village Forests'. As stated earlier, Village Forests are carved out of Reserved Forests and assigned to village communities for conservation and sustainable use (Chapter III, Section 28 of the Indian Forest Act, 1927). This provision should be imaginatively implemented to provide timber and other forest produce as well as pasture rights to the local communities. The state FDs, by an executive order can convert reserved forests to village forests (or community forests) and assign them to the local communities. This does not require any amendment in the Act. The state governments can frame rules for the administration of these community forests and the rules so framed should give sufficient powers to the local communities to manage the forests independently with minimum state control.

Local self-rule is now a constitutional right under the 73rd and 74th Amendments. Under Article 243G, the state legislature should devolve powers and authority to the rural *panchayats* to enable them to function as institutions of self-government with particular reference to matters listed in the Eleventh Schedule of the Constitution. This Schedule consists of 29 subjects including: soil and water conservation, and social and farm forestry. This is further emphasized in the Panchayat (Extension to Scheduled Areas) Act, 1996. There is enough scope for the local communities to serve as stewards of local natural resources. What is lacking is the political will.

There are certain ongoing positive efforts in the context of CBC in India, e.g., the Van Panchayats of the Central Himalayas and the Joint Forest Management (JFM) Programme of West Bengal (Maikhuri et al. in this book; Sarin et al. in this book). There are other ongoing positive attempts. The Van Gujjars of Rajaji National Park have drafted proposals for Community Forest Management (CFM) in the PA, with the help of some NGOs. In a similar attempt the tribals of Nagarhole National Park have prepared a people's plan for community management of their habitats. Possibilities are emerging, though they do not yet have the stamp of legislature.

The way forward

Through the historic Earth Summit declarations, the issue of traditional communities' right to manage their natural resources has received greater global attention. Agenda 21 (UNCED 1992a), The Rio Declaration on Environment and Development (UNCED 1992b), Forestry Principles (UNCED 1992c) and Convention on Biological Diversity (UNCED 1992d), are some of the major achievements. All these major international instruments refer to community participation in the conservation of natural resources.

For instance, Agenda 21 forthrightly urges governments to 'adopt or strengthen appropriate policies and/or legal instruments that will protect indigenous intellectual and cultural property and the right to preserve customary and administrative systems and practices'.

India is under international obligation to transform these principles into effective legislation. As one step, operationalizing the National Forest Policy, 1988, becomes absolutely imperative. Social scientists and concerned citizens should strive for mobilization of social forces for the implementation of the Forest Policy since it would mean replacement of both the Indian Forest Act and the Wild Life (Protection) Act by enactments that are more humane and equitous.

There are other possibilities too in the sphere of human and constitutional rights. Social movements and legal struggles have forced recognition of a number of rights, like right to customary law and practice, right to environmental integrity, right of access to the restricted local catchment area, right to common resources, right to indigenous knowledge, minority rights, and right to habitats of the indigenous community. In an emerging situation pregnant with various possibilities, this list is only illustrative. These rights are now collectively referred to as 'bundle of rights' of the indigenous or local community. Though they are not codified by any international instrument as such, the legal community across the spectrum is not unaware of their explosive potential. The 'bundle of rights' rightly assumes that the conservation and sustainable use of biological diversity wholly hinges on local communities having secure rights to access and use of biological resources, and the local knowledge associated with them. Community conservation is not a complete circle in itself; it is one of the three corners of the triangle, the other two corners being sustainable use and equitable sharing of benefits (the objectives of the CBD). And the subject lies at the tri-junction of human rights law, biodiversity conservation and economic development.

The two Conventions of the International Labour Office (Convention 107 of 1957 and Convention 169 of 1989) with respect to indigenous peoples assume significance in this context. These two conventions as well as the UN draft charter of the rights of indigenous people have emphasized that the right to maintain group identity and to conserve and enrich their physical environment should be respected. Article 11 of Convention 107 declares: 'The right of ownership, collective or individual, of the members of the populations concerned over the land which these populations traditionally occupy shall be recognized.' Article 12.1 of the Convention bars governments from evicting these populations from their habitual territories without free consent. India has ratified this Convention, and hence its provisions are enforceable in India. To put it differently, international law protects local livelihood systems as human rights and any violation can be challenged in a court of law (SCC 1993).

There are many international instruments which deal with rights of individuals, but there are no declarations to recognize collective human rights. Community conservation rights are collective human rights, alternately titled Traditional Resource Rights (TRR) (Posey 1996). The livelihood security of these traditional communities are hinged on their local resource base. Therefore the TRRs are their 'right to life', a right considered fundamental, both under the international human rights law as well as the Constitution of India.

In India, in the absence of specific constitutional provisions in this regard, Public Interest Litigation holds the key. Observes Justice A.M. Ahmadi (1996), former Chief Justice of India:

> In the post-Emergency era, the Apex Court, sensitized by the perpetration of large scale atrocities, during the Emergency, donned an activist mantle. In a series of decisions, starting with Maneka Gandhi Vs. Union of India, the Court widened the ambit of constitutional provisions to enforce the human rights of citizens and sought to bring the Indian law in conformity with the global trends in human rights jurisprudence. Simultaneously, it introduced processual innovations with a view to making itself more accessible to disadvantaged sections of society giving rise to the phenomenon of Social Action Litigation/Public Interest Litigation. During the nineteen eighties and first half of nineties, the Court has moved beyond being a mere legal institution; its decisions have tremendous social, political and economic ramifications.

After analyzing the expanded powers of the Supreme Court and comparing it with other constitutional courts from across the globe, the noted academician, Professor Upendra Baxi, came to the conclusion that the Supreme Court is the most powerful court in the world

(Ahmadi 1996). The apex court then can and should rise to restore to communities their lost habitats and reiterate their right to life.

REFERENCES

Ahmadi, A.M. 1996. Judicial Process: Social Legitimacy and Institutional Viability. Public address.

AIR. 1928. Bhiku vs. Sheoram, Nagpur 87.

———. 1931. K.R. Ramasamy Iyer vs. Secretary of State, Madras 213.

———. 1951. SC 210, Ram Dhanlal vs. Radha Slam.

———. 1952. SC 231, Gokalchand vs. Parvin Kumari.

———. 1958. Rev. For. Joseph Valamangalam vs. State of Kerala, Kerala 290.

———. 1965. SC 314.

———. 1980. SC 707.

Berkes, F. 1993. Traditional Ecological Knowledge in Perspective, in J.T. Inglis (ed.). *Traditional Ecological Knowledge: Concepts and Cases*. Canadian Museum of Nature and International Development Research Centre, Ottawa, Canada.

CSE. 1996. *Protection of Nature Parks: Whose Business?* Proceedings of a debate. Centre for Science and Environment, New Delhi.

Durning, A.T. 1993. Supporting Indigenous Peoples, in L. Brown, A. Durning, C. Flavin, H. French, J. Jacobson, N. Lenssen, M. Lowe, S. Postel, M. Renner, L. Starke and P. Weber. *State of the World*. Worldwatch Institute, Washington D.C., and W.W. Norton and Co., New York.

Indian Forest Act, 1927. Act No. 16 of 1927. Government of India, New Delhi.

Krishnan, B.J. 1996. Legal Implications of Joint Management of Protected Areas, in A. Kothari, N. Singh and S. Suri (eds). *People and Protected Areas: Towards Participatory Conservation in India*. Sage Publications, New Delhi.

Kumarappa, J.C. 1945. *Economy of Permanence*. Sarva Seva Sangh Prakashan, Varanasi.

Narsayya Vs. Sami – 12 – Madras Page – 43. 1889.

Posey, D.A. 1996. *Traditional Resource Rights: International Instruments for Protection and Compensation for Indigenous Peoples and Local Communities*. IUCN—The World Conservation Union, Switzerland.

PRIA. 1993. Report on National Workshop on Declining Access to and Control over Natural Resources in National Parks and Sanctuaries. Society for Participatory Research in Asia, New Delhi.

Roy–Burman, B.K. 1993. Quoted in Report on National Workshop on Declining Access to and Control over Natural Resources in National Parks and Sanctuaries. Society for Participatory Research in Asia, New Delhi.

Saxena, N.C. 1996. Forests and the Poor in India. Keynote address at the workshop on National Forest Policy, Dehradun. October 1996. Organized by NCPCLR, Dharwad.

SSC. 1993. Nilabati Behera vs. State of Orissa and others. *Supreme Court Cases*. p. 746.

Stebing, E.P. 1922–27. *The Forests of India*. Vol.I–II–III. John Lane. The Bodley Head Limited, London.

UNCED. 1992a. Agenda 21. Adopted at the UNCED, Rio de Janeiro, Brazil. 3–14 June 1992, U.N.Doc.A/Conf.151/5.

UNCED. 1992b. Rio Declaration on Environment and Development, Adopted at UNCED. Rio de Janeiro, Brazil, 3–14 June 1992, UN.Doc.A/Conf.151/5.

———. 1992c. Non-legally Binding Authoritative Statement of Principles for a Global Consensus on the Management, Conservation and Sustainable Development of All Types of Forests, 1992. Adopted at UNCED. Rio de Janeiro, 3–14 June 1992, U.N.Doc A/Conf.151/6/Rev.1.

———. 1992d. Convention on Biological Diversity 1992. Adopted at UNCED, Rio de Janeiro 3–14 June 1992, U.N.Doc.151/26.

PART 4

case studies

18

Self-initiated conservation in India's protected areas: The cases of Kailadevi and Dalma Sanctuaries

Priya Das and K. Christopher

This paper is based on action research conducted in Kailadevi Wildlife Sanctuary, Rajasthan (western India) and Dalma Wildlife Sanctuary, Bihar (eastern India) as a part of the project 'Towards Participatory Management of Protected Areas', conducted by a team at the aegis of the Indian Institute of Public Administration (IIPA).[1] This project aimed at exploring the possibilities of shifting from the conventional centralized, top-down management of wildlife habitats which emphasizes separation of wildlife and human populations, towards a more participatory process which integrates wildlife and human livelihood concerns.

One of the aspects studied in these two protected areas (PAs) has been the significant conservation efforts which have been largely self-initiated by the resident rural communities, and the relationship of these efforts to official attempts at wildlife conservation. Even though the community-based attempts in both the cases aim at the same objectives, set in two different social contexts, there are important operational and institutional differences. And yet, both the cases bring out similar cardinal issues.

Case 1: Kailadevi Wildlife Sanctuary

Introduction

The Kailadevi Wildlife Sanctuary (KWS) forms the buffer zone of the Ranthambor Tiger Reserve, one of India's 21 tiger reserves. KWS is spread over an area of 674 sq. km in district Sawai Madhopur, the state of Rajasthan.

Falling within the biogeographic zone 4 (semi-arid zone) and biotic province 5B (the Gujarat Rajwara Province) of the Wildlife Institute of India classification (Rodgers and Panwar 1988), the vegetation is mainly of the dry deciduous type with a predominance of *dhok* or *Anogeissus pendula* (Champion and Seth 1968). According to the Forest Department (FD), given the size of the area, the wildlife population is relatively low. However, in the past this area is known to have nurtured a rich faunal life, including the Tiger (*Panthera tigris*), Leopard (*Panthera pardus*), Blue bull (*Boselaphus tragocamelus*), Sloth bear (*Melursus ursinus*), Indian porcupine (*Hystrix indica*), Striped hyena (*Hyaena hyaena*), and a wide variety of birds. These species still exist here, but in small numbers. The Sanctuary also has several settlements of predominantly pastoral people, of the Gujjar (OBC) and Meena (ST) communities.[2]

In the past, KWS was influenced by several external pressures including the hunting activities of the imperial rulers and of the Bargi community; forest coups; illegal felling; mining activities; and the advent of the Rabari community, the migratory graziers in that region.

Community-initiated forest protection committees

Reasons for formation

Community-initiated conservation efforts at Kailadevi are more than anything motivated by the threat that declining forest resources pose to the livelihood of local communities. From oral evidence it appears that until the recent past, villagers had evolved a lifestyle in relative harmony with their habitat. Predominantly pastoralist, the forests were a critical source of fodder and sustenance. Though agriculture has spread, dependence on the forest for fuel, fodder and timber continues to be critical.

Sustainability started breaking down due to the gradual increase in population, several external pressures (felling by the FD and private sources, overgrazing by migrant graziers), and some new internal socio-economic factors (changes in lifestyle, etc.). With increasing shortage of fodder and fuel caused by deforestation, villagers realized the significance of conservation and the quality of life that available forests can ensure. This feeling appears to have manifested itself in the formation, some time in the mid-1980s, of local village forest protection committees (FPCs). Since their concern for their cattle is pre-eminent, their highest incentive for conserving the forests is the sustained availability of fodder followed by their requirement for fuelwood. In Lakhruki, a village inside Kailadevi Sanctuary, they also directly linked the increase in rainfall to the increase in their regenerated forests.

It is significant to note that while the villagers do not harbour any hostility towards wild animals, neither do they seem to attribute much significance to them in their day-to-day existence. Their value for wildlife is derived more from their religious realm and their basic reverence for nature.

The immediate trigger for the formation of the FPCs was the surmounting fodder shortage caused by the entry of huge flocks of livestock belonging to the Rabaris. The body of the 12 villages (Baragaon ki Samiti) that had been initially formed in 1986 to take stock of the Rabari problem, gradually started taking up the responsibility of protecting the forest area. Thus, in 1990, the 'Baragaon ki Panchayat' came to be synonymous with 'Kulhari Bandh Panchayat of Baragaon' (the *panchayat* of 12 villages that prevents the carrying of axes into the forests).

Constitution

FPCs in the villages are constituted entirely by the people, and operate at two levels: (*a*) At the level of the individual villages. This body is constituted of members of the village, with a small decision-making body whose members represent the village at the apex level; (*b*) The apex level or at the level of joint sessions of a number of villages included under a single *panchayat*. The apex body, subject to circumstances, may also include villages outside of this *panchayat*.

'Baragaon ki Panchayat' was the first apex body and included 12 villages of the Lohra Panchayat, including all their hamlets. This body is generally summoned if there are inter-village disputes among the member villages over resource use, or if a particular village tends to break the prescribed norms (see the section on functions of FPCs later). Similar apex bodies have been established in a number of areas.

The *panch patels*, the handful of village elders, are primarily responsible for enforcing the various norms and regulations of the committee. In case of an offence committed, matters are first taken up at the village level generally by the *patels* and only in the event that it cannot be settled by them, are they taken up at the higher level. In some villages (e.g., Kased), the *panchayat* exists in isolation of other villages, and is rather volatile, breaking up on minor issues). Thus it appears that a village-level unit is probably more stable when it is part of a larger conglomerate like the apex bodies, than when it functions in isolation.

In some villages, residents categorically stated that even though they vest the decision-making rights in the hands of the *patels*, the *patels* at no point of time can function in an autocratic manner or take arbitrary decisions. So far as the *panchayat* is concerned, functionally each and every villager has an equal standing and the right to decision-making. While enlisting members at the village level it is ensured that almost all families are represented. The main objective of this is to ensure that the responsibilities of the FPC are equally shared by everyone in the village.

What is remarkable is the representative nature of the committees. Most villages in this area are multi-caste villages. FPCs, both at the village level and thus eventually at the apex level, comprise members from almost all caste/communities residing inside the village. The number of representatives is in proportion to their strength in the village. Even the *panch patels* are representative of the various communities in a particular village.

However, though some older women are quite vocal, gender discrimination, an integral part of local culture, prevents women from active involvement in the FPCs. Involvement of women could be crucial. Since this area is a politically reserved constituency for women, and most of the members of the formal political *panchayat* are women, their participation in the village FPCs would enable them to effectively participate in the formal *panchayat*, which at present is severely lacking.

Functions and rules

The FPCs have laid down certain rules pertaining to use of forest resources: no one is allowed to carry an axe into the forest; villagers are allowed to collect only dead and dry wood for fuel and that too only for personal use; and timber for household purposes can only be brought with the FPC's consent and in only the amount fixed by the FPC. Members of the FPC are meant to keep a constant vigil in their area and dutifully report any untoward happening with regard to their area. The committees affix certain fines for offences committed which may vary from Rs 11 to Rs 11,000. The money collected from the various fines is treated as community resource and is put to use for community services. In Lakhruki, a *talai* (pond) has been built from its pooled resources.

These rules have apparently been reached by common consensus. It is interesting to note that most of these rules, other than that of timber extraction, are along the same lines as those set by the FD (it prohibits the extraction of timber from this region completely). No action is taken by the committee against anyone unless witnesses as well as evidence are produced in the *panchayat* meetings.

The people feel that in cases that lack concrete evidence and a person is held on the basis of suspicion alone, their religious sanctions against such acts are far more effective. Irrespective of whether there is a witness or not, a person refrains from lying after having taken the oath of honesty ('Ganga *uthana*') in the name of the goddess Ganga.

Meetings are summoned either at regular intervals or only in the case of a serious eventuality. In some villages, minutes registers are maintained. Conflicts over resources between villages are generally settled by the exchange of letters between the *patels* of the concerned village. These letters have a great social bearing. The inter-village relationships are subject to the manner in which these letters are written and responded to.

One criticism of many government-initiated JFM programmes across the country is that the rules framed by them are too rigid, not allowing for situation-specific innovations. The Kailadevi FPCs seem to be different. In some instances we noticed that villagers who had been representing the FPC in all meetings, carried their axes into the forest area. On inquiry, we found that the axes are used only to chop the dry wood into collectable sizes, because especially for older people it is almost the only way of collecting wood. Thus while the FPC may not object to a person carrying in an axe, they would certainly take him to task if the person came back with green wood or chopped an excessive amount. Communities allow for some flexibility in the rules only to facilitate their day-to-day existence, so long as the basic principles are not compromised.

Relations with other organizations

Village *panchayats*: Even though the FPCs remain distinct from the administrative village *panchayat*, the decision-making powers in both cases are vested with the *patels*. A distinction is effected in the two institutions because they are not only structurally (with a defined membership of sorts) but also functionally different. The FPCs are more active and meet more frequently as they have to keep a check on forest-related activities on a day-to-day basis.

***Panchayats* at the level of clusters of villages:** At this level only the involvement of the *sarpanch*, the head of a formal *panchayat* that includes a cluster of villages, has been of any significance.

Such involvement has had both positive and negative results. Apparently, the FD has vested some powers in the *sarpanch* to take to task offenders. This is done by lodging complaints and putting on trial any local person who commits a forest-related offence. However, this has been of little use to the FPCs because as per the people, this power has been always abused to serve the vested interest of the corrupt *sarpanch* and has made it more difficult for the FPCs to function. On the other hand, as seen in the case of village Rahar, the *sarpanch* has played a significant role in facilitating the involvement of most of the villages under his jurisdiction in the government-initiated FPCs. In some cases (e.g., village Lakhruki), the ex-*sarpanches* assume an important role in the decision-making body of the FPC.

The Forest Department and its Van Samitis: Van Suraksha Samitis (VSS) are FPCs officially constituted by the FD at the village level, under the JFM policy of the FD. According to local sources, the scheme had been initiated as far back as 1985, but nothing was done on the ground. At present there are only five VSSs in the Sanctuary.

The VSS is essentially constituted on very simple and workable principles. In the Sanctuary, most of the VSSs have been constituted in villages where community-initiated FPCs already existed, and are based roughly along the same lines.

In village Rahar we witnessed the constitution of such a committee. The Assistant Conservator of Forests (ACF), along with the Forester, arranged for a meeting in the village on 21 May 1996. About 35 members, representing the various lineages and all the three communities, were registered as members of the committee, by common consensus. A record was maintained both by the forest officials as well as the village primary school master. The villagers who had been thoroughly disillusioned with the FD for its inaction on earlier occasions, agreed to constitute the committee after much cajoling, coaxing and assurance by local forest officials.

From amongst them, the villagers chose a spokesperson, who would officially be known as the *adhyaksh* or the head. A few rules (same as those applicable in the community-initiated FPCs) regarding fines were laid down and it was made mandatory for all the registered members and the FD representative to be present at every meeting that was to be convened at regular intervals. It is worth mentioning that at this level the formulation of the body was essentially done by the villagers at the behest of their *patels*. The FD only endorsed it.

We were also able to attend one of their meetings (convened every 15 days) on 21 June 1996. The committee, after much deliberation, was able to extract a fine of Rs 11 each, from all those members who had remained absent without intimation in the previous meeting. The villagers had also made sure that action was taken against the forest guards who had failed to turn up at the previous meeting, which none of the officials had attended. Throughout the meeting the FD officials only offered suggestions and answered queries directed at them.

The VSS does not operate as efficiently everywhere. In village Maramda, the residents complained that their VSS has never existed in reality. About three years back, a forest guard had registered their names individually. But till date no meetings had been held. Forest protection activities are being carried out by the community-initiated FPC in complete isolation of the VSS.

The exact nature of involvement of the FD with the VSS remains ambiguous. While in Rahar they were a part and parcel of the system and their attendance in every meeting was made mandatory by the villagers, in Lakhruki the involvement of the FD has become absolutely marginal, with no official attending meetings organized by the villagers.

This inaction is rather unfortunate, as both forest officials and villagers feel that the FD could provide much needed official recognition to villagers' efforts, which could help prevent internal caste conflicts since all castes would be equally subject to the same set of laws, as also help villagers to resist illegal activities by outsiders (see below).

Conservation effectiveness

Owing to the involvement of practically every family in FPCs, everyone keeps a vigil on each other. Consequently, the extent of illegal felling by the local people has gone down drastically. This fact is substantiated by the extensive regeneration of the forest cover that one witnesses in the area. The FPCs have not only been effective in checking themselves and some infiltrators from illegal felling, but have also accosted some FD officials, tried them in the FPC meeting, and levied appropriate punishments! The FPCs have also been assisting the FD in keeping a check on illegal activities. In a number of cases the FPCs have taken the offenders to the FD following their own trial. This however happens mostly in cases where they haul a truck carrying out timber and other such large-scale offences. The constant vigil of the villagers has also to a certain extent pulled up the forest guards who may at times tend to play truant or are lax about their work.

What is most interesting to note is the transition from protecting their resource base for survival to getting involved in the broader issues of conservation, becoming aware of their position vis-à-vis other stakeholders, and their 'right' to a say in decision-making regarding resources.

Issues regarding the functioning of FPCs

Lack of official recognition: The point of conflict between the two (the FPC/VSS and the FD) is over the nature of involvement that each envisages for the other. Legally and administratively, the FD is the sole authority in the KWS; the FPCs only regulate people's activities by imposing social and religious sanctions.

The FD's concept of participation of FPCs in management of the forests is limited to seeking cooperation for the policies and rules that it wishes to implement, and does not extend to any devolution of powers. The FPC on the other hand, with the gradual breakdown of the community's social and religious ties, feels constrained without any officially sanctioned powers. Members feel that empowering them to a certain extent will enable them to take to task errant forest officials, avail of benefits offered by the FD's schemes, and more effectively and extensively check illegal activities in the forest area.

Villagers with the help of local NGOs have defined the exact nature of empowerment they seek at the FPC level:

- In the formation of FPCs or VSSs, the members should be elected by the people in a village.
- These bodies should be registered and legally recognized by the government.

- In order to check illegal felling, the FPCs should be empowered with the powers that are with a Forest Ranger.
- All development work relating to the Sanctuary and its resources should be channelized through the FPC.
- Where a case cannot be resolved by the FPC alone, it should be jointly arbitrated by the FPC and the FD, and 50 per cent of the fine levied should go to the FPC.
- In cases of losses suffered by the villagers at the hands of the wild animals, the report of the affected person, if endorsed by the FPC, should be considered valid and should be accepted by the FD (thereby avoiding the delays and harassment of having to get official inspections conducted).
- The FPC should be adequately empowered to check the entry of livestock from outside the Sanctuary.

Local pressures and disputes: Notwithstanding their commendable efforts, the functioning of the FPCs is marred by a number of internal pressures:

Caste conflicts: Caste-based politics and differences often cause dissension and render some FPCs ineffective. In Rahar, the earlier existing FPC disintegrated due to this. At Kailadevi village, the Baragaon Panchayat had not been able to stop the rampant illegal felling and lopping of fuelwood; the Jatavs, a low-caste community noted for illegal selling of fuelwood, said that they were discriminated against both by the Meenas as well as the FD, as most officials also belonged to the Meenas. The Meenas on the other hand complained that it was extremely difficult to take any action against the Jatavs, as at the slightest opportunity, they threatened to slap a case of caste-based discrimination. Interestingly, in some villages the rules were relaxed for Harijans. Since they are mostly landless, and make a living out of basket-making and other such activities, they are allowed to lop a limited quantity of green branches.

Favouritism: Some villagers complained that nepotism and favouritism by the decision-makers has caused people to lose faith in the system of FPCs. This was specially the case with the kith and kin of the *patels* and other influential villagers.

Outside influence: Villagers who have started settling outside the village are no longer threatened by social sanctions against them, and sometimes indulge in illegal activities without fear of reprisal.

Attitudes towards joint management

Notwithstanding the differences in perceptions between the FD and the people, the 'need' for some form of joint management was supported by both. It has now become apparent to officials that without public support, KWS cannot be saved. The people, especially in the villages where community FPCs are in operation, are fully aware of their significant contribution in conserving and protecting the forest cover in that area, but also of their inherent limitations.

The forest officials feel that the legal powers should continue to be vested with them. They feel that since the local communities are mostly illiterate, it may not be feasible to just empower them without actually preparing them for it. However they were most welcoming to the idea of involving them in the local or micro-level management.

According to the villagers, the FPC should be allowed to handle all management problems within the village boundaries. They would prefer least involvement of the FD in their internal matters. However, they want that the FD should recognize their authority and give their committee a legal status, including treating their decisions and complaints with seriousness. They also feel the need for the FD's intervention to handle external pressure, like that of the migratory graziers, mining, illegal felling, and to intervene in situations that cannot be handled within the village, like serious caste and inter-village conflicts over resource use.

Existing institutional structures at the village level as well as the official level offer ample opportunities for joint management. At the village level, the smallest unit of management suggested was the existing community-initiated FPC. Villagers hope to achieve the following from a joint management system:

- A platform for open dialogue and interaction with the FD.
- An opportunity to counter-check the malpractices of the FD as well as a process by which their FPCs can further be strengthened.
- An opportunity for eliciting full cooperation of all the villagers as well as a process where there would be an equal distribution of responsibilities for protecting the Sanctuary.

Towards joint management?

As part of its project, IIPA in association with the Society for Sustainable Development (SSD), Karauli, organized a workshop entitled 'Kailadevi Sanctuary: Prospects for Conservation', on 6–7 December

1996. The workshop was aimed at initiating constructive dialogue between villagers living inside the KWS, wildlife officials and NGOs. Unfortunately, while 60 villagers and three NGOs (SSD, WWF-India and IIPA) participated, the FD was represented by only a solitary Forest Guard. All senior officials kept away, despite repeated assurances that they would come.[3] Workshop participants issued a joint resolution on these issues, containing recommendations for specific measures:

- On livelihood and employment, it was recommended that water availability be urgently enhanced, the productivity of their limited land and livestock be improved, and suitable sources of employment be provided. In addition, adequate and quick compensation for the damage done to crops and livestock by wild animals should be paid. The villagers rejected any attempts at forcible displacement, and stressed that they would ensure conservation of the forest while meeting their livelihood requirements in their existing locations.
- On external pressures, it was recommended that cattle camps set up by villagers from outside the Sanctuary not be permitted once adequate water/fodder arrangements for the resident villagers were available inside; that migratory Rabaris with their sheep herds not be permitted into the entire area; and that illegal felling from outside be tackled by giving more powers to the village-level committees.
- On people's participation in the management of the Sanctuary, it was recommended that the FPC should be legally registered, and should consist of all caste/ethnic communities of the village. At least two-fifths of the members should be women. The FPCs should have the powers of a Forest Ranger, should be involved in arbitration in cases of forest offences, and should get 50 per cent of all fines levied in such cases.

Case 2: Dalma Wildlife Sanctuary, Bihar

Introduction

The Dalma Wildlife Sanctuary (DWS) is situated in the East and West Singhbhum districts in the Chhotanagpur Division of South Bihar, a few kilometres north of the steel-producing city of Jamshedpur. The DWS covers an area of 193.22 sq. km, with a core zone of

55 sq. km and a buffer zone of 138 sq. km. The highest point is Dalma hill (926 m).

The vegetation consists of Northern Tropical Dry Deciduous Forest (5B/C 1C) and Peninsular Sal (*Shorea robusta*) Forest (5B/C2), in Champion and Seth's (1968) classification. The mammal species include the Asian elephant (*Elephas maximus*), Sloth Bear (*Melursus ursinus*), Wild boar (*Sus scrofa*), Indian giant squirrel (*Ratufa indica*), Barking deer (*Muntiacus muntjac*), Mouse deer (*Tragulus meminna*), Blacknaped hare (*Lepus nigricollis*), Jackal (*Canis aureus*), Indian porcupine (*Hystrix indica*).

There are 132 villages in and around the Sanctuary. Of these, 88 have their settlements and fields located outside the Sanctuary but their common lands and village forests inside, and one village is totally within the Sanctuary area. The total population in these 132 villages is 32,000, with a cattle population of around 41,000. The numerically dominant communities are the Mahato, Bhumij and Santhal, the last two being Scheduled Tribes. Agriculture is the main occupation and is often supplemented by the manufacture and sale of wood and non-timber forest produce (NTFP).

Dalma's forests provide food, drink, timber, fuelwood and various other products to local people. Many of the forest products are sold in markets, especially in Jamshedpur. There is a great demand for fuelwood and timber, a significant proportion of which is extracted, illegally, from the neighbouring forests of Dalma. A number of illegal *mahua* liquor distilleries are operating inside the Sanctuary, using a large quantity of fuelwood. Quarrying, stone-crushing, and brick-making activities at the edge of the Sanctuary are additional pressures. The construction of the Suvarnarekha irrigation canal through a part of the Sanctuary, begun in the late 1980s, has also resulted in a lot of forest clearing.

A tribal ritual known as the *akhand shikar* (The Great Hunt) or the *sendra*, takes place in the Sanctuary in the month of May each year. This event is of great religious, social and cultural significance to the local tribal population. Around 20,000 people participate in it every year, even though it violates the provisions of the Wild Life (Protection) Act, 1972.

The FD has so far not been able to deal with these various issues, due to non-availability of funds, shortage of adequately trained staff, lack of proper equipment, internal inefficiency and corruption, lack of coordination with other government agencies, and hostile relations with local communities.

Community-initiated forest protection committees

Reasons for formation

Over the last few years, villagers have initiated significant forest regeneration and protection measures, mainly out of concern over depletion of livelihood resources. Emphasis has been on the protection of timber species and fruit trees. The villagers explain that due to large-scale felling first by the FD and later by the timber mafia, many areas had become devoid of trees. Most protection efforts date from the early 1990s. Around that time, the passing of the Bihar Joint Forest Management (JFM) Resolution, information of which was spread by NGOs and the FD, greatly encouraged the creation of a number of village FPCs or Van Samitis. These FPCs were primarily self-initiated, with some encouragement by NGOs. The success of the first few FPCs in Sari and Nutandih villages served as models for others to follow. Religious and cultural sensibilities also had an important role to play in encouraging forest protection. Certain parts of the forests in the area, and a number of tree species (e.g., *sal*), are considered sacred.

Constitution

The FPCs are informally constituted bodies and enjoy no legal or official recognition. Each FPC usually consists of a general body consisting of all the residents of the village, and a smaller core group. The general body meeting, once every three–four months, is attended by all the villagers. The core group, which meets more frequently, consists of those who are taking a special interest in protection work in the area. In most cases, these persons are the educated youth of the village and belong to the more privileged communities in the village. The office bearers, including the President and Secretary, are elected at the time of the formation of the FPC. Maintaining of registers and minutes of meetings is followed in a few cases.

Decision-making in the FPCs is by no means fully participatory. Women attend the general body meetings, but rarely play any role. If they have any views to be conveyed, they ask their husbands to speak for them. The underprivileged communities of the area (such as the Sabar/Kharia and Paharia, notified as 'primitive tribes' by the Government of India), often do not attend (e.g., in Koira village). In most villages, it is these communities and other 'poor people' who are usually accused of cutting wood from the forest and selling it in the market. In Gobarghusi and Kherua villages however, it is these very communities who are playing a vital role in forest protection, as their homes adjoin the forest area.

Functions and rules

The function of the FPC is to ensure the protection of a particular forest patch demarcated for protection. The areas chosen for protection usually fall within the territorial boundary or *mauza* of the village. This area is usually well demarcated by physical features such as ridges of hills or streams. In certain areas such as Sari and Nutandih, signs have been painted on rocks by the FPCs declaring that tree-felling is prohibited in the area. The FPCs work therefore includes formulating policy and rules, holding meetings to discuss issues such as illegal felling, coordinating the protection activities of the various *tolas* (hamlets) of the village, trying and punishing offenders, exhorting villagers to aid in forest protection, and representing the villagers during interactions with the FD and NGOs.

The majority of rules regarding forest protection are decided by popular consent. There are a variety of approaches followed with regard to forest protection. In Mirzadih, Nutandih and Sari villages, cutting of trees from the protected patch is completely banned, but another part of the forest is earmarked to fulfill resource requirements. This area is referred to as *chhada*. In villages such as Gobarghusi and Koira, the approach to protection is less strict. In Gobarghusi, only certain species which are of economic or food value are protected and there is no restriction on the collection of 'dry wood'. However, the term 'dry wood' invariably tends to be given a very liberal interpretation and includes poles and dry saplings. In both these villages, the Committee may, in certain cases, grant permission for the cutting of even protected species for self-consumption. The emphasis is on curbing large-scale commercial exploitation of the forests.

Systems for enforcement of rules and imposing punishments vary from place to place. In most cases however, first-time offenders are let off with a warning. The severity of the offence as well as the economic status of the offender are taken into account while levying fines. Offenders are first tried by the FPC. If the FPC is unable to arrive at a decision or if the offender wishes to appeal against its decision, the matter is taken to the court of the *panchayat sarpanch.* In the case of severe offences, the offenders may be handed over to the FD.

Relations with other organizations

Local Bodies: The lowest statutory local body is the *panchayat* which includes a number of villages. The *mukhiya* is the *panchayat* head. The *sarpanch* sees to legal matters and dispute resolution at the *panchayat* level. The *panchayat* heads do not exert much authority nowadays as the *panchayat* system is not functioning very effectively in the area. This is partly because the last *panchayat* elections were held over 18 years ago. The FPCs are autonomous of the control of the *panchayats.*

Non-governmental Organizations: A number of NGOs, such as Shramjivi Unnayan, Tata Steel Rural Development Society, Tribal and Harijan Welfare Cell, and Gram Vikas Kendra, have been crucial in encouraging and sustaining the functioning of FPCs in the area.

Forest Department: An atmosphere of distrust marks the relations between the villagers and the FD. Each holds the other party to be responsible for degradation of the areas' forests. While forest officials have recognized the recent contributions made by the FPCs in forest conservation, they have not been willing to accord them official recognition or registration, or coordinate forest conservation work with them. In Sari village, FPC members felt that they were doing a better job of protecting their forests than the FD had done and that they did not require the FD's assistance in this work. In Gobarghusi, in contrast, it was felt by the FPC that the assistance of the FD was very essential, especially for tackling outside timber poachers. In both these villages, it was reported that ever since the villagers had begun the work of forest protection, the Forest Guards had become lax in their work and had stopped visiting the areas regularly.

Conservation effectiveness

One of the biggest pressures on the DWS is that of illegal wood extraction, both for timber and fuelwood. Local requirements are met almost entirely from the DWS. No serious thought seems to have been given to searching for alternatives to resource use from the

Sanctuary. It is believed by the villagers that wood extraction for local consumption alone does not result in forest degradation. But far more serious is the issue of extraction for commercial purposes. Sale of firewood to Jamshedpur and the weekly village markets at Baram, Gobarghusi and Patamda, forms an important part of village economy as most families cannot produce enough by farming alone.

While extraction for this purpose is generally banned from an FPC-protected patch, there is no restriction on the cutting of wood from other parts of the forest. There are also many instances of wood extraction by the Jamshedpur-based timber mafia. The mafia is also involved in running a number of illegal liquor distilleries in the villages around Dalma. While the villagers find it difficult to face the powerful mafia head on, in many instances, in Pagda and Gobarghusi villages, for example, they have resorted to blocking or digging up of roads to stop the plying of timber trucks. Many villagers feel that the FD does not extend them enough assistance in such situations, and that even when offenders who have been caught are handed over to the FD, no action is taken against them.

Despite these problems, community-based protection over the past five years or so has resulted in the regeneration of *sal* forests in Sari, Nutandih, Punsa, Dalapani, Mirzadih and Kherua villages.

One development which has had an adverse consequence on forest conservation in the area has been a Bihar government notification of 1994, which allowed the felling of 10 species of trees, mostly fruit bearing, on private land, and their subsequent sale and transportation without a transit permit. This move, reportedly done at the behest of the World Bank, has resulted in large-scale cutting of species which play a very important part in the village economy.

As this cutting is mainly of trees on private land, and has been given legal sanction, the Committees have not taken any action. In fact, in Nutandih village, some of the Committee members have themselves been involved in this activity. In many villages, cutting of fruit trees has been reported from forest areas also.

It must be noted that the success of the Committees in dealing with these various pressures has not been uniform. Success has been contingent upon various factors:

Geographical factors: Remoteness of the area, such as in the case of Sari, has contributed to successful protection. Conversely, in the case of Punsa village, the proximity of the village to the city means the availability of employment, in turn meaning that dependence on the sale of firewood for livelihood is low.

Ecological factors: Community protection has been most successful in areas which have traditionally had a good *sal* forest in the past, such as Sari, Nutandih, Kuarama and Kherua.

Economic factors: The availability of fertile, well-irrigated land in Kherua and Nutandih has meant low dependence on sale of firewood for livelihood, helping in forest protection. In Bataluka and Amdapahari, the fact that most households depend on the sale of wood has led to the failure of community protection.

Political factors: The level of inter- and intra-village factionalization as well as the effectiveness of dispute resolution mechanisms, has affected the success of community conservation.

The FPCs have generally not tried to restrict either the Annual Hunt, as the event is a customary feature, or the occasional hunting for wild boar and birds for personal consumption. In fact, hunts are sometimes organized by the Committees as a means of patrolling the area they are protecting as well as symbolically asserting their authority over the area.

Issues regarding the functioning of forest committees

Lack of recognition: One of the most critical issues facing FPCs is the lack of official and statutory sanction for community forest protection activity. So far, the FD has not granted any form of official recognition to them. Even FPCs such as that of Kherua village, which is outside the Sanctuary, have not been recognized as official JFM committees. As they have no statutory authority, they find it difficult to take action against offenders, especially those from other areas, over whom it is impossible to enforce any form of social sanctions. It must be noted here that those FPC members who are aware of the Bihar state JFM notification find the guidelines for the constitution of village committees to be extremely restrictive. For example, the notification stipulates that the village committee must be headed by the village *mukhiya*; further, the local forester, schoolteacher and 'defeated *mukhiya*'—the person coming second in the *panchayat* elections for the *mukhiya*—as well as three other persons nominated by the *mukhiya* are to constitute the rest of the committee. Such stipulations do not allow much freedom of choice for the villagers in choosing their committee representatives.

Absence of benefit-sharing mechanisms: Under the existing system of wildlife laws in the country, the system of sharing of benefits such as timber and NTFP, as envisaged by the Bihar JFM resolution, cannot be practised inside a Sanctuary. This may not have been well understood by some of the local NGOs and even the FD. NGOs went ahead and promised villagers a one-third share in the profits earned from harvesting the forest area (inside the Sanctuary) protected by their village as an incentive. However, problems soon arose. FPCs began permitting a controlled harvest of wood from their protected patches, purely for local use. As this felling was against the Wild Life (Protection) Act, the FD took action against the villagers. In many villages, cases were registered against members of FPCs. With benefit-sharing in terms of harvesting of timber not being permitted inside the Sanctuary, many became disinterested in the ongoing protection work.

However, many villagers still hope that someday the FD will permit them to cut and sell some of the timber they have been protecting. It is essential that some incentive or assistance is provided by the Department to the FPCs in return for their work. The villagers desperately want their user rights to the areas they are protecting to be recognized.

Local pressures: It seems to be a common pattern here that while villagers protect their own patches of forest, to fulfill their resource demands they regularly encroach upon the areas claimed by other villages. Villagers generally say that they are powerless to deal with 'encroachers' from a larger village or one that is strategically located (e.g., near a town). Such conflicts also take place between *tolas* of the same village.

Local disputes: Village-level disputes often occur. These often take on an ethnic or communal tinge as villages often get factionalized along caste/tribal lines. In Dangdung and Bonta villages, such disputes have resulted in the disintegration of the FPCs. It has been suggested that an apex body, consisting of a number of FPCs, both at the *panchayat* level and at the Block level could serve as an effective dispute-resolution mechanism. This has been attempted recently in the Koira Panchayat, but it is too early to judge its effectiveness.

Possibilities for joint management

Considering the local initiatives described above, as well as the nature of the various pressures on Dalma Sanctuary, there is a need as well as significant scope for a more participatory approach towards management of the PA.

The villagers are keen to obtain some legal recognition from the FD. This, they feel, will guarantee them some rights over their forest patches as well as some statutory authority. While the FD has recently begun to recognize the contributions of the FPCs, there is still much disagreement over the nature of recognition and rights to be granted. The FD insists that due to the various provisions of the Wild Life (Protection) Act, 1972, it is not possible to grant any access to forest produce to the FPCs. The FD favours an eco-development approach; DWS is in fact one of the sites identified by the Wildlife Institute of India, Dehradun, to be covered by a World Bank sponsored forestry and eco-development project, and the Sanctuary authorities feel that this will help solve the problem of pressure due to livelihood issues. On the other hand, local villagers insist that any form of benefit-sharing must be based on forest resources. Any exclusion of the villagers from access to these resources will severely affect their lifestyle and economy.

One recent positive development has been the increase of interaction between the FD and the local people. This has been a result of a number of workshops and meetings held over the past year. The Department has also been conducting village surveys as part of the process of formulating a management plan for the Sanctuary, making it more familiar with the needs of the local villagers.

Towards joint management?

As part of its project, IIPA organized, along with the Tata Steel Rural Development Society, a workshop on 'Dalma Wildlife Sanctuary: Prospects for Conservation'. This workshop, conducted at Jamshedpur in August 1996, was attended by FPC and NGO representatives as well as senior FD officials. The participants issued a joint resolution (perhaps the first one for a PA in India in which both the FD and villagers signed), with the following recommendations pertaining to joint management:

- A suitable government notification should be passed for the purpose of granting recognition to FPCs operating in and around the Sanctuary, as Forest Protection and Eco-development Committees.
- Any work involving rural development or which is forest related should be given to these Committees.

- Any externally initiated projects being undertaken on village land or forest land being protected by village-level Committees, should be undertaken after securing the full agreement and participation of these Committees.
- Collection of any forest produce, which is not harmful to the forest or its wildlife, should be allowed to continue inside the Sanctuary. However, it must be ensured that the main beneficiaries of this extraction are the local people. The Committees must be paid the market rate for these forest products, and this revenue should go towards rural development work in the village.
- The cultural significance of the Annual Hunt should be recognized, and efforts made to make it less destructive by reviving traditional restraints on hunting.
- Full efforts should be made to find alternatives to the firewood demands of Jamshedpur, and to stop other external pressures on the Sanctuary, while finding alternative livelihood sources for villagers dependent on these.
- The management planning exercise for the Dalma Wildlife Sanctuary (DWS) should be fully consultative, involving all the affected villages and concerned NGOs.

The emphasis of the Committees' efforts are directed towards those species of immediate use to them only and not towards biodiversity in general. There is a need to involve the Committees in the protection of fauna as well.

Conclusion: Issues emerging from the two cases

Both Kailadevi and Dalma Sanctuaries offer several possibilities for moving ahead. FPCs in both have demonstrated the ability of people to organize themselves around a common issue, evolve complex mechanisms of control, and sustain themselves at least to some extent. However, villagers do reach their limits in handling external and some internal factors, for which they feel governmental intervention to be necessary.

The following are the broad issues that need to be addressed:

- Meeting people's livelihood requirements is crucial to the sustaining of conservation activities in PAs; relocation of so many people is neither desirable nor possible, but integration could be achieved with some innovative thinking.

- To achieve the above, villagers' own approaches towards conservation and resource use need to be understood and built upon, rather than displaced by alien, official approaches; in turn, villagers need to be made aware of outside concerns regarding biodiversity conservation.
- It is imperative that the FPCs be accorded official recognition and legal status, but not be bound down by rigid homogeneous rules which would stifle local innovation and flexibility.
- State intervention may be necessary in situations of intense inter- and intra-village conflicts and inequities.
- FD and villager cooperation is absolutely necessary to tackle external pressures on the resources of the PAs, with help from concerned NGOs.

Past and ongoing experience in these PAs suggests that there is considerable scope for some form of participatory or joint management in the future, but that this will require sustained dialogue and negotiation on the part of the potential partners, especially villagers and wildlife officials.

NOTES

1. A more detailed discussion on these two areas is in Das (1997) and Christopher (1997), and on the project as a whole (including research methodology), in Kothari et al. (1997). As a guiding principle, the definition used in the project for joint management of protected areas (JPAM) was: 'JPAM is the conceptualisation, planning, and management of protected areas and their surrounds, with the objective of conserving natural ecosystems and their wildlife, while ensuring the livelihood security of local traditional communities, through mechanisms which ensure a partnership between these communities, government agencies, and other concerned parties' (Kothari et al. 1996).
2. OBC (Other Backward Castes) and ST (Scheduled Tribes) are terms used in India to denote ethnic groups who have been given special constitutional privileges to overcome traditional disprivileges.
3. Subsequently, at a follow-up meeting organized by SSD and the villagers, a couple of middle-level officials of the KWS came briefly. Further attempts at strengthening the dialogue were continuing at the time of going to press.

REFERENCES

Champion, H.G. and **S.K. Seth.** 1968. *A Revised Survey of the Forest Types of India.* Manager of Publications, Government of India, New Delhi.

Christopher, K. 1997. Dalma Wildlife Sanctuary: Prospects for Joint Management, in A. Kothari, F. Vania, P. Das, K. Christopher and S. Jha (eds). *Building Bridges for*

Conservation: Towards Joint Management of Protected Areas in India. Indian Institute of Public Administration, New Delhi.

Das, P. 1997. Kailadevi Wildlife Sanctuary: Prospects for Joint Management, in A. Kothari, F. Vania, P. Das, K. Christopher and S. Jha (eds). *Building Bridges for Conservation: Towards Joint Management of Protected Areas in India*. Indian Institute of Public Administration, New Delhi.

Kothari, A., N. Singh and **S. Suri** (eds). 1996. *People and Protected Areas: Towards Participatory Conservation in India*. Sage Publications, New Delhi.

Kothari, A., F. Vania, P. Das, K. Christopher and **S. Jha** (eds). 1997. *Building Bridges for Conservation: Towards Joint Management of Protected Areas in India*. Indian Institute of Public Administration, New Delhi.

Rodgers, W.A. and **H.S. Panwar.** 1988. *Planning a Wildlife Protected Area Network in India*. Vol. 1 and 2. Wildlife Institute of India, Dehradun.

19

People's protected areas: Alwar district, India

Rajendra Singh

Profile of Alwar district

Location and climate

Alwar district is located in the northeast of Rajasthan, western India, between 27°15′ and 28°15′ N latitudes and 76°15′ and 77°0′ E longitudes. Temperatures of the area fluctuate from 0°C during some cold winter nights to as high as 49°C during the summer season. Mean relative humidity during the fall weather is 63 per cent while average annual rainfall is around 620 mm. Ninety per cent of the rainfall occurs during the monsoon months (July to September). Topographically, the Alwar tract may be divided into two zones: (*a*) hilly area comprising Thanagazi, Rajgarh, Bansur and parts of Mandawar, Behror and Alwar-Sadar tehsils and (*b*) the remaining parts having a more or less plain-like appearance with very small and low hill-like terraces or plateaus. The well-known Sariska Tiger Reserve encompasses about 866 sq. km area in Thanagazi, Rajgarh and Alwar-Sadar tehsils. The total area of district Alwar is 8,380 sq. km with a population of 2.3 million living in its 1991 villages and five towns.

Biodiversity of the district

Given the low rainfall and arid conditions, the forests are deciduous in nature, most trees having a shrub-like appearance with low height, small leaves and thorny branches and stems so typical of desert vegetation.

The biological diversity of district Alwar is one of the most significant in Rajasthan and perhaps in India, with several thousand species of flora and fauna found in the area. A number of biologically rich areas of the district, such as Sariska, have not yet been completely surveyed and explored. It is not only diversity which is significant, but its uniqueness too. Many species of plants as well as animals (especially reptiles), found here are believed to be endemic. Also, a great range of medicinal plants and shrubs are found in the district.

Present state of conservation

Biological diversity in the district is declining rapidly, due to a number of factors. Currently, at least 3 per cent of the recorded wild flora and a somewhat larger fraction of wild fauna, are on the threatened list. For example, among trees, the most important species *Chandan* (*Santalum album*) and *Kadamb* (*Anthocaphalus indicus*) are severely over-exploited; many species of mammals, such as the Tiger (*Panthera tigris*), Leopard (*Panthera pardus*), and Four-horned antelope (*Tetracerus quadricornis*) are threatened. And this should not surprise one, for in the last few decades, this region has lost about 70 per cent of its forests. This has happened due to abandoning of traditional practices of resource conservation and management. Policies and laws introduced and enforced by the government play a significant role in this regard, in particular those governing developmental activities. Floods, droughts, desertification, etc. also affect and degrade biological diversity. Habitat destruction through encroachment, hunting, wood stealing, over-exploitation, poisoning by pesticides and chemical fertilizers, excessive zoological and botanical collections, displacement of indigenous species by exotics, mining and quarrying have all contributed to this colossal destruction. Almost all government policies, rules, regulations and schemes end up working in a negative direction, especially at ground level. Today, even in traditional communities, village institutions and traditional systems of natural resource conservation and management have seriously deteriorated. New value systems (especially commercialization) and enforced state control over land have shoved communities aside, and away from management of natural resources and conservation of flora and fauna.

Since there is no adequate record of wild biodiversity in the past, no one can say for sure how many species we have already lost. There might be hundreds of species of wild flora and fauna which have gone forever, unrecorded because we do not even know that they existed. The process continues for a number of species and sub-species, as habitats which have scarcely been explored continue to be destroyed.

Ethnic features

Most of the ethnic groups and sub-groups of the Indo-Gangetic plains occur in this district also, to a lesser or greater extent. But two major communities, Gujjars and Meenas, are prominent in this tract. The hamlets of Gujjar tribes are found more in hilly regions because their main occupation is animal husbandry. Meena is a Scheduled Tribe of the area and is primarily engaged in agriculture, with animal husbandry as the secondary occupation. Gujjars are nature lovers and prefer wilderness for their habitat. The average landholding per household is between 1 and 1.5 hectares, while the number of cows and buffaloes ranges between 10 and 15 per household. Both Gujjars and Meenas are directly dependent on forest resources including for grasses, fuelwood, leaves, honey, etc. Both communities are known for their interest in conservation and their traditional skills, strength and self-discipline.

Government efforts at conservation

History

In princely states forests and other natural resources were managed by a complex mix of practices and traditions including hunting reserves. Formal official activities regarding natural resource management and conservation started only after independence. In 1956, soon after the formation of the present-day Rajasthan state, Sariska, which had already been declared as a *shikargah* (hunting reserve) by Maharaja Mangal Singh in 1885, was converted into a Sanctuary. Subsequently, Sariska became a Tiger Project area and a National Park in the 1970s, the latter under the Wild Life (Protection) Act of 1972. Thus the level of official efforts, laws, policies, activities, boundaries, etc. in the context of Sariska have changed from time to time. This has had severe implications for the villages inside and in the surrounding areas, as well as on other natural resources of Alwar district.

While there is a lot of talk of securing community support for, and participation in, ecological conservation, the current laws, government policies and official attitudes are incompatible and in fact opposed to public cooperation and are bound to fail. They ignore community interests and rather than evolving conservation programmes with community support, ask people to blindly support official programmes, which of course, does not happen.

Community reaction to government efforts

The official policies and efforts for conservation, which have given the state full control over forests and wildlife, have caused considerable conflict between people (especially forest dwellers) and the official machinery. These laws and policies are entirely against nature's law under which both humans and wild biodiversity are creations of one and the same 'God' and have equal right to share the common habitat. Government laws are based on distrust, and have promoted separation of the people from nature; the result is that people are losing their love for nature. In place of love, hostility is generated among the villagers, resulting in frequent clashes and conflict.

Another reason of conflict is the decreasing number of tigers in the forests, due to poaching by outsiders and loss of habitat, resulting in an increase in number of prey species like the Blue bull or *Neelgai*, deer and wild boar. Large herds of these herbivores damage agricultural crops of farmers around Sariska.

Traditional forest management and conservation in Rajasthan

As people living in arid and semi-arid areas, communities of Rajasthan were traditionally well aware of the importance of trees, forests and biodiversity and had a well-established and efficient forest management and conservation system appropriate for their region. The indigenous systems of management of forests in this area were evolved long before colonial times. Some practices traditionally adopted in Alwar region are briefly mentioned below:

- *Kakad Bani*—the forests on the common geographical boundary of two or three villages. The responsibility for protecting this area was shared between these villages. This area was used for controlled grazing and harvesting 'minor forest products', but no felling, logging or even lopping was allowed except by a joint decision.

- *Rakhat Bani*—the forest area that belonged to a single village independently. This was an exclusive common resource for this village and access was regulated by the village itself. This was essentially a reserve forest, basically a buffer area used only during periods of scarcity, famine and droughts.
- *Dev Bani* and *Oran*—forest patches dedicated to gods, or sacred groves attributed to the common deity of the villages. These were religiously protected.
- *Rundh* or *Balbani*—forest area protected by the king, *jagirdar* or *zamindar* (landlords) for the purpose of hunting. Access of others was highly restricted but not totally forbidden. Villagers were allowed to collect fuelwood and fodder in periods when there was no hunt.
- *Dharadi Partha*—a traditional system to conserve forests in which different castes and tribes were prohibited from cutting trees for use as fuelwood, and each different *gotra* (clan) was to protect particular species.

Stone monuments observed in dense forests established to demarcate the usage of that particular patch still reveal the old practices. Most of the classification, enforcement and management of the above was done by the communities themselves. Even when the king or his agent was involved, community opinion was given ample weightage. The systems were aimed at regulating extraction of forest products and systematizing intra- and inter-village utilization.

Culture and religion played a major role in developing, managing and protecting natural biodiversity and common resources. Later, as foreign and modern cultures diluted the respect for traditional practices, the indigenous and traditional natural resources management systems declined. Even after independence the Indian government continued to follow the forest policies initiated by the colonial regime prior to them. These policies further led to the breakdown of the traditional systems. For instance, in Bhaonta village, the disappearance of village forest boundaries in 1947 (after the government takeover of the surrounding forests) meant also that village controls broke down. Extraction of wood by the nearby urban settlement and by the Forest Department (FD) also led villagers to start cutting their own forests, some of which were traditionally not cut even in times of greatest need, 30–40 years ago.

TBS experience of community-based conservation

During the last few years, efforts were made by the NGO Tarun Bharat Sangh (TBS) to revive the past traditions and create environmental awareness to involve local communities in the area. Special efforts have been made for the conservation and management of natural resources.

Restoration of the traditional symbiotic relationship of forest dwellers with their surroundings and natural habitat is a continuous process where collective action towards conserving, regenerating and managing basic resources becomes the concern and pride of the local community. By the same process, the local community's conscience is awakened towards protection of wildlife against the constant threat from outside encroachers, commercial poachers, tree-fellers and migrant hunters.

Motivation for and the process of formulation of village *samitis* (committees) and *gram sabhas* (village councils) took place through extended discussion between TBS and the villagers, e.g., in three camps at Suratgarh (1985–87). This reintroduced the idea of common property and village self-reliance. Most often, all adult members of the village are members of the *sabha*.

Mobilization of community organization and effort often takes place over the question of water and its short supply. This area faces serious water shortage problems; thus *johads* (water tanks) were constructed in the village as a solution, with inputs from the local villagers themselves. In Gopalpura, the villagers assessed topography, helped design the checkdam system with local skills and traditional technology, and contributed one-fourth of the cost, aside from giving *shramdaan* (free labour).

A separate *mahila sangathan* (women's organization) was organized in Gopalpura as it was realized that though the women are more sober and long-sighted than the men, they are not included in the *gram sabha*. Each adult woman is made a member of the women's organization. This organization makes decisions by consensus and often before the *gram sabha* meets on an issue. Their decisions are then carried to the *sabha* through their menfolk, who have been encountered and convinced at home.

The rules laid down by the *gram sabha* in the forests surrounding the villages follow a basic pattern. Grazing is controlled, and takes place only in demarcated *gochars* (grazing lands). Graziers are not allowed to carry axes and in some villages are given the duty of guards. The cutting of green branches is regulated on the basis of either thickness or permission from the *gram sabha*. Hunting is prohibited to both villagers and outsiders. Fines/punishments are laid down, usually more severe for one neglecting to report a crime than for the defaulters themselves.

In Gopalpura and Suratgarh villages, the *gram sabha* was formulated keeping equity and self-management as the primary goals. All decisions are taken by consensus. Adherence to the rules has been ensured by *satyagraha* (peaceful resistance or struggle) and even social boycott. It is said that in the villages, these rules are accepted without any illwill because they are formulated by the village itself.

The successful construction of *johads* and the immediate results (more water in the dry season, improved soil moisture and plant growth) provided the confidence and the will to move on to the protection of forests. One village took the decision to re-green its hills to prevent excessive runoff. Other activities undertaken for conservation include a *parikrama* (circumambulation) of the forest in Bhaonta, as a message to conserve healthy trees and wildlife. The same village also declared the Bhairudev Public Wildlife Sanctuary on 1,200 hectares of regenerated forest.

These measures have shown encouraging results for forests and wildlife. Regeneration due to restrictions on lopping and grazing, increased soil moisture, and massive plantation (e.g., in Gopalpura) has resulted in green hills. The Bhairudev Sanctuary now has *Neelgai* (*Boselaphus tragocamelus*), Leopard (*Panthera pardus*), Jackal (*Canus aureus*) and Indian hare. Another result, e.g., in Gopalpura, is that though water is available for irrigation, cultivated area has not increased. This is because contact with TBS has impressed upon the villagers the need for forests and grazing land to remain as such. Hamirgarh village has also started using biological fertilizers instead of chemical ones.

The contention of TBS, that development of environmental consciousness amidst the people alone can bring about recognition of their responsibility in ecological conservation and management, has proved correct. Even the initial (during 1986–90) attitude of hostility on the part of government authorities (especially the forest staff), has given way to appreciation and endorsement, at first tacit (during 1991–92) and finally formal (during 1993 and later), of the positive role played by the non-formal, non-official grassroot-level bodies in the villages. Since 1993, 15 village committees (out of about 40 existing ones) inside the Sariska Tiger Reserve have been accorded the legal status of 'Forest Protection Committees' by the Field Director of Sariska.

Other forms of joint management are emerging. In Bhaonta village, offenders are sometimes brought to the *gram sabha* by the FD, when it finds itself unable to punish them. At Suratgarh, an official guard is stationed at a special request from the village, though there is no provision for an official post there. The guard is paid jointly by the village and the FD, and is treated as part of the village.

Within and around Sariska National Park, as also in other working areas of TBS, more than 100 forest protection committees (FPCs) have been established which form their own village rules for the conservation and management of forest and wildlife. Through their effective efforts, conservation and regeneration of forest and pasture lands is being continuously promoted. In this connection, the establishment of the Bhairudev Public Sanctuary by the people of village Bhaonta is an interesting and significant example of community-based conservation (CBC) of biodiversity (see Box 19.1). Revival of traditional practices as well as evolution of new ones is now well established in these villages. Unfortunately, efforts by TBS and the villagers to bring in a system of joint management of the Tiger Reserve

Box 19.1

A People's Wildlife Sanctuary

Bhaonta and Koylala,villages consisting of a few dozen households each in the Alwar district of Rajasthan, have set a remarkable example for the rest of India by establishing a 'public sanctuary' called Bhairudev 'Sonchiri' (bird of gold). Over the last few years they have protected a regenerating forest of 1,200 hectares, nestled in a valley at the head of which the villagers have built a checkdam for irrigation and drinking water purposes. Realizing the importance of protecting the catchment of this reservoir, and wishing to revive traditions of wildlife protection, the villages have set up a Forest Protection Committee and established strict rules regarding gazing, fuelwood collection, etc. in the forest area. A small temple dedicated to a new deity Bhartari Baba, seen as a protector of forests, has been constructed to emphasize the importance of conservation. Anyone caught violating the rules (regular patrolling is done by graziers who are allowed limited access to the area) is fined or socially ostracized; in the past, this has included one of the most powerful leaders of the village itself, who had transgressed limits by inviting his relative with livestock into the area for grazing. Hunting is strictly banned, and villagers are proud of the revival of herbivore and carnivore (leopard) populations.

as a whole, initiated during a workshop on the subject in Delhi (Sarkar et al. 1995), have not seen much positive response from the government so far.

TBS has been instrumental in the revival of the *lok samitis* (people's committees), which consist of at least one member of each community; and at least 70 per cent of households of the village are represented. However, this traditional system does not always conform with the village-level official institution (*panchayat*), a traditional system revived under the recent amendments in the Indian Constitution. This sometimes leads to a conflict situation.

The response of local communities to TBS's attempts at involving them in ecological conservation have mostly been positive. Campaigns and public meetings have played an important role in awareness building and improving the understanding of villagers about their roles, rights and responsibilities in conserving natural biodiversity.

Sometimes caste differences, or vested interests of a few corrupt officials and selfish villagers do appear as irritants, e.g., in the village Suratgarh, the higher caste (*brahmins*) do not cut trees themselves, but to meet their needs buy these from the lower-caste poor, for whom it is a source of income generation. Yet the higher caste who dominate the village FPC make rules regarding not cutting the wood from the forest, and thus these poor headloaders end up paying the fines imposed by the *gram sabha*. Another incidence is from the village Gopalpura, where the *sarpanch* of the *panchayat*, moneylenders and local politicians saw the *gram sabha* as a threat to their privileges, and managed to instigate a split in it. However, subsequently, they were won over through discussions, when it was explained to them that the smooth functioning of the *sabha* was to everyone's benefit. A second split due to trouble from the *sarpanch* was also dealt with in the same way. It is these irritants which create the need for the continuous presence of some committed individuals or an NGO in the area.

Often, the *mahila sangathans* also play an important role in the resolution of these conflicts, by keeping in check their menfolk, who are otherwise easily led astray by vested interests.

Another traditional conflict resolution institution that has been revived in many villages is that of the *thain*. In this system, a coordinator or *panch* is elected by the *sabha*. Decisions on matters of dispute are then made by this *panch* while seated on a stone slab called the *thain*, or the seat of justice. The decisions of the *thain* are accepted by all.

Constraints and opportunities for CBC

Rigid policy or law, snobbish behaviour of officials and the serious lack of mutual trust and understanding between local communities and government officials, result in seriously discouraging community efforts. For instance, Gopalpura villagers at the time of building their *johad* had sought technical assistance from the Block Development Officer, and the junior engineer in the Irrigation Department. The State Irrigation Department, however, responded by declaring this dam illegal, claiming that since all tanks, streams and drains are government property, construction of water harvesting structures on them is not allowed. Notice was issued to TBS to pull down the dam or face criminal charges. Only later, on witnessing the soundness of the structure and the commitment of the village, did they relent. The government also imposed a fine of Rs 5,000 on the village for planting trees and grasses on government land. Sariska National Park forest guards also began to extract a grazing fee after regeneration had taken place in Dewri village because of the villagers' efforts.

Local people often have to fight against the government itself to safeguard their forests. In 1989, a decision was taken by the government to settle some landless people on Gopalpura's forest land, with highly politically motivated intentions. This decision was strongly opposed by the local villagers. The district administration finally had to halt any further resettlement in that area.

Another instance which shows the attitude of the government is from Hamirpur village. A successful revival of its water sources, including a medium-sized *johad*, resulted in an increase in the population of fish in the mid-1990s. Suddenly, the Fisheries Department began showing interest, and without any consultation with the villagers, issued licences to fish contractors. When the residents of Hamirpur got to know, they angrily decided not to allow any fishing; contractors were stopped from entering the area, and Department officials were told to take back the licences. Several governmental threats followed, but the villagers remained firm till the Department had to cancel the permits (Patel 1997).

To conserve natural biodiversity, continuous dialogue and discussion at different levels between the government, NGOs, scientists (natural and social), journalists and the local community, are a must to develop understanding and trust.

In addition, caste/class/gender/other differences and inequities within and between communities need to be addressed. It has been noted in many areas that apart from the lower castes as mentioned above, the women very often get adversely affected by the conservation rules as they have to work harder to get fuelwood. But in these areas this situation is less problematic, especially where women work more with the livestock (which benefits from forest protection), and alternative fuel sources are available.

The recent nation-wide stress on JFM and CBC are encouraging signs. All one needs is a respect for the traditional system, a trust among various parties involved, the common good sense of the people, and a flexible approach to demands for significant changes in laws and policies of the country emphasizing communities' interests, rights, roles and responsibilities. It is worth mentioning here that, on a positive note, the Government of Rajasthan has decided to lease out land to village *panchayats* for regeneration.

Suggestions for the future

The modified approach to ecological conservation should include:

1. Intensive dialogues and discussion with local peoples including forest dwellers.

2. Intensive campaigns to revive environmental awareness amongst local people, and the belief that the forests, biological diversity, and other natural resources are their own and can help them satisfy their need to live happier lives.

3. Ensuring through education that whatever villagers take from natural resources, they reimburse the same in return, e.g., by protecting regenerating vegetation, and by additional plantation.

4. A prohibition on forced displacement of people from PAs, as direct working experience in these areas clearly shows such a move to be counter-productive. Where displacement has already taken place, optimum options for rehabilitation and prosperity be provided to the displaced communities.

5. Creating awareness, skills, proficiency and worthiness by conducting various training programmes.

6. Enabling policies to facilitate the full expression of people's knowledge, practices, institutions, and to empower them to manage resources. These policies and laws should also consider or legitimize people's own land use patterns.

7. Formation of *lok samitis* in each village and giving them legal status would be necessary.

8. Conversion of the role of the Forest and other related Departments to one of facilitators and reporters, rather than owners. Hundred per cent authority and decision-making must be with local communities. However, it must be mentioned that this is relevant only in the situations where it is clear that the community has the organizational ability to manage their resources and people are mobilized to conserve biodiversity, while fulfilling the basic needs of the community equitably. The examples of villages Bhaonta and Suratgarh, mentioned above, are instructive here.

9. Legally prohibiting activities like mining, quarrying, encroachment, industrialization, urbanization, poaching, and green-tree felling, and ensuring that the prohibition is implemented effectively.

10. Conceptualizing and planning the process of conserving and managing natural resources at the level of local people, not at the governmental level, to start with.

CBC is the only way likely to succeed in protecting natural resources and biodiversity. The approach to this shall need the trust of people, and decrease in the bureaucratic authority and snobbishness of officials. Once awareness and trust are generated, the local folk will contribute heartily to conservation through self-discipline and community decisions taken in *gram sabhas*. Of course, the responsibility for generating awareness, encouraging community organization and conducting training programmes, demonstrations and exposures will have to be shared by NGOs. The process of CBS will not only mean sharing of the benefits by the communities but also the responsibility to protect and manage the resources sustainably.

REFERENCES

Patel, Jashbhai. 1997. *Story of a Small Rivulet Arvari: From Death to Rebirth.* 2nd Edition. Tarun Bharat Sangh, Alwar.

Sarkar, S., N. Singh, S. Suri and **A. Kothari.** 1995. *Joint Management of Protected Areas in India: Report of a Workshop.* Indian Institute of Public Administration, New Delhi.

Traditional community conservation in the Indian Himalayas: Nanda Devi Biosphere Reserve*

R.K. Maikhuri, Sunil Nautiyal, K.S. Rao, K.G. Saxena and R.L. Semwal

Introduction

Emphasis on law enforcement to achieve the goal of conservation, ignoring the dependence of the local people on natural resources of the protected areas (PAs) for their subsistence needs, is aggravating the conflicts between indigenous people and reserve authorities all over the world, particularly in developing countries (Azeez et al. 1992; Johnsingh and Panwar 1992; Kiss 1990; Kothari et al. 1996; Misra et al., 1992; Schultz 1986; West and Brechin 1991).

Many studies have shown that an exclusive ecological perspective towards conservation may have detrimental impacts on the subsistence system (Grossman and Underwood 1971). It is being increasingly realized that the success of PAs depends upon the extent of support from local people towards such establishments (Heinen 1993; McNeely and Miller 1984; UNESCO 1995). The importance of integrating PA management with rural development schemes is also being increasingly recognized, as is the importance of including people in the management process (Bandaratillake 1992; Dasmann 1984; McNeely 1984; Ramakrishnan 1992; Sharma 1990). PA planning should consider cultural, political, socio-economic and ecological issues in a holistic perspective. The old concept of shielding

* The authors are very grateful to the Director, G.B. Pant Institute of Himalayan Environment and Development for providing facilities, and the help of MacArthur Foundation/UNESCO for the financial support. The help of Mr R.P. Sati for typing the manuscript is gratefully acknowledged.

parks from outside human influences must gradually evolve to adopt changing socio-economic realities while still fulfilling the primary objective of nature conservation (McNeely 1984). Efforts should be made to educate managers and policy-makers to discard the fortress mentality of reserve protection, the misconception that setting aside areas is the best approach for conservation.

An integrated study in the Nanda Devi Biosphere Reserve (NDBR) was undertaken because of the complexity of management conflicts and unique biological wealth and ecological fragility of the area. The broad goal of the study was to develop and test sustainable natural resource management models through peoples' participation in NDBR for achieving the objective of conservation. The Forest Panchayat system and the Chipko movement, the people's initiatives in forest conservation and people-conservation policy conflicts are discussed in this paper.

Profile of Nanda Devi Biosphere Reserve

The NDBR comprises a unique combination of ecosystems including mixed temperate forests, alpine meadows, a number of high mountain peaks and glaciers. The Reserve, located in the northern part of the western Himalayas, consists of a central core zone (624.62 sq. km), surrounded by a buffer zone (1,612.12 sq. km). The core zone has the legal status of a National Park since 1982. The Reserve includes Reserved Forests, Civil Forests, Panchayat Forests and farmland. A total of 17 villages are situated in the buffer zone, of which 10 villages fall in Garhwal (district Chamoli) and seven villages in Kumaon (districts Pithoragarh and Almora). Two villages in district Pithoragarh and two in district Almora are presently uninhabited. There are 2,580 inhabitants in the buffer zone of NDBR. In the study area (i.e., in the Chamoli district part), the population is 2,253. From the geomorphological point of view, the buffer zone occupies the whole Rishi Ganga catchment (a tributary of the Dauli Ganga) which is encircled by higher Himalayan peaks, among which is India's second highest peak, Nanda Devi, which flanks the northern part (Map 20.1).

Biological significance of the area

The NDBR region is one of the most biologically diverse areas of the western Himalaya. The variation in altitude, topographical and climatic conditions, has resulted in a variety of vegetation types. The

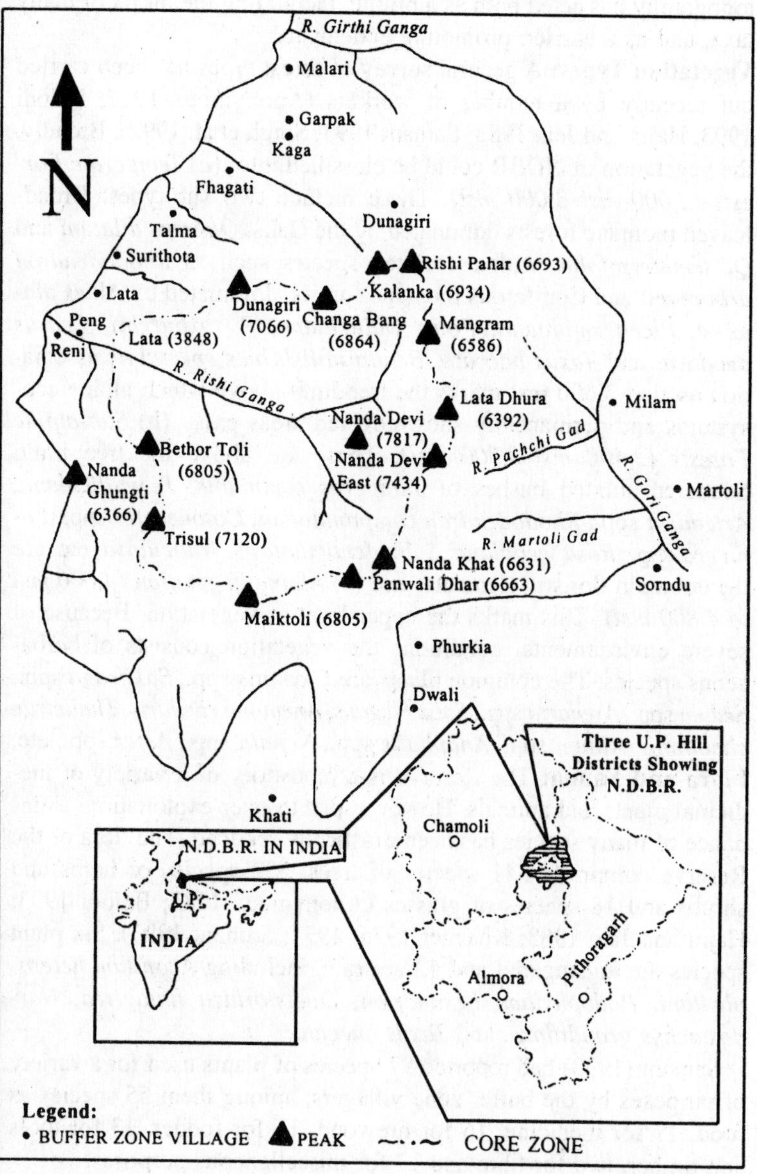

Map 20.1: Nanda Devi Biosphere Reserve with core and buffer zone

topography has acted both as a bridge, facilitating the influx of many taxa, and as a barrier, promoting endemism.

Vegetation Types: A general survey of forest types has been carried out recently by a number of workers (Anonymous 1987; Balodi 1993; Hajra and Jain 1983; Samant 1993; Singh et al. 1992). Broadly, the vegetation of NDBR could be classified into: (*a*) *Temperate Forests (2,000 msl–3,000 msl)*. These include two sub-types: Broad-leaved montane forests dominated by the Oaks, *Quercus dilatata* and *Q. semicarpifolia*, with associated species such as *Rhododendron arboreum*; and Coniferous montane forests, dominated by *Abies pindrow*, *Picea smithiana*, *Pinus wallichiana*, *P. roxburghii*, *Cedrus deodara*, and *Taxus baccata*. *Betula utilis/Abies spectablis* association around 3,600 msl marks the tree limit above which alpine ecosystems and permanently snow-covered areas exist. (*b*) *Sub-alpine Forests (3,400 msl–3,500 msl)*. These are above the tree limit. Scattered stunted bushes of *Juniperus communis*, *J. wallichiana*, *Artemisia* spp., *Rhododendron campanulatum*, *Cotoneaster* spp., *Lonicera* spp., *Rosa webbiana*, *Salix denticulata*, *S. fruticulosa*, etc. are the common floristic elements; and (*c*) *Alpine Vegetation (4,000 msl to 4,800 msl)*. This marks the upper limit of vegetation. Because of severe environmental conditions the vegetation consists of herbaceous species. The common plants are *Clematis* spp., *Saxifraga* spp., *Sedum* spp., *Arenaria* spp., *Salix elegans*, *Anemone rivularis*, *Thalictrun chelidonii*, *Allium* spp., *Anaphalis* spp., *Nepata* spp., *Aster* spp., etc.

Flora and Fauna: The Reserve is a repository of a variety of medicinal plants and animals. However, due to over-exploitation, abundance of many species has been drastically reduced. The flora of the Reserve comprises 341 species of trees, 552 species of herbs and shrubs and 18 species of grasses (Anonymous 1987; Balodi 1993; Hajra and Jain 1983; Khacher 1976; 1977; Samant 1993). Six plant species are endangered and 12 are rare, including *Aconitum heterophyllum*, *Podophyllum hexandrum*, *Dactylorhiza hatagirea*, *Nardostachys grandiflora*, and *Taxus buccata*.

Samant (1993) had reported 97 species of plants used for a variety of purposes by the buffer zone villagers; among them 55 species as food, 17 for medicine, 16 for firewood, 15 for fodder, 13 for tools and timber, two for fibre and 17 for miscellaneous purposes.

The occurrence of very rare fauna such as the Snow leopard (*Panthera uncia*) and Brown bear (*Ursus arctos isabellinus*), Blue sheep (*Pseudois nayaur*), Himalayan tahr (*Hemitragus jemlahicus*), Musk deer (*Moschus chrysogaster*), Monal pheasant (*Lophophorus impejanus*), Himalayan snowcock (*Tetraogallus himalayensis*) and Snow patridge (*Lerwa lerwa*) etc., makes NDBR an area of immense conservation value. Eighty-six species of mammals, 534 species of birds, and 54 species of reptiles and amphibians are reported from NDBR (Anonymous 1987; Lamba et al. 1981; Lamba 1987; Reed 1978; Sathya Kumar 1993; Tak and Kumar 1983). Of these, seven mammals and eight birds are endangered (Lamba et al. 1981).

Apart from biological wealth, NDBR is also critical as a large sub-catchment of river Ganga, which sustains the life of a dense population in the north Indian plains, and as the birthplace of the Chipko movement and a number of people's institutions for forest management. It also has great potential for tourism which could help alleviate the economic condition of the local people.

Forest categories in NDBR

Three kinds of forests exist in the area: (*a*) Civil/Community forests, developed by the Forest Department (FD) on community lands. These forests remain under the direct control of the village communities, which enjoy extensive rights to utilize their resources without needing permission from the FD. Forest cover in this type of forest remains very poor; (*b*) Van Panchayat Forests, carved out from the Reserved Forests and local communities through the leadership of the *sarpanch* and *van panchayat* members. The Sub-Divisional Magistrate supervises them. These parties collectively look after the conservation and management of these forests. Forest resources can be taken from Van Panchayat Forests (VPFs) only after a resolution in this regard, which is sent to the FD for final permission. Generally, permission is granted but the beneficiaries have to pay a defined price for the resource to the *van panchayat*. Generally, the forest cover remains fairly good in these forests; (*c*) Reserved forests under the control of the FD. Resource extraction of any kind is not permitted without prior permission. In certain cases, restricted permission is given for collection to the local community, called the *ramana* system. This system is usually operational on a time-bound basis.

NDBR: History of its conservation

Almost all the reserve forests within the NDBR fall in the Chamoli district. All the areas of the NDBR falling in the Almora and

Pithoragarh districts are civil areas outside the limit of the Reserved Forest. These areas came under the control of the British after they defeated the Gurkhas in the year 1815. However, government controls on the forest resources could commence only in the last two decades of the 19th century. In 1911, the reservation process of these forests began, and in 1912, technical management was brought under the purview of the Conservator of Forests, Kumaon circle (Mohan and Bhadauria 1993).

The scientific management of forests here came into existence in 1960. Fir, Spruce and Kail working circles were demarcated during 1973–83. However, timber exploitation on a large scale could not succeed because of the famous Chipko movement which took off from village Reni, in which villagers successfully protested against tree-felling by contractors.

The NDBR also has a long history of wildlife conservation. This area was first approached by the famous English mountaineers Eric Shipton and Tilman in 1934, who explored the difficult sage route to Nanda Devi peak. They were surprised to see herds of Blue sheep, locally known as *Bharal* (*Pseudois nayaur*), gazing as if they had never seen humans before. Realizing the wildlife value of this pristine area it was declared a Wildlife Sanctuary in 1939. The post-independence era saw a rush of mountaineers into the catchment in pursuit of high peaks like Nanda Devi, Trishul, Dunagiri, etc. This led to serious damage and destruction of both flora and fauna of the area, which forced the government to declare the whole catchment a National Park in 1982. Entry in the Park was banned except for the purpose of ecological research, patrolling staff and local inhabitants. On 18 January 1988, taking a cue from UNESCO's Man and Biosphere (MAB) programme, Nanda Devi National Park (NDNP) was given the status of NDBR. In 1992, NDBR got the recognition of a World Heritage Site.

Socio-economic profile of the buffer zone of NDBR

The larger buffer zone of NDBR contains several human settlements. The people in these villages belong to two ethnic groups, viz., Indo-Mongoloid (Bhotiya tribals) and Indo-Aryan. The Bhotiya people living in Chamoli district belong to Tolchha sub-community whereas those living in the Pithoragarh district belong to Marchha sub-community. People living in Almora district are largely non-tribal.

Except the residents of Reni and Peng villages, all Tolchha Bhotiya households have two permanent dwellings—one at high altitudes between 2,400–3,500 msl and the other in the lower valleys between 800–1,500 msl. They stay in high altitudes during summers and in low altitudes during winters. The migration from one dwelling to the other usually takes six–20 days. With the advent of modern transportation facilities, some members, usually the old persons, travel by bus or truck and cover the distance in a day, while the majority walk the distance with livestock.

Agriculture: In the 2,000–3,500 msl zone, rainfed cultivation on terraced slopes is the common land use. Almost all households are involved in agriculture, but members of 11 per cent of all households are also involved in business, 31 per cent in services and the remainder either work for others or are daily wage labourers. The farming system of this region is representative of the so-called solar power agro-ecosystem. The system consists of four different interlinked components: support area or forest land, cultivated land, livestock and humans. Major agricultural operations/activities are generally performed by the women folk. Bullocks are used for drought power.

The farmers are subsistence farmers. Most of their crops are prospective cash crops. They could fetch considerable income if sold in the market down in the plains. However, farm produce in excess of the household needs is marketed locally in monetary terms or bartered for commodities like rice, sugar and salt, not produced locally. People thus consider local-production-based food security more important than a cash-market-based food security system.

Livestock population: Altogether 1,105 cattle, 2,066 sheep, 2,857 goats, 82 horses, 19 mules and 32 poultry birds were reared by the people in the buffer zone. The average number of animals per family was 18.6. After the establishment of the National Park, livestock population, particularly that of sheep and goats, declined substantially due to the drastic reduction in the alpine pasture area accessible for grazing.

Forest: All the households of the buffer zone villages depend entirely on forests for fuel, fodder, timber and organic leaf manure. Wild resources of plant and animal origin make a significant contribution to food security. Wild edible fruits, seeds and leaves often provide food during critical periods when staple foodgrains are in short supply. Many plant species are used in the traditional healthcare system. Temperate bamboos support the handicraft cottage industry.

Non-timber forest products not consumed domestically are marketed for cash. In some cases money earned from wild plant products is the principle source of cash in the household, but in others it is merely a bonus (see Table 20.1).

Major threats to biodiversity

Forests and alpine meadows of the region provide subsistence to the local inhabitants. Traditional resource use and management systems aimed for sustainable supply of natural resources in the geographically isolated and ecologically fragile area. In recent times, improvement in accessibility through road construction by the government has brought cultural changes and penetration of market forces and monetary considerations; and consequently commercial exploitation of the natural resource base.

Segregation of Panchayat Forests, and the use of the government-owned Reserved and Civil Forests for economic and ecological benefits to a wider national community, have resulted in the following adverse effects: (*a*) alienation of local communities from government-owned forest land; (*b*) unsustainable exploitation of government forests by outsiders whose prime objective was to maximize profits rather than to maintain sustainable yields; (*c*) inability of government agencies to ensure desirable balance between exploitation and regeneration, because of the vast area, dissected terrain, inaccessibility, and limited manpower and financial resources.

The region has gradually become a supplier of timber and NTFP to pharmaceutical, cosmetic and timber industries. Grazing pressure in government forests and pastures has increased to levels beyond the carrying capacity because of an influx of livestock into the buffer zone from outside villages. This has rendered some parts of the catchment prone to top soil loss and landslides. Ineffective administrative controls and management practices paved the way for unsustainable extraction activities by outsiders on the one hand, and promoted a sense of resentment among the local inhabitants on the other. There are instances when the local people themselves use government forests unsustainably. This has resulted in the decline of the biological resource base.

Although the commercial exploitation of NTFP from the area has been banned since 1982 (when it was declared a National Park), yet

Table 20.1: Important medicinal plants (along with value and uses) collected by the villagers in the buffer of NDBR

Scientific Name	*Local Name*	*Part Used*	*Cure for*	*Rs per Kg*	*Total Quantity Collected (kg or litre per year)*	*Monetary Equivalent (Rs)*
Dactylorrhiza natagirea	Hathazari	Tuber	Fever, wounds, cuts	450.00	24.00	10800.00
Saussurea costus	Kut	Tuber	Stomach pain, fever	25.00	324.00	8100.00
Picrorhiza kurrooa	Katuki	Root	Jaundice, stomach pain, fever, dysentry, dyspepsia	25.00	73.00	1825.00
Nardostachys grandiflora	Jatamashi	Root	Rheumatism, burn	30.00	30.00	900.00
Pleurospermum angelicord	Choru	Root	Fever, stomach pain, dysentry	25.00	711.00	1775.00
Angelica glauca	Chippi	Root	Rheumatism, burn, joint pain	25.00	542.00	13550.00
Aconitum heterophyllum	Atis	Root, Tuber	Stomach pain, diarrhoea, dyspepsia	550.00	73.00	40150.00
Morchella esculenta	Guchhi	Fruiting body	Cold and cough	1800.00	904.00	16272.00
Paeonoia emodi	Chandra	Leaves	Dysentry, blood purifier	15.00	3075.00	40125.00
Megacarpia polyandra	Barmoa	Leaves	Stomach pain	15.00	8195.00	122025.00
Bergenia ligulata	Shilpori	Leaves	Cold and cough	60.00	66.00	3960.00
Smilacina purpurea	Puyanu	Leaves	Dysentry	15.00	4596.00	68940.00
Cedrus deodara	Devdar	Wood	Skin disease, wounds and cuts	45.00	548.00	24660.00
Taxus baccata	Thuner	Bark	Cold and cough	35.00	2077.00	72625.00
Betula utilis	Bhojpatra	Resin	Cold and cough	45.00	97.00	4365.00
Prunus persica	Kirol	Seed kernels	Joint pain, cold and cough	40.00	392.00	15680.00
Relium australe	Dholu	Whole plant	Boils, wounds and cuts	30.00	38.00	1140.00
	Doom	Leaves	Dysentry	35.00	10932.00	382620.00

the extraction continues because of practical difficulties in enforcing the ban. A similar situation exists in the case of poaching of wildlife. A huge number of Nepalese labourers depend upon wage earning in the area, to secure their livelihood. These people, because of their familiarity with mountain resources, are being hired by contractors for illegal commercial exploitation of resources in government forests. The advantage of engaging Nepalese labour is that the reserve officials hesitate to institute legal proceedings for punitive action against these foreign nationals. It is also practically impossible in legal terms to identify the contractors as the prime culprits.

Probably the greatest threat to the region's biodiversity lies in the indifferent attitude of government officials charged with the responsibility of protecting the environment and resources. While the local communities have taken an active role in managing the Van Panchayat Forests, they are indifferent towards the government forests because they do not have any legal responsibility for conservation of public resources.

Institutional structures for conservation

The Uttar Pradesh hills have a longer history of officially sanctioned local peoples' participation in forest management than any other part of the country. The *van panchayat* system is a village-level institution and it has considerable potential for involving local communities in forest management and conservation. There are about 4,804 *van panchayats* in the U.P. hills covering an area of about 2,44,800 hectares (Saxena 1995). Some *van panchayats* have a good understanding among their members and are functioning well, but many are not functioning.

Van panchayat and its functioning in the buffer zone villages of NDBR

Presently there are two local institutions in the buffer zone, the *van panchayat* responsible for local management of forests, and the *gram sabha* responsible for all other rural development works. *Van panchayats* headed by the *sarpanch* cater to more than one village, while the *gram sabha* headed by the *pradhan* caters to only one village. The *gram sabha* essentially functions under the Block Development Agency, over which the District Magistrate has overall control. Panchayat Forests are the areas where land rights vest with the Revenue Department, but resource use rights vest with the peoples' institution. People are empowered to decide resource use practices within the framework of the forest policy. Technical support to people in the management of forests is the responsibility of the FD.

A notification to declare an area a Panchayat Forest is issued by the Commissioner when the villagers ask for it. The location and area of the Panchayat Forest are decided by the Commissioner in consultation with the FD. There is practically no say of villagers in decisions on which location or what extent of area is notified. Following notification of Panchayat Forests, the village community elects members of the *panchayat* (five–10 members) and the chairman is called *sarpanch*.

The *van panchayat*'s income is derived from grazing fees, fuelwood collection fees, and other income from the forest resources. It is empowered to decide on subsistence use of the forest resource but has virtually no powers on the commercial exploitation of forest resources. The *van panchayat* acts as a forwarding agency for the proposals of commercial exploitation for grant of permission by government agencies. Because of distances and costly communication means, this commercial exploitation often proceeds illegally by a select few powerful individuals. The powers of *van panchayat* to penalize an individual caught defying the rules are limited (maximum fine imposed is Rs 500), whereas income from illicit extraction is enormous. Further, whatever income is generated by the *van panchayat*, is deposited in the government account and this income can be utilized only with the permission of revenue officials. Because of practical difficulties in approaching the government officials, the *panchayat* often adopts unofficial ways of income generation and expenditure.

Membership of *gram sabha* is more prestigious and lucrative than that of the *van panchayat* because the former has a strong political influence and receives development grants from the government. Separate institutions for forest management, i.e., *van panchayat* and *gram sabha* do not seem to be an appropriate proposition when one looks at the inter-dependency of forests and rural development in these landscapes. A single institution for a given village could perhaps be a better institutional arrangement.

In the Chamoli district part of NDBR, three *van panchayats* (Tolma part I & part II, Lata–Reni and Peng) and five *gram sabhas*

(Tolma, Lata, Reni, Dunagiri and Malari) exist. The area under *van panchayats* is not clearly known. According to *panchayat* authorities the total area of three Panchayat Forests is about 4,737.7 hectares, while as per the map used by the NDBR authority the total area comes to around 9,351.4 hectares. Out of the total forest area of the buffer zone (34,384 hectares), 13.75 per cent of the forest is under *van panchayats*, only 1.1 per cent under civil/community forests, and 85.15 per cent under reserved forests.

As per the *sarpanch* of Lata village, the *van panchayat* concept was established during 1964 in this area. All these *van panchayats* have full control over protection of the forests and have been utilizing resources on a sustained basis. They have large numbers of huge wind-blown dead logs, lying in Panchayat Forests and are demanding their removal to earn revenue for the *panchayat*, but are unable to get approval from government authorities. All *van panchayats* have appointed watchmen, but as the area is quite large the watchmen are only able to protect or keep vigilance on the accessible areas. Since most of the settlements receive heavy snowfall during the winters, collection of fodder and fuel is allowed without any restrictions during summers and the rainy season. The *van panchayats* have limited funds in their account. However, sometimes, they utilize some funds for plantation and other community works also.

The equity in distribution of benefits is maintained in all the three *van panchayats* studied. All the members of *van panchayats* enjoy equal status, access and usufruct rights. There is seldom any discrimination in forest utilization based on socio-economic status. Nevertheless, the nature and quantum of resource use differs among households due to wide variations in need, ability to utilize, accessibility, and forest quality determined by aspect, altitude and soil type. Lata–Reni *van panchayat* of the three above mentioned *panchayats* has three women members.

Traditional methods of resource use such as allowing people in groups and social sanctions allowing for a higher level of bamboo use by weaker sections of the society created a consciousness for sustainable use and equity.

Afforestation in degraded areas carried out by different government departments (Civil Soyam, Soil Conservation, Block Office and FDs) have largely failed due to extreme ecological stress. Lack of people's participation in decision-making on the choice of plantation species, and in soil and water management practices, is another factor contributing to the failure of these plantations. People participate in these programmes for wages and not with any aspiration for improvement in forest resources. Government officials are also not committed because their work efficiency is evaluated in terms of physical and financial targets instead of the survival and growth of the plantation. Such programmes cause people to become disillusioned and disinterested in the *van panchayat* because the majority feel that their elected representatives are, in one way or the other, involved with such unproductive activities. Often the members (*panchs*) behave indifferently because they think that the *sarpanch* connives with the development officials for instituting such programmes. Consequently, there has been a decline in the frequency of *van panchayat* meetings in the last few years, particularly in Tolma and Peng *van panchayats*. The electoral processes in *van panchayat* systems are uncertain. Elections have not been conducted for the last 10–12 years in Lata-Reni and Peng *panchayats*.

The problems in managing Van Panchayat Forests of NDBR could be summarized as:

- Insufficient staff to manage/keep regular watch on vast inaccessible areas.
- Status of *van panchayat* being subordinate to the status of *gram sabha*.
- Poor/no budgetary support from the government to Panchayat Forests.
- Procedural complexity in commercial exploitation of Panchayat Forests.
- Lack of cooperation and coordination among the members of *van panchayats*, government officials and the people.
- Low level of formal education of elected representatives.
- Lack of training support to create awareness among the villagers about environment and resource management.

Chipko movement and its impact on the buffer zone of NDBR

In the initial stages, commercial exploitation of forests for timber by government agencies or their agents did not attract the attention of the local people because indigenous resource uses were largely based on NTFP and people enjoyed unrestricted customary rights of NTFP collection. Gradually, people realized that economic exploitation by the government essentially means marginalization of their long-term economic interests and socio-cultural disintegration.

People of Reni village, Chamoli district, were amongst the first to realize this, and decided to oppose such exploitation. Women were at the forefront of this move, being the first to feel the impact of the reckless deforestation. The government had auctioned several forests in the late 1970s, and humble informal pleas from the villagers to stop the auction fell on deaf ears. One of the forests was close to the village of Reni in the ecologically fragile catchment area of Alaknanda river. In order to save the forest, villagers organized themselves to oppose the fellings, and some political parties supported them. The government, angered by opposition to the proposed fellings (a source of revenue), tried subterfuge. On the day of the proposed felling, they called the prominent activists to a meeting and offered compensation to affected villagers in far-away Chamoli. With most of the menfolk gone, the contractors assumed that felling would go smoothly. However, the women of the village, led by the late Gaura Devi, confronted the lumbermen, threatening to hug the trees if they attempted to cut them (hence the term Chipko, which in Hindi means 'to hug'). Despite threats and rude behaviour the women refused to budge, and prevented the fellings. Other important forests saved by Chipko activists include Advani, Salet, Loital and Malgadi in the same region (Bhatt 1985; Tiwari and Ravindran 1993).

Thus, what began as a protest against government policies became a highly successful grassroots movement that achieved its objectives through non-violent and peaceful protests. Chipko raised questions against the government's developmental policies, like spending huge amounts on constructing roads, setting up industries, etc., at the expense of forests. The Chipko movement was the first of its kind in independent India. It helped create environmental awareness throughout India and forced the government to address environmental problems seriously, taking into consideration the people–forest dependency. It protested against policies which favoured the urban minority. It was truly an indigenous environmental movement (Jain 1984). When the rest of the world was just beginning to think about the role of the environment in the economy, the people of Garhwal Himalaya were actually fighting a war against the FD for protection of their forests, culture and regional economic independence.

Chipko helped initiate grassroots activity particularly among women of this region. Today women of the buffer zone of NDBR are well aware of the value of forests and are actively taking part in protecting forests. They do not hesitate to protect trees even if such actions incur the displeasure of male members of their own families. Today there are many women organizations in the region known as Mahila Mangal Dals that protect women's interests in a male dominated society (Tiwari and Ravindran 1993).

Biosphere Reserve and people conflicts

The main issue of conflict in NDNP and Biosphere Reserve is over the right of the people to use the forest resources which they had traditionally been using. Declaration of the area as a National Park has led to the curtailment of peoples' rights in the area. The government authorities have not provided any alternate livelihood means, leading to increase in the people–government conflict.

Conflicts because of livestock grazing

In the past, the communities of the area were mostly on the move, trading with Tibetans during the summers, and with lowland people in the *terai* region during the winters. Grazing pressure was diffused. Alpine pastures were used during the summer season and the *terai* forests during the winter season. They took their livestock along as they went about their strenuous journey. This had two positive impacts. First, the damaging effects on natural resources were limited to narrow corridors which formed the main routes. Second, their constant movement allowed sufficient time for resource regeneration through natural processes.

The creation of NDNP in 1982 and subsequently NDBR in 1988 created problems for the local people, because they had to reduce the livestock holdings for want of adequate grazing area. Alpine pastures were denied access to the people in the name of conservation. People in Lata, Tolma, Reni and Peng villages were most affected because all their grazing grounds fall in the core zone where entry is strictly prohibited. For this reason, they were forced to graze their livestock in the alpine meadows of other villages, and therefore have to pay grazing tax (Rs 10 per sheep and Rs 25 per cattle and horse). Considerable expenditure is incurred on arranging *Anwals*, a district social group who earn their livelihood by taking care of the livestock of one village which is grazing in another village. *Terai* forest grazing lands traditionally used by people during the winter season, have in most cases been closed. The shepherds have to obtain a temporary grazing permit annually, from the state capital Lucknow which is approximately 500 km away. Instances of theft of the sheep and goats are getting more and more frequent in the *terai* region. The overall result is that, on the one hand, reduced availability of grazing land causes overgrazing in available pastures, and on the other hand, local people are reducing their livestock holding because of grazing problems (e.g., the sheep and goat population has declined by two-thirds each). This in turn has resulted in a reduction of wool production and wool-based traditional handicraft.

Ban on NTFP extraction

Before the creation of NDNP, residents of the buffer zone villages could earn by way of marketing surplus medicinal and aromatic products. Once in every five–six years (compartment-wise) the FD used to issue permits to the contractors through Bhesaj Sangh (a government institution for medicinal plants) to collect the medicinal plants. Now, with the exception of mushrooms, the collection of medicinal and aromatic plant products is strictly prohibited. An important source of income has thus been blocked.

Ban on expedition/mountaineering on the Nanda Devi and other peaks

Over 90 per cent of the young men of Reni, Lata, Tolma and Peng villages used to work as tour guides for expeditions to high peaks such as Nanda Devi and Trishul till the early 1980s. After successful completion of expeditions, foreigners used to gift many household goods to the locals, apart from the wage payments. The per capita income per year from tourism during the 1960s and 1970s in the above-mentioned four buffer zone villages, which had the advantage of being located at the trek start point, was estimated to be about Rs 2,014.00, and about Rs 1,455.00 as average for all the buffer zone villages. Despite inhabiting remote and far flung hills, people did not feel isolated from the outside world because of regular visits of nature lovers to the area. A ban on tourism to the core zone has eliminated an important source of income for the people. The locals have been showing strong resentment and pressurizing the government to withdraw the ban but have not succeeded so far.

Ban on removal of dead logs from the Van Panchayat Forest

A large number of huge dead logs are lying in these forests. *Van panchayats* have applied for permission to remove these logs to earn revenue for strengthening the institution and for other development activities. Since the permission has not been granted, a feeling that conservation policies are anti-people has been generated.

Damage by wildlife

Bee-keeping is one of the oldest traditional activities in all the villages. The intensity of activity measured in terms of number of hives maintained per household is high in Tolma, Peng, Lata, Reni and Phagti, as compared to other villages. But damage is done by wildlife, particularly the bear, to the hives during the night. Sometimes in search of beehives, bears damage houses also.

It was reported by the locals of the buffer zone villages that every year a large number of livestock (a total of 875 heads in the period 1988–96) are killed by wild animals. The highest number of reported losses include sheep and goats. Large animals, i.e., horses, sheep and goats are killed by leopards, birds (chickens) by small cats, and calves by jackals and foxes. The management plan of NDBR has a provision for a fair compensation for such killings. In the beginning when the National Park/NDBR was created, disbursement of compensation was a quick and fair affair as per the people's experiences. Gradually, people felt that it became difficult for them to get the claims settled quickly and fairly. Limitations of the budget for the purpose, and a tendency of putting up fake claims by the people are serious drawbacks in the claim settlement process. It could also be possible that PA managers, in order to earn credibility, took interest in the initial stages of establishment and subsequently started adopting a callous attitude. Of course before creation of Nanda Devi Biosphere or National Park, killing of livestock did occur but since the Wild Life (Protection) Act was not in force, locals generally did not hesitate to kill the wildlife through collective efforts.

Crops, fruits and horticultural plants damaged by wildlife

Crop damage by wildlife (ungulates, monkeys, wild boar and bear) has badly affected the subsistence economy of the people. In three villages (Lata, Peng, Tolma), the annual loss was worth Rs 25,000 each. Often people are forced to reduce cropping intensity and even abandon agriculture. Dominant horticulture crops, apple and peach are also damaged to a great extent. It was also reported that huge manpower is wasted every year for protection efforts.

Instructions are issued to the contractors at the time of licence distribution by the Bhesaj Sangh for employing only locals who are likely to be more committed and conscious about sustainable levels of extraction. However, the motive of maximization of profit in contractors leads to employment of Nepalese labourers who work on cheaper wages and exploit the resource base to a higher intensity. Locals of the buffer zone villages complain that despite plenty of forest-based employment opportunities, they are being deprived of these by the government machinery. At present neither Bhesaj Sangh nor local people are allowed to collect these products.

Unemployment in the buffer zone villages

One of the burning problems of the people in this region of Garhwal is the exodus of the able-bodied workforce. Every year a considerable number of young people leave their homes in search of jobs outside. Likewise, fewer people are undertaking seasonal migration between their winter and summer settlements annually. Although there are many reasons for this change, the lack of adequate opportunities for horizontal expansion of productive activities locally, is the most prominent. Agricultural expansion has reached its saturation point. Sheep and cattle rearing has become non-profitable on account of new environmental policy changes. Traditional woolcraft and bamboo handicraft have declined. Given these circumstances, the present out-migration trend is naturally likely to continue. Radical reforms are required, otherwise this community will soon lose its identity and rich cultural heritage.

Another reason for unemployment in the region is that the government developmental schemes have not been able to reach the people for various reasons.

Conclusion

The local communities in and around NDBR are dependent on the reserve for a variety of resources including building materials, fuel, fodder, wild edible products, and occasionally (despite regulations), for medicinal and aromatic plants. Such is the case in many areas of the South Asian countries. To date, there have been very few major studies made in any of the Indian reserves, which addressed the sustainability of forest produce extracted legally/illegally by the local people. Such studies are critically needed for this and other areas; Ramakrishnan et al. (1994) reported that even long-term subsistence use of forest produce, such as fodder and fuelwood in the Himalaya, can degrade the resource base over time.

Based on the present study, the following suggestions emerge for improvement in *panchayats*:

- The *sarpanch* must have good leadership qualities and also have the capacity to interact with government officials.
- *Panchayats* should be given more autonomy in respect of forest resource-based socio-economic development.
- Government agencies, viz., FD and the Rural Development Department should provide technical and financial backup to people's initiatives, but not interfere with them.
- Single village *panchayats* function better than a multi-village *panchayat*.
- The hill villages are so small that the separate institutions of *gram sabha* and Forest Panchayat should be merged.
- Instead of people going to government officials, the latter should come to people for participatory decision-making.

REFERENCES

Anonymous. 1987. *The Nanda Devi Biosphere Reserve* (Project Document 3). Programme of Man and Biosphere (MAB). Department of Environment, New Delhi.

Azeez, P.K., N.K. Ramachandran and **V.S. Vijayan.** 1992. The Socio-Economics of the Villages of Keoladeo National Park, Bharatpur (Rajasthan), India. *International Journal of Ecology and Environmental Sciences* 18: 169–79.

Bandaratillake, H.M. 1992. Managing the Buffer Zone in Sinharaja World Heritage Forest. *Parks* Vol. 3: 15–19.

Balodi, B. 1993. Expedition to Nanda Devi: Floristic Analysis. The Corps of Engineers' Scientific and Ecological Expedition, Nanda Devi. Unpublished.

Bhatt, C.P. 1985. A Movement to Conserve Forest Wealth. *India and Foreign Review* 2 (9): 2–10 and 29.

Dasmann, R.F. 1982. The Relationship Between Protected Areas and Indigenous Peoples, in J.A. McNeely and K.R. Miller (eds). *National Parks, Conservation, and Development: The Role of Protected Areas in Sustaining Society.* Smithsonian Institution Press, Washington, D.C.

Grossman, D.R. and **B.A. Underwood.** 1971. Technical Change and Caloric Costs. *American Anthropology* 73: 725–40.

Hajra, P.K. and **S.K. Jain.** 1983. *A Contribution to the Botany of Nanda Devi National Park*. Botanical Survey of India, Howrah.

Heinen, J.T. 1993. Park-People Relations in Kosi Tappu Wildlife Reserve, Nepal: A Socio-economic Analysis. *Environmental Conservation* 20 (I): 24–34.

Jain, S. 1984. Women and People's Ecological Movement. A Case Study of Women's Role in Chipko Movement in Uttar Pradesh. *Economic and Political Weekly* 13 October 4: 1788–94.

Johnsingh, A.J.T. and **H.S. Panwar.** 1992. Elephant Conservation in India: Problems and Prospects, in P. Wegge (ed.). *Mammal Conservation in Developing Countries*. Proceedings of a workshop held at the 5th Theriological Congress, Rome, Italy, NORAGRIC. Occasional Papers Series C. Agricultural University of Norway, As, Norway.

Khacher, Lavkumar. 1976. Nanda Devi Sanctuary: A Naturalists Report. *Himalayan Journal* 19: 191–209.

———. 1977. The Nanda Devi Sanctuary. *Journal of the Bombay Natural History Society*, 75 (3): 868–87.

Kiss, A. (ed.). 1990. *Living with Wildlife Resource Management with Local Participation in Africa*. World Bank, Washington, D.C.

Kothari, A., N. Singh and **S. Suri** (eds). 1996. *People and Protected Areas: Towards Participatory Conservation in India*. Sage Publications, New Delhi.

Lamba, B.S. 1987. *Status Survey of Fauna: Nanda Devi National Park*. Occasional Paper No. 103. Zoological Survey of India, Calcutta.

Lamba, B.S., M.L. Narang and **P.C. Tak.** 1981. A Preliminary Report on the Status of Endangered and Threatened Species of Wildlife in Nanda Devi Sanctuary. Unpublished.

McNeely, J.A. 1984. Introduction: Protected Areas are Adapting to New Realities, in J.A. McNeely and K.R. Miller (eds). *National Parks, Conservation and Development: The Role of Protected Areas in Sustaining Society*. Smithsonian Institution Press, Washington, D.C.

McNeely, J.A. and **K.R. Miller** (eds). 1984. *National Park Conservation and Development*. Smithsonian Institution Press, Washington, D.C.

Misra, H.R., C. Wemmer, J.L.D. Smith and **P. Wegge.** 1992. Biopolitics of Saving Asian Mammals in the Wild: Balancing Conservation with Human Needs in Nepal, in P. Wegge (ed.). *Mammal Conservation in Developing Countries: A New Approach*. Proceedings of workshop held at the 5th Theriological Congress, Rome, Italy, NORAGRIC Occasional Papers Series C., Agricultural University of Norway, As.

Mohan, D. and **R.S. Bhadauria.** 1993. *Management Plan for Nanda Devi Biosphere Reserve for the Period 1993–94 to 1997–98*. Wildlife Preservation Organization, Forest Department, Uttar Pradesh.

Ramakrishnan, P.S. 1992. *Shifting Agriculture and Sustainable Development of North-Eastern India*. UNESCO MAB Series, Paris, and Parthenon Publishing, UK. (Reprinted by Oxford University Press, New Delhi, 1993.)

———. 1996. Conserving the Sacred: From Species to Landscapes. *Nature and Resources* 32: 11–19.

Ramakrishnan, P.S., A.N. Purohit, K.G. Saxena and **K.S. Rao.** 1994. *Himalayan Environment and Sustainable Development*. Indian National Science Academy, Diamond Jubilee Publication, New Delhi.

Reed, T.M. 1978. A Contribution to the Ornithology of the Rishi Ganga Valley and the Nanda Devi Sanctuary. *Journal of the Bombay Natural History Society* 76 (2): 275–82.

Samant, S.S. 1993. Diversity and Status of Plants in Nanda Devi Biosphere Reserve. The Corps of Engineers' Scientific and Ecological Expeditions, Nanda Devi. Unpublished.

Sathya Kumar, S. 1993. Status of Mammals in Nanda Devi National Park. The Corps of Engineers' Scientific and Ecological Expedition, Nanda Devi. Unpublished.

Saxena, N.C. 1995. *Towards Sustainable Forestry in the U.P. Hills*. Centre for Sustainable Development, Lal Bahadur Shastri National Academy of Administration, Mussoorie, U.P.

Schultz, B. 1986. The Management of Crop Damage by Wild Animals. *Indian Forester* 112 (10): 133–44.

Sharma, U.R. 1990. An Overview of Park-People Interaction in Royal Chitwan National Park. *Nepal Landscape and Urban Planning* 19: 133–44.

Singh, S.P., R.P. Singh and **Y.S. Rawat.** 1992. *Patterns of Soil and Vegetation and Factors Determining their Forms and Hydrologic Cycle in Nanda Devi Biosphere Reserve, Final Technical Report*. Department of Environment and Forests, Government of India.

Tak, P.C. and **G. Kumar.** 1983. Nanda Devi National Park: The Home of Several Animals and Birds. *Science Reporter*. Oct.–Nov., 1983: 569–74.

Tiwari, D.D. and **V. Ravindran.** 1993. An Analysis of Chipko: A Socio-Political, Economic, Cultural, and Ecological view. Working paper No. 1115, Indian Institute of Management, Ahmedabad.

UNESCO. 1995. Seville Strategy for Biosphere Reserves. *Nature and Resources* UNESCO—Parthenon Publishing 31: 2–17.

West, P.C. and **S.R. Brechin** (eds). 1991. *Resident People and National Parks*. University of Arizona Press, Tucson, Arizona.

21

An NGO-initiated sanctuary: Chakrashila, India

Soumyadeep Datta

Introduction: A profile

On the 14th of July, 1994, a unique forest patch of Dhubri district of Assam was declared as Wildlife Sanctuary by the gazette notification of the Assam government. This area has been named the Chakrashila Wildlife Sanctuary (CWS). It is the youngest sanctuary of Assam, with an area of 4,500 hectares. Chakrashila is unique because of the presence of the Golden langur (*Presbitis geei*) found only along the Assam and Bhutan border. CWS also harbours other rare species of plants and animals.

CWS lies between latitudes 26°15′ and 26°26′ N and longitudes 90°15′ and 90°20′ E. It is in the district of Dhubri, the western-most region of Assam. It is 68 km from the district headquarters, Dhubri.

The terrain is hilly, covered with dense forest, mostly semi-evergreen and moist deciduous, with patches of grasslands and scattered bushland. There are several small springs for quenching the thirst of wild animals; the two major springs are Howhowi Jhora and Bamuni Jhora. Climatic conditions are like that of the temperate zone, with dry winters and hot summers followed by heavy rains. Annual rainfall is between 200 to 400 mm, the maximum being in July (25–50 cm) and minimum in January (less than 5 cm). The temperature throughout the year varies between 8°C and 30°C.

The diverse ecosystems of the CWS support various mammals like the Tiger (*Panthera tigris*), Leopard (*Panthera pardus*), Golden langur (*Presbitis geei*), Leopard cat (*Felis bengalensis*), Gaur (*Bos gaurus*), Crabeating mongoose (*Herpetes urva*), Porcupine (*Hystrix indica*), Chinese pangolin (*Manis pentadactyla*), and Flying squirrel

(*Petaurista* spp.). The only documentation on the local fauna available is that prepared by the local NGO Nature's Beckon.

The two internationally recognized wetlands of Dhir and Deeplai are a part of the Chakrashila ecosystem, though not yet a part of the notified Sanctuary. Hopefully they will soon be included.

Tribal communities of Chakrashila

The fringe villages of Chakrashila are inhabited by people belonging mainly to the Rabha (or Rava) and Bodo tribes. Their roots lie with the Indo-Mongoloid family (Assam–Burmese linguistic section). They are the original inhabitants of Assam. Scattered among these communities are also a limited number of Garo and Rajbanshi families, and of late some Muslim people.

Most of the villagers are agriculturists, paddy being the main crop. Potatoes and green vegetables are grown for their own consumption. Most families have cows, pigs and hens, and own looms for weaving their own clothes.

They construct houses with wood, bamboo and thatch, which they obtain from the forests. They are also dependent on the forest resources for making their looms, equipment for agriculture and fishing, carts and household furniture, and for fuel and fodder. Substantial protein intake also comes from the forest, in the form of fishes, snails and insects. For irrigation and potable water they are dependent solely on the perennial springs of the forest. Even most of their musical instruments are made of forest products like bamboo, wood and cane.

Unfortunately, due to the rapid destruction of the forest, these indigenous cultures are fading away fast. The socio-economic conditions of the tribal peoples have also been subjected to great strain due to deforestation. Consequently, they are facing starvation which was unheard of till the recent past. Many young people have left their villages in search of jobs for supporting their families; however, due to non-availability of jobs in nearby towns, they have been forced to take labourer's jobs in faraway places like the coalfields of Meghalaya.

The present plight of the tribals has been exploited by political parties; tribal youth have been misguided to take up arms to improve their status. Unless this situation is controlled, the forests and local cultures will soon be finished.

Fortunately, most of the tribal youth of Chakrashila have been assisted by Nature's Beckon to raise their socio-economic status, and insurgency has been contained in the area.

Problems in the area

Today, CWS is remarkable in that it is guarded round-the-clock by villagers, and not a single FD official has been posted here. However, about 12 years back, the situation was quite serious. Everyday a large number of *sal* (*Shorea robusta*) trees and other valuable trees were felled and smuggled out in trucks and hand-pulled carts. Hunting was a favourite pastime of the rich and the poor as well. The forest was ransacked by all kinds of exploiters.

One of the lucrative options for many merchants was trading firewood extracted from Chakrashila. The poor villagers worked as labourers due to dire poverty. They participated in the destruction of their own forests upon which they are solely dependent for their daily requirement of fuel, fodder, house-building materials, medicines and even food. They did not realize that they would soon be deprived of their minimum daily requirements. This unfortunate situation left more than five kilometres of once thick forest along the periphery of Chakrashila completely denuded. Consequently, the villagers dependent on the forest for their daily requirements moved inside the forest with their huts and hearth. This resulted in a drastic shrinkage of the forest area.

Some members of Nature's Beckon who frequented the then lush green forests of Chakrashila during their boyhood days were utterly dismayed to see its miserable state in the course of their bird-watching trips in the 1980s. They began to contemplate on the ways and the means to save Chakrashila and to conserve the still remaining resources of this hill reserve.

Chakrashila has several villages on its periphery, inhabited by the Rava and the Bodo tribes. Most of them are very poor and have been grossly exploited by the affluent timber merchants and poachers. Some of the indigent villagers could understand the unfair business dealings and the nefarious mode of operations of the merchants. But they failed to retaliate due to the lack of resourcefulness, self-confidence, courage and unity. Above all, none could give the right type of leadership that was needed by their community. Inspite of all these handicaps they had many good qualities that remained unspoilt due to the inhibitions acquired through generations.

Beginnings of Nature's Beckon

Members of Nature's Beckon came to realize that the conservation of Chakrashila is next to impossible, unless the villagers living on the periphery stop extending a helping hand to the exploiters of their forest resources. This realization called for an action plan to make villagers aware of the importance of conservation.

The first step of educating the people is to win their confidence, love and respect. This demands a lot of patience, perseverance and the faculty to maintain aplomb in the face of adversities. We realized that these indigenous people were very knowledgeable in many respects and their traditional practices were very useful for conservation and sustainable development.

Since communication and transport between Dhubri town, the headquarters of Nature's Beckon, and Chakrashila was very poor, we set up a temporary hutment at Jor-nagra village on the periphery of Chakrashila. The hut was strategically built next to the house of Longthu Rava, whom members of the organization had been friendly with since their childhood days.

Longthu introduced us to a handful of local youth and village elders and acted as our guide initially. Due to our active birdwatching and trekking through the forest—identifying plants and mammals, we soon had a number of village youths who developed an interest in our activities and enrolled themselves as members of Nature's Beckon. They were each given the organizations' badge and identity card. This gave them a sense of identity and purpose.

For most of the time during our stay at Jor-nagra we went about the forest preparing checklists of plants and animals and studying their habitat. Gradually, the local tribals were convinced about our intentions and began to gather around our hutment. In the evenings, over a comfortable bonfire, we discussed the various aspects of our environment and the interdependence of plants and animals. In the course of time, we began to talk to them about the natural and environmental problems that were besetting Chakrashila. We started convincing them that only they could save Chakrashila.

Some of the intelligent youths realized their problems and agreed that their days were getting progressively harder due to the dwindling of the forest resources. But they did not know how to stop influential merchants and poachers from exploiting the forest resources. We convinced these youths that smugglers and poachers were involved in illegal activities punishable under the law of our country and they had no moral grounds to fight us with, if all of us remained firm about our decision and stood together against them.

The youths were ready to take direct action immediately. But we stopped them. We knew that the time was not yet ripe for taking drastic action against the timber smugglers and the poachers. Before we ventured to a confrontation with the timber smugglers, it was necessary to garner the full support of all the villagers. Every one of them had to understand our intentions. It was also necessary for them to realize that we were organizing them to take severe action against the timber smugglers and poachers, only for their benefit.

So in the next phase we started visiting every house of Jor-nagra village. We discussed issues related to Chakrashila in detail with the elders in the village. We also examined how the problems of education, health and poverty could be gradually overcome through the sustainable development projects. The importance of women in environmental management and development was also taken up for discussion and the need for their involvement was emphasized. It took about one year to gather the total support of the villagers of Jor-nagra.

We selected the month of November for direct action against the smugglers and poachers. This was because the harvesting season was over and sufficient food was available with every household. Moreover, November was a period of recess/rest for the villagers.

There were clashes in the beginning. Some of the youth were injured, yet, we did not complain to the police or the FD, as that would have made the villagers dependent on these agencies. Our opponents being guilty, could not lodge any complaint either, but tried their best to frustrate our conservation efforts by influencing the police and the local administration.

We never sought the help of the FD to detect the timber smugglers. But after driving them out from the forest, we handed over the seized saws, axes, choppers, chain and rope to the officers of the FD. Mr D. Jaman, the Divisional Forest Officer of Dhubri, in his written comments on our work said, 'It is only with the help of such organizations that the FD will be able to achieve its *prima facie* objective of preserving and improving the forest.'

Once the villagers had a confrontation with a large gang of smugglers. They were successful in driving the smugglers out of the forest area, and in seizing a huge quantity of their saws, axes and other tree-felling equipment. In appreciation of this achievement, the Principal Chief Conservator of Forest of Assam, Mr R.N. Hazarika, wrote to us: 'Your organization has done a commendable job in detecting timber smugglers and seizure of saws used for illegal tree felling in Chakrashila forest, for which I congratulate you as well as the members of Nature's Beckon and offer thanks for such services for the cause of environment and preservation and protection of wildlife in the state.'

On one occasion the villagers, with their bows and arrows, surrounded a gang of poachers inside the forest and seized four guns with ammunition, and handed over the weapons to the then Deputy Commissioner of Dhubri.

In appreciation of the dedicated work done by the villagers and the members of Nature's Beckon for the protection of wildlife of Chakrashila, the state government rewarded them with Rs 5,000 from the Chief Minister's Relief Fund. This was a great encouragement for the villagers and their morale was further raised.

Programmes in the village

To teach a good lesson to the timber thieves, villagers once burnt a truck which entered the forest for the purpose of smuggling trees. This was a big jolt to the smugglers. After this incident they never dared bring a truck inside Chakrashila.

During this time, we came to realize that constructive work cannot be done successfully in a hostile environment. Peace, development and environmental protection must go hand in hand. We were very eager to bring about complete peace in Chakrashila. The discovery of the Golden langur at Chakrashila and our prolonged association with the villagers, strengthened our demand for its all-round development. Destructive outside pressure also subsided to a great extent. This helped in the regeneration of the denuded forest along the periphery of Chakrashila.

Fortunately, the periphery is a *sal*-dominated area, so the regeneration was quite fast for both the shrubs and the medicinal plants as well. Regeneration was further accelerated because of round-the-clock vigilance by the villagers.

Apart from the protection of the wildlife of Chakrashila, our main concern was about the villagers upon whose well-being the sustainable protection and development depended. We have since then constantly worked for the economic upliftment of the poor villagers. Since we had very little funds, we concentrated our activities in managing and developing the forest resources. By selling NTFP like thatch, bamboo, and grass, we somehow supported the villagers. We encouraged them to cultivate their traditional foods like wild flowers, edible roots like tapioca, and to raise edible insects. We also encouraged them to eat their traditional foods like snails, field rats and crabs.

Then we helped them to raise kitchen gardens and supplied them with various vegetable seeds. They were also encouraged to raise poultry and pigs. The products of these kitchen gardens, poultry and pig farms not only sustained them but also helped them earn a little extra money. Weaving, which was a vital source of income for the tribal families, was also started anew in many poor families with some support from Nature's Beckon.

In this way, slowly and steadily we helped them to improve their economic condition to some extent. They got back their self-respect and their forest for sustainable development.

The achievement of the villagers of Jor-nagra encouraged the villagers from adjacent villages like Abhyakuti, Bandarpara, Kaljani, Damodarpur, Banshbari, etc., and they volunteered to join our group. We started working together for the development of the whole area.

The organization

When the other villages around the Chakrashila Hills Reserve joined us to support our cause, we felt the need for a central meeting point. We converted our small hutment into a big hut, around which trees were planted and a well was dug. This space was then used as both an office and a training centre for the youth and women of Chakrashila; it came to be known as Tapoban. It became an active centre for interaction among the villagers and greatly increased the cohesiveness within the village.

Now, Tapoban is a vital centre of learning for the children, youth and women of Chakrashila. Tapoban also offers hospitality to naturalists and enlightened tourists from far off places. These visits of educated people have enhanced the status of the unknown villages of Chakrashila. The villagers are thus very happy and want to do more for the development of their area. It needs mentioning here that Tapoban came into reality at no extra cost, as most of the work and resources necessary for its construction came spontaneously from the villagers.

One of our popular slogans in the early days of the development of Chakrashila was 'Bring the forest to your village'. The villagers were taught to plant trees, shrubs, medicinal plants, edible roots, fruits and flowers, fast-growing fuelwood trees, thatch and bamboos, in their own village plots so that they could be spared the drudgery of collecting those from deep inside the forest. Mass-scale plantation was taken up on the peripheral area of the forest to meet the fuel and fodder requirement of the villages.

The one single small project which doubled the production of crops at Jor-nagra village was digging of furrows to connect the cultivable lands of Jor-nagra with a perennial source of a water spring called Mauria Jhora. Most of the villages around Chakrashila became self-sufficient to a great extent and the dependence of the villagers on the forest products became negligible.

The sanctuary

During our developmental work at Chakrashila, we also started surveying the forest areas, identifying the birds, mammals, reptiles, plants, etc., and preparing checklists of these species. No faunal survey of Chakrashila was ever done by the FD. We took the initiative for this, and to the surprise of many naturalists and FD officers, it was discovered that the Chakrashila forest is not only the home of the Golden langur but also the habitat of many other endangered mammals, reptiles and birds. It was hence logical to seek the upgradation of this Hill Reserve into a Wildlife Sanctuary under the Wild Life (Protection) Act.

Permanent protection of Chakrashila would also bring immense benefit to the villages lying in the periphery. If the forest cover remains intact, the perennial water sources which irrigate the cultivable lands and supply water for drinking and other household purposes will not dry up. The biological wastes of the forest will continually enrich the fertility of the lands in and around Chakrashila. The upgradation of Chakrashila into a Wildlife Sanctuary will now also provide more scope for the socio-economic development of the villagers living on the periphery, through eco-development projects.

Tribals are known to prefer living on the periphery of forests, where they lead a healthy and productive life in harmony with nature. Since human beings are at the centre of concern for sustainable development, it is essential that the forest around which they live must remain intact. This could be best achieved by converting the forest into a Sanctuary.

It must also be kept in mind that it would be a gross injustice if the forest which the villagers have been protecting since the last 12 years with all their love and might, sacrificing many of their pleasures, goes into the hands of timber merchants or encroachers with the connivance of greedy politicians and bureaucrats.

Keeping all these perspectives in view, we appealed to the state and central governments to upgrade the Chakrashila Hill Reserve into a Wildlife Sanctuary. When the state government remained silent on the issue, we created public pressure on it through repeated appeal and media coverage. We received unstinted support in our endeavour from the public, NGOs, and prominent personages, who wrote to the Chief Minister of Assam and the Ministry of Environment and Forests, Government of India, justifying our demand for immediate upgradation of Chakrashila to a Wildlife Sanctuary.

During this period of dialogue with the government, we utilized our time for the management of Chakrashila. We started plantation in the denuded areas, prevented the poachers and the smugglers from entering into the forest, planted various food trees for the Golden langur and other wild animals. Haphazard burning of the forest land was also prevented, and artificial salt licks were created for the animals inside the forest. The villagers volunteered to clear the weeds and guard the forest. Signboards for the preservation of wildlife were also installed at various points.

Meanwhile, we were able to gather the support of the people of Assam to upgrade Chakrashila into a Wildlife Sanctuary and the issue was debated in the state assembly in our favour. The media also supported our cause. Ultimately, on 14 July 1994, the Governor of Assam, by notification number FRW-6/93/63 under section 26A(I)(b) of the Wildlife (Protection) Act 1972, upgraded the area to a Wildlife Sanctuary.

After this announcement by the government our responsibility has increased manifold, to vindicate the claim that the Sanctuary will be best managed by NGOs and villagers with the benevolent control and support of the government.

Constraints and opportunities of community-based conservation

The tribals of Chakrashila traditionally believe in the concept of the community-based approach in agriculture, fishing, in the construction of their houses, in cultural and religious festivals and in settling any disputes or for punishing a culprit.

Although each family has a separate plot of land, the entire village gathers together to offer prayers to the village deity, before the cultivation is started. They then together plough the fields. Harvesting too, is done jointly, thereby providing help to one another. The whole community comes forward for the construction of any house. This system is known as '*Shouri*' in the village. These communities have common wetlands for fishing as well as common grazing lands and lands for growing thatch, bamboo and cane, etc. Judging from these strong traits of their temperament, it is needless to emphasize that CBC will be best suited for such tribal communities.

CBC imposes accountability on every villager and thus checks injudicious use of forest resources, as unjustified individual opinion cannot prevail over the general will of the villagers and forest officials. CBC generates a sense of closeness among the villagers and forest officials; many problems are solved amicably, under the good leadership of the officials.

It is also to be noted that with the present strength of staff and resources of the FD, it will be rather difficult to conserve the biodiversity of our forests unless the communities living on the fringe areas are taken into confidence.

Of course, some constraints will always be there in this approach of conservation, as no administrative system can claim to be perfect. But these constraints are manageable and many of them can be removed with timely remedial actions.

Some of the constraints of CBC at Chakrashila are the total absence of infrastructure for the management of the biodiversity of this PA, uncertain rights of communities, and lack of knowledge of government policies and laws about the PAs among the villagers living on the periphery of the PAs.

Proposals for the future

Natural resources are not infinite. As such, people must utilize these resources judiciously to reap their benefits generation after generation. This realization should not remain confined to the educated people only but should also be spread among the grassroots. So, educating the vast multitudes of people living on the peripheries of PAs, about CBC, is crucial.

Such educational programmes can be planned and implemented by NGOs with support from the government. Statutory laws relating to PAs, and the rights of the people vis-à-vis these PAs, should be made transparent to local communities. These communities should also have a reasonable say in policy formulation.

Conservation efforts will remain ineffective unless the socio-economic upliftment of the extremely poor villagers is taken up with right earnest. Healthcare, sanitation, and drinking water are basic needs which should be fulfilled first of all. Next, infrastructure to develop cottage industries based on NTFP should be provided, and farmers helped to upgrade their skills. Suitable eco-development projects to grow medicinal plants, firewood, bamboo, thatch, fruit trees and indigenous food plants, can be promoted in the surrounds of every family holding. Finally, the development of eco-tourism can also pay rich dividends to villagers, if properly managed.

Benevolent leadership on the part of NGOs and government officials will go a long way in solving many intricate conservation problems amicably.

22

Conservation by local communities in north-east India

D. Choudhury

Introduction

In the last few years, policy-makers have gradually come to accept the view that in conservation efforts, a participatory approach is essential for success, rather than one based solely on government initiatives (Malhotra and Poffenberger 1989; Poffenberger 1990b). The success of Joint Forest Management (JFM) in West Bengal and later in Haryana and Gujarat, has strengthened this acceptance (see McGean 1991; Poffenberger 1990a and 1990b; Sarin et al., and Raju in this book).

Although JFM has attracted both national and international acclaim, it has also drawn a fair amount of criticism. The critics level their reservations at the lesser degree of control the villagers have in decision-making and benefit-sharing (in comparison to the authorities), and question the level of equity ensured among members, especially against a backdrop of caste, class, and gender pressures within the village (Rizvi in Sarkar et al. 1995; Sarin et al., in this book). Despite the criticisms against JFM, the results have been appreciable enough to induce a section of policy-makers and academia to laud such efforts (Poffenberger 1990a and 1990b; Gadgil and Guha 1995) and explore the possibilities of extending the concept to the management of PAs (Kothari et al. 1996; 1997).

In the euphoria over JFM's success, both conservationists and policy-makers have tended to overlook traditional peoples' initiatives at common resource conservation evolved over generations, and nurtured and managed under traditional customary laws. While it is true that most of the traditional management practices have degenerated over time in most parts of India (Gadgil and Guha 1995; Gadgil, in this book), some are still in practice in several parts of the country, particularly in the North-East (N-E) (Gadgil and Guha 1995; Ramakrishnan, in this book). Unfortunately, the examples from the N-E have rarely been highlighted and therefore, remain little known to both conservationists and policy-makers (however, see Gadgil and Guha 1995; Choudhury in Sarkar et al. 1995).

North-east India, especially the predominantly tribal areas where traditional management practices still exist, is known for its notoriously high degree of deforestation, attributed to shifting cultivation and uncontrolled timber felling (FSI 1993). A large section of forest officials argue that the fact that forests in most parts of the N-E are under traditional management systems or the District Councils, and not under the direct control of the state, contributes significantly to this situation and therefore calls for greater government control. This opinion has become so ingrained that it has succeeded in negating all the merits of resource management traditionally carried out by the tribals of the region, and in shifting the focus away from the need to examine the socio-economic factors responsible for giving rise to a situation where large-scale deforestation has become routine. There is an urgent need therefore, for an assessment of real factors contributing to wanton destruction of forest resources in areas in the N-E that used to have very good forest cover and a strong community-based management system.

Evolution of community-based conservation in the North-East

North-east India is an area of high rainfall spread through most of the year, high humidity and temperature, thus accounting for lush vegetation. The region represents the transition zone between the Indian, Indo-Malayan and Indo-Chinese biogeographic regions, as well as a meeting place of Himalayan mountains and the Peninsular landmass of India. It therefore acts as a biogeographic gateway. About 8,000 species of flowering plants are estimated to grow here. This region is considered a sanctuary of ancient angiosperms as a number of primitive flowering plants are found here (Takhtajan 1969; Rao 1994). Several groups of orchids, rhododendrons, ferns, bamboos, zingibers and lichens have expressed their maximum diversity in this zone. Over 7,000 endemic species are found in India, of which 3,000 alone are found in the N-E. The region is considered one of the centres of origin of rice and citrus and a secondary centre of origin of maize. The region is also known to have a great treasure of medicinal plants and animals. Over 2,500 plants and innumerable animal species are used for medicinal purposes by the area's tribals (SAARC 1992; Changkija 1997).

The rich floral diversity along the altitude gradient from the flood plains of the Brahmaputra river to the snow-capped peaks in the N-E, supports diverse terrestrial and avian fauna. These faunal elements have evolved, over millennia, in intricate association with the communities living in the mountains and hill slopes. Animals comprising a total of 11 orders, 21 families, 86 genera, 148 species and 186 sub-species (BHCP 1993) have been reported from this region. A total of 50 per cent of the mammalian species are endemic to the region (SAARC 1992). Of the 16 species of primates found in India, 11 species have been reported from here. The state of Arunachal Pradesh alone has reported nine species of primates (Jerdon 1874; Pocock 1939; Prater 1948; Roonwal and Mohnot 1977). Out of 71 species of birds which are globally threatened and found in India, as many as 55 species are found in the N-E region (Bhattacharjee 1996). Most of the endemic avian and mammalian species of the region are rare or on the edge of being lost.

That the isolation and insularity of the local communities was significant in past generations, is evidenced by the fact that each village was equivalent to a tiny republic. Inter-village rivalry was intense, even within the same tribe, over most of the N-E highlands (Hutton 1921a; Mills 1929). This may have naturally resulted in tribal communities of the uplands evolving survival strategies based on local resources available, and developing a thorough, intuitive understanding of their immediate surroundings.

Resource management practices among the tribals of north-east India are deeply interwoven in their day-to-day lifestyle, and forest-related livelihood pursuits (Ramakrishnan 1992; 1995; Choudhury, in press). Their understanding of the environment is also reflected in their philosophy which embodies a host of traditional beliefs and practices and acts as a cultural means of conservation. Numerous examples of environmental conservation through such traditional beliefs and practices can be seen among tribal societies of the N-E region. For instance, among the Nagas, destruction of forests in close proximity to the villages is prohibited as it is believed that this will result in a loss of prosperity and disease outbreak. Behind traditional practices like conserving certain trees while clearing *jhum* fields

(Changkija 1996a) and a taboo on hunting during the mating season of animals, lies the belief that if such rules are broken, it will bring ill-luck to the people of the villages. The traditional Naga practice of shifting (*jhum*) cultivation in association with the alder tree (*Alnus nepalensis*), which is nitrogen fixing, is also an ingenious innovation towards sustenance and soil conservation.

Another important aspect of this region is the existence of sacred forests in villages. Such sacred groves are usually strategically located and collectively owned by the community. It is here that all the important religious rituals and ceremonies are performed. Such forests are believed to be the abode of spirits. It was taboo to cut trees, hunt game, or even collect herbs from here. Children were strictly restricted from entering such a forest for fear that the rage of the spirits would be invoked if the forest was disturbed in any way (Changkija 1997).

Traditional customary laws governing these tribes ensure resource conservation and management in the larger interests of the community. They also ensure inbuilt social security mechanisms addressing the needs of each member of the community. These mechanisms have helped nurture the community feeling to such an extent that individualistic traits favouring personal gain over larger community welfare are discouraged and rarely entertained. This singular fact has helped in retaining community control over resources and contributes largely to containing private efforts, influenced by market forces, in encashing natural resources into individual largesse.

Many examples of conservation through community efforts among these societies can be found in the traditional systems of landholding and tenure, forests and forest resource management, water management and even in their agricultural practices. A discussion on all these aspects would be beyond the scope of this paper and therefore the focus here will be confined to community efforts in the conservation of agricultural diversity and forest resources.

A brief comment on the landownership systems prevalent among tribal societies of the N-E, especially in areas which are under traditional management and not under the direct control of the state, is pertinent here in order to emphasize the true community spirit of resource management and the inherent intuitive ecological understanding these communities have. Although customary laws governing landownership and tenure vary with each tribe and sometimes even within tribes and among villages, some fundamental principles are common. Some exceptions do exist, notably among the Semas and Konyaks of Nagaland, and the Khasis, Jaintias and Garos of Meghalaya. Landownership is vested with the *Gaonbura* (chieftain) among the Semas, the *Ang* (Chief) among the Konyaks, the *Dolois/Raja* among the Jaintias, the *Syiem* (King) and his *Mintries* (Council of Ministers) among some Khasi groups and the *Nokma* (Headman) among the Garos. Even among these tribes, the spirit of community welfare is upheld in varying degrees ranging from that of a benevolent feudal overlordship (Semas/Lothas) to one closely resembling the community ownership reflected elsewhere, and executed primarily through elaborate tenureship rights (Gurdon 1914; Hutton 1921b; Tiwari et al. 1995).

The people's right to land is protected under the provisions of the Naga Hills Jhumland Regulation Act of 1946, supplemented by the Naga Jhumland Act of 1970, which gives latitude to the Nagas' customary rights. Unlike in other parts of the country, the land customarily belongs to the people and is controlled by them, and not by the government. Each tribe has a well-defined territory, within which each village is well demarcated. The inter-tribal and inter-village demarcations are permanent and indisputable. The people's customary rights to the land and its resources find ample protection under Article 371A of the Indian Constitution, enforced by the 13th Amendment Act of 1962. Thus, in the state of Nagaland, 80 per cent of the land is owned and controlled by the village/individual. This could be why efforts at conservation by government agencies could not be carried out successfully (Changkija 1997).

Land assets, including forests, amongst most N-E tribal communities can be broadly categorized as community land, clan land and private land. Community lands are common property resources, accessible to all community members, for usufruct rights only, except in *jhum* lands where even the ownership is vested with the community, demarcated clanwise. All matters relating to such land are under the jurisdiction of the traditional village council and ownership is non-transferable.

Clan lands, as the name suggests, belong to a particular clan, the ownership and usufruct rights of which are vested only among members of the clan; ownership transfers are not uncommon within the clan after a thorough discussion by clan elders, but are never given to non-tribals. Clan members are the sole arbitrators in all matters related to such land and the village council has little control except when clan decisions conflict with community interests in a major way. The binding factor in the management of such land is one of kinship and this, at least among the tribes of Nagaland, is extremely strong.

The third category of landholding is private and, owned by individual families or members of families. Management of such land and the natural assets within these are sole family concerns and traditional councils have little say in such matters, except in the case of land transfer, which is generally restricted to members within the tribal group, or when the land is required for community welfare and benefits. The ownership and associated rights for such cases, however, remain vested with the individual.

In the case of Naga tribes, the three categories of landholdings (individual, clan and village) are each further classified into forest land, and *jhum* or cultivable land. *Jhum* lands are cultivated on a rotational basis. The area under forest land is usually preserved for minimal foraging and extraction of resources for house building and firewood. It is the reserve area into which cultivation may be extended if required, and if so sanctioned by the village council. Irrespective of whether the forest land falls under individual, clan, or community holding, no individual or single clan can encroach upon it for cultivation purposes without the sanction of the council. The village council is also empowered to earmark any plot of land within the village boundary and conserve it as a Reserved Forest. For example, the village council of Changki village in Mokokchung district of Nagaland, has taken certain conservation-oriented measures to be followed by the community. Some of the important measures include:

- Conservation of about 24 sq. km of land surrounding the village as Reserve Forest
- A ban on the use of fish poison (both chemical and indigenous herbal)
- Not allowing the trapping of nesting birds
- Strictly prohibiting the cutting of edible, wild fruit trees
- Prohibiting hunting during the breeding seasons of animals, or of female animals

An excellent account of the evolution of land ownership and administration among communities practising shifting cultivation is discussed in detail by Spencer (1966) and as pointed out by him, a thorough understanding of such systems is necessary in order to avoid any misconceptions when taking decisions. Similar accounts of land ownership amongst various tribes of the N-E have been dealt with by others, notable of whom are Gurdon (1914) for the Khasis, Hutton (1921a; 1921b) for the Semas and Angamis, Mills (1921; 1929) for the Lothas, and Ao (1980) for the Aos.

Community efforts at conservation of agricultural biodiversity

The predominant agricultural practice among the tribal communities of north-east India is shifting cultivation, popularly called *jhum*. The exceptions are the Apatanis of Arunachal Pradesh who are better known for their wet terrace paddy-cum-fish cultivation (Ramakrishnan 1992) and the Angamis of Kohima area who also practise terrace cultivation, and some *jhumming*. Shifting cultivation has been much maligned, but as pointed out by earlier workers (Spencer 1966; Ramakrishnan 1992; Choudhury, in press), there is much more to *jhumming* than has been highlighted by its detracters. This is probably the only form of agriculture that allows the soil to replenish its nutrients and regenerate over a long period of time. Even today, the village of Changki, mentioned above, allows for a fallow period of up to 50 years. The shortest cycle in this village is presently over 20 years (Changkija 1996b). It is only in recent years, due to a variety of reasons, that the fallow phase in many other places has been drastically reduced to an average of 10 years or less.

The activities associated with *jhumming* involve community participation during all the phases, starting from site selection to harvesting, including land development, sowing and weeding. The upland communities in the N-E have little access to resources that are commonly available to their counterparts elsewhere in India, and have little cash to spare even if the resources were available. Thus they have to depend on their surrounding forests or *jhum* fields for their day-to-day requirements, and their resource base has to be large enough to cater to their needs. Consequently, the crop range in their fields is composed of a rich array of species. Spencer (1966) lists a total of over 243 species of crops grown by shifting cultivators. Ramakrishnan (1992) and his co-workers have reported between eight to 35 crop species grown in *jhum* fields. During our own studies in Mokokchung district, Nagaland, the crop range has varied from 25 to 43 (Choudhury, in press). In addition to the major cereals like rice, the Naga tribes also grow several indigenous secondary crops in these fields. A survey in Changki village showed that 69 species (belonging to 19 families) were used as secondary crops. Inter-specific diversity therefore remains high in shifting agriculture, whatever its other demerits. Intra-species diversity is also high, especially in paddy, yam, taro, beans and chillies. In a study of Chuchuyimlang village and its environs in Nagaland, we recorded a total of 11 landraces of paddy still in use, and an additional 11 landraces of paddy were reported to be in use in some villages of Tuensang, samples of which were collected by us. Similar diversity for other crops was also observed. Villagers have been cultivating paddy and other crop landraces over generations and their knowledge about the different attributes of these varieties (pedalogical, slope and water adaptability) is remarkable (Choudhury, in press). A more detailed survey may yield greater information on the crop diversity of the N-E, thus aiding in a more composite picture of shifting cultivation in the area.

Community efforts at conservation of forest resources

Community efforts at forest resource conservation among the N-E tribes have evolved due to two reasons—the recognition that forests provide a substantial part of their day-to-day requirement for supplementary food, medicine, energy, and other livelihood resources (Choudhury, in press), and religious beliefs and superstitions. Most tribal villages all over the N-E uplands have forest conservation as part of their traditional customary laws and landholding systems.

Like in most mountainous regions, the natural forest is an indispensable component in the traditional agro-systems of the north-east India (Changkija 1996b). People engage in collecting, grazing, hunting and selective logging activities in the forest to supplement their cash income. Plants collected in the forest include edible fruits, seeds, flowers, leaves, tubers, mushrooms, bamboo shoots and stems. In addition, about 100 medicinal plants, fibres, weaving and dyeing material, are also collected. Hunting provides animal protein to the community, and cash by sale of fur and certain animal parts that may be used for medicinal purposes. Trees are felled selectively only in specific areas. Collection is controlled and does not result in forest degradation. Since local needs are limited, only small amounts and certain plant parts are harvested/collected (Changkija 1997).

An examination of most tribal customary landholding systems reveals that each village or community has forested land set aside for community purposes, some categories of which are open for specified use by members to meet everyday requirements, while others remain out of bounds for all community members, and any transgression results in severe consequences. The latter fall into what may be termed sacred groves/forests. Tiwari et al. (1995) have documented in detail the various aspects associated with sacred groves in Meghalaya, including customary laws governing the establishment and governance of such systems, their present status, and the factors responsible for the gradual depletion of these forests. These and other authors (e.g., Gadgil and Guha 1995) have also highlighted the rich diversity that remains conserved in such systems due to traditional efforts.

Other examples of CBC can be seen in village forests of different categories, either clan- or community-controlled (as described in an earlier section), in most tribal villages of Arunachal Pradesh, Manipur, Meghalaya and Nagaland. These forests are managed and protected by community efforts under the control of traditional tribal institutions such as the Village Councils, however, they are dwindling in size in most parts of the N-E. The novelty of such forests is that management and protection are through oral law and no formal protection force is necessary as in conventional state-managed forests, or even in JFM. While resource use for the day-to-day requirements of the villagers is permitted, any forest resource removal for personal gain is severely strictured and cannot be carried out without the knowledge of the traditional institutions managing these forests.

The threat to traditional conservation systems

With rapid changes overtaking most societies and the advent of modern consumeristic values, aspirations of people, even in the relatively isolated communities of the N-E highlands, are undergoing transformation. Although communication in most parts of the N-E is rudimentary even today, and most areas remain cut off during the monsoon months, the infusion of information technology, especially the electronic media and television, is widespread. Consequently, even these societies have started experiencing the effect of market forces, and it is not surprising therefore, that rural communities have vigorously started looking for opportunities to enhance their cash flow.

Such attempts are mostly seen in the communities' ready acceptance of income-generation schemes introduced by various government agencies, especially those related to horticulture, cash crop promotion, or afforestation. The efforts of these agencies have met with limited success in cash flow enhancement. The reasons for failure are not due to the lack of effort by the villagers, but because of the near absence or poor quality of backup services (especially technical), storage, and product processing, transportation and market linkages.

An elaboration of the shortcomings at different levels of linkages is beyond the scope of this paper. However, it is sufficient to mention here that the result is economic insecurity brought about due to the shift in land use the villagers undertake when they adopt such schemes with the hope that their income will improve. Attempts at homogenization have several pitfalls and ultimately result in increased external dependency for communities which, hitherto, were self-dependent. The increased exogenous nature of market driven economies ultimately ensures the gradual erosion of traditional management systems (Choudhury, in press).

The promotion of income-generation schemes, affecting a change in landuse patterns, has also resulted in a competition for land resources which up to now could be put to agricultural use. In areas with shifting cultivation, the conversion of fallow lands (classified as wastelands by government agencies) to permanent plantations, especially of cash crops, has severely reduced land hitherto available for *jhumming*. Such reductions have also, ironically, resulted from the establishment of Sanctuaries and National Parks (as in Garo Hills, Meghalaya).

The competition for land for conservation needs, income-generation mechanisms (horticultural and cash crops), industry (coal mining) and agriculture is so severe in Garo Hills, it has reduced agricultural land availability to the extent that *jhum* cycles have been reduced to as low as two years. The resultant deterioration in socio-economic conditions has created resentment and consequently worked against conservation efforts (Choudhury and Marak 1996). Such drastic reductions in the *jhum* cycle spell disaster for crop productivity and the local ecology. The effect is seen in the deterioration of the socio-economic conditions of the community. Economic compulsions arising out of such conditions together with the changing market variables compel the villagers to convert natural resources into cash. A significant factor has also been the flow of central government funds to this region, gradually affecting consumption patterns to such an extent that traditional customs have been compromised. For the subsistence farmer, where survival is at stake, and opportunities for income generation absent, the lure of a high wage in logging operations is hard to resist.

With declining economic conditions in the village and poor returns from *jhum* fields, villagers, especially the youth, migrate to the urban localities looking for opportunities for capacity-building, and employment. This slow migration robs the rural populace of able-bodied labour, and with the fast regenerative capacity of vegetation in these areas, leads to labour shortage in *jhum* fields, especially during weeding. The effect is a dramatic reduction in productivity (Kitzer 1996) and ironically leads to a further deterioration in the socio-economic condition of the villager, leading to further depletion of their purchasing power.

Although conversion to Christianity and a resultant erosion in traditional beliefs is attributed as a cause for the degradation of sacred groves (Tiwari et al. 1995), the main factor responsible is the gradual transformation in value systems, catalyzed by the increasing influence of market forces. Although no statistics can be provided, deforestation appears to be more prevalent in clan and private forests (e.g., regenerating *jhum* fallows), than in those controlled by the community. It is in rare cases that timber logging is done in community-owned forests. A point to note is that in most of the hill states, government control and management of forests through the FD is confined to less than 10 per cent of the total forests. In most cases, the forests are actually under the ownership of clans or individuals. While it cannot be denied that deforestation rates are increasing in the N-E (whether due to shifting cultivation or logging) the fundamental reason compelling the communities into such action is undoubtedly their socio-economic condition. Any attempt to arrest this trend has to begin with measures that ensure a sustained food and economic security.

Conclusion

The operation of cultural sanctions and mechanisms maintained and sustained the richness of the N-E region's biodiversity in the traditional system. However, various forces such as urbanization, indiscriminate logging, unplanned road construction, population explosion, and other anthropogenic activities such as hunting game, uncontrolled harvesting of minor forest products, expansion of unplanned and unscientific agricultural systems, and so on, are exerting heavy pressure on the natural resources. The traditional *jhum* system of cultivation is becoming maladaptive to the concept of conservation. The practice of *jhum* has existed for thousands of years and has moulded the lifestyle of the region's people, and influenced their ecosystems as well. Culture and traditions are closely linked with the *jhum* cycle. Therefore, if the people are to be motivated for conservation, the *jhum* system and its cycles need to be stabilized and intensified.

While urbanization and commercialization are symbols of modernization, forces of which cannot be stopped, the prospect of establishing a CBC system by harnessing the traditional system of village administration and land use pattern offers hope and needs further investigation.

For evolving suitable conservation strategies for various species at risk, it would be essential to catalogue them to begin with. Detailed biological conservation studies should follow the initial process of identification of species and their habitat and threats to their survival. The government agencies or NGO efforts at conservation should aim at motivating and involving people at grassroots level. Strategies for conservation should be formulated on the basis of an understanding of the local situation. Since the forest resource management system is a participatory one, direct communal involvement should be invited and encouraged. As far as possible, conservation programmes should aim at utilizing and strengthening local enforcing bodies like the Village Councils. This will ensure the commitment of the community towards conservation efforts.

REFERENCES

Ao, T. 1980. *Ao Customary Laws*. Annada Printing House, Jorhat.

Bhattacharjee, P.C. 1996. *Wildlife: The Boon for North-eastern Region*. Proceedings, Wildlife Festival of Nagaland, Kohima.

BHCP. 1993. *Status Report. Biological Diversity in Eastern Himalaya*. Biodiversity Hotspots Conservation Programme, WWF-India, New Delhi.

Changkija, S. 1996a. Personal communication with S. Changkija, GB Pant Institute for Himalayan Environment and Development, Dimapur.

———. 1996b. *Biodiversity of North-east India*. Proceedings, Wildlife Festival of Nagaland, Kohima.

Changkija, S. 1996c. Ethnobotanical Folk Practices and Beliefs of the Ao Nagas in Nagaland, India. *Journal of Ethnobotany*, 8: 13–17.

———. 1997. Community Based Conservation of Biodiversity in North-east India. Paper presented at the Regional Workshop on Community-based Conservation: Policy and Practice, New Delhi, 9–11 February 1997.

Choudhury, D. and **T.T.C. Marak.** 1996. Biosphere Reserves in North-east India: A Status Report of Namdapha and Nokrek Biosphere Reserves, in P.S. Ramakrishnan, A.N. Purohit, K.G. Saxena, R.S. Rao and M.K. Maikhuri (eds). *Conservation and Management of Biological Resources in Himalaya*. Oxford IBH, New Delhi.

FSI. 1993. *The State of Forests 1993*. Forest Survey of India, Dehradun.

Gadgil, M. and **R. Guha.** 1995. *Ecology and Equity. The Use and Abuse of Nature in Contemporary India*. Penguin Books India Ltd., New Delhi.

Gurdon, P.R.T. 1914. *The Khasis*. Macmillan, London.

Hutton, J.H. 1921a. *The Angami Nagas*. Macmillan, London.

———. 1921b. *The Sema Nagas*. Macmillan, London.

Jerdon, T.C. 1874. *Handbook of Mammals of India*. Mittal Publications, Delhi.

Kitzer, S. 1996. Personal communication with S. Kitzer, Jt. Director, Agriculture, Government of Nagaland.

Kothari, A., N. Singh and **S. Suri** (eds). 1996. *People and Protected Areas: Towards Participatory Conservation in India*. Sage Publications, New Delhi.

Kothari, A., F. Vania, P. Das, K. Christopher and **S. Jha** (eds). 1997. *Building Bridges for Conservation: Towards Joint Management of Protected Areas in India*. Indian Institute of Public Administration, New Delhi.

Malhotra, K.C. and **M. Poffenberger.** 1989. *Forest Regeneration through Community Protection: The West Bengal Experience*. Ford Foundation, New Delhi.

McGean, B. 1991. *NGO Support Groups in Joint Forest Management: Emerging Lessons*. Sustainable Forest Management Working Paper Series, Working Paper 13. Ford Foundation, New Delhi.

Mills, J.P. 1921. *The Lotha Nagas*. Macmillan, London.

———. 1929. *The Ao Nagas*. Macmillan, London.

Pocock, B.L. 1939. *Fauna of British India*. Mammalia Vol. 1. Today and Tomorrow's Printers and Publishers, New Delhi. 1985 Reprint.

Poffenberger, M. 1990a. *Joint Management for Forest Lands: Experiences from South Asia*. Ford Foundation, New Delhi.

———. 1990b. *Forest Management Partnerships: Regenerating India's Forests*. Workshop on Sustainable Forestry, New Delhi. Ford Foundation, New Delhi.

Prater, S.H. 1948. *The Book of Indian Animals*. Bombay Natural History Society, Bombay.

Ramakrishnan, P.S. 1992. *Shifting Agriculture and Sustainable Development. An Interdisciplinary Study from North-eastern India*. MAB Series 10. UNESCO and Oxford University Press, New Delhi.

———. 1995. *Ecological Function of Biodiversity: Human Dimensions*. INSA-IUBS Seminar on 'Biodiversity—Genes to Ecosystems: Towards Sustainable Management'. INSA Diamond Jubilee Celebrations, New Delhi.

Rao, R.R. 1994. *Biodiversity in India: Forest Aspects*. Bishen Singh Mahendra Pal Singh, Dehradun.

Roonwal, M.L. and **S.M. Mohnot.** 1977. *Primates of South Asia: Ecology, Socio-biology and Behaviour*. Harvard University Press, Massachusetts.

SAARC. 1992. *Regional Study on the Causes and Consequences of Natural Disaster and the Protection and Preservation of the Environment.* South Asian Association for Regional Co-operation, Kathmandu.

Sarkar, S., N. Singh, S. Suri and **A. Kothari** (eds). 1995. *Joint Management of Protected Areas in India: Report of a Workshop.* Indian Institute of Public Management, New Delhi.

Spencer, J.E. 1966. *Shifting Cultivation in South-eastern Asia.* University of California, California, U.S.A.

Takhtajan, A. 1969. *Flowering Plants: Origin and Dispersals.* Oliver and Boyd, Edinburgh.

Tiwari, B.K., S.K. Barik and **R.S. Tripathi.** 1995. *Sacred Groves of Meghalaya: Status and Strategies for their Conservation.* Report submitted to the North-eastern Council, Shillong.

23

Community enterprise for conservation in India: Biligiri Rangaswamy Temple Sanctuary

Sharachchandra Lele, K.S. Murali
and Kamaljit S. Bawa

Introduction

The debate over the goals and means of biodiversity conservation has reached a stage where it is becoming increasingly accepted that, for ideological reasons or instrumental ones, the involvement of the local community living in and around protected areas (PAs) is critical to the success of the conservation effort. This acceptance of 'community-based conservation (CBC)' is, of course, at a very general level; the devil is in the details of what is meant by involvement, what incentives are necessary and sufficient to obtain local involvement, what rights and responsibilities can be and should be devolved to the community, what process and institutions will ensure broad-based participation and sustainability of the arrangements as well as of the ecosystem, and so on.

One particular approach being experimented with by several agencies is rooted in the belief that if communities living in and around PAs are to be willing partners, they must develop a *direct and substantial economic stake* in the biodiversity of the area. If such an economic stake is created (either by strengthening an existing biotic resource-based enterprise or creating a new one), the community will have the incentive to regulate its own activities vis-à-vis the protected ecosystem/area. In this paper we describe the preliminary results of an ongoing 'enterprise-based conservation' effort taken up by us in southern India. This work is being carried out under the aegis of and financial support from the Biodiversity Conservation

Network (BCN, a programme funded by the US Agency for International Development; see also Bhatt in this book).

The BRT Sanctuary and the local community

The Biligiri Rangaswamy Temple (BRT) Wildlife Sanctuary, located in the Mysore district of Karnataka state, lies at the confluence of (and hence includes biota from) both the Western and Eastern Ghats regions. Its 540 sq. km of forests, spread over an undulating terrain ranging from 600 msl to 1,800 msl, can be categorized into five broad types: scrub thorn forest (28 per cent of total area), deciduous forest (61 per cent), evergreen forest (7 per cent), high altitude grassland (3 per cent) and high altitude stunted cloud forest (*shola*, 1 per cent). The Sanctuary harbours many large mammals like the Elephant (*Elephas maximus*), Tiger (*Panthera tigris*) and Leopard (*Panthera pardus*), apart from a variety of small mammals and amphibians.

The earliest inhabitants of these forests are a tribal community called Soligas, of whom about 4,500 still live in 25 settlements scattered in or on the fringes of the Sanctuary. The Soligas traditionally engaged in shifting agriculture and trapping. They also collected a wide range of non-timber forest produce (NTFP), initially for their subsistence needs, but later for forest contractors as well. Shifting agriculture has been discouraged since the late 19th century, and with the declaration of much of the area as the BRT Wildlife Sanctuary in 1974, shifting agriculture and hunting were completely banned. The Soligas were allocated small pieces of land[1] where they could practise settled agriculture. However, the extraction of NTFP continued under the aegis of tribal cooperatives, called Large-scale Adivasi (tribal) Multipurpose Societies (LAMPS).

The existing situation with respect to NTFP harvesting and marketing in BRT Sanctuary is shown in Figure 23.1(a). The Soligas harvest NTFP and sell them to the LAMPS. The LAMPS, created in the late 1970s–early 1980s as vehicles for tribal development, are state-controlled tribal forest cooperatives (one in each taluka, a sub-district-level area) that are given two-year renewable leases for NTFP extraction in the forests of that taluka by the Forest Department (FD). Their main objective is to ensure maximum prices for NTFP like the fruits of *amla* or *nellikai* (*Phyllanthus emblica*) and gallnut or *aralekai* (*Terminalia chebula*) to the tribal collectors through pooled marketing of the raw produce.

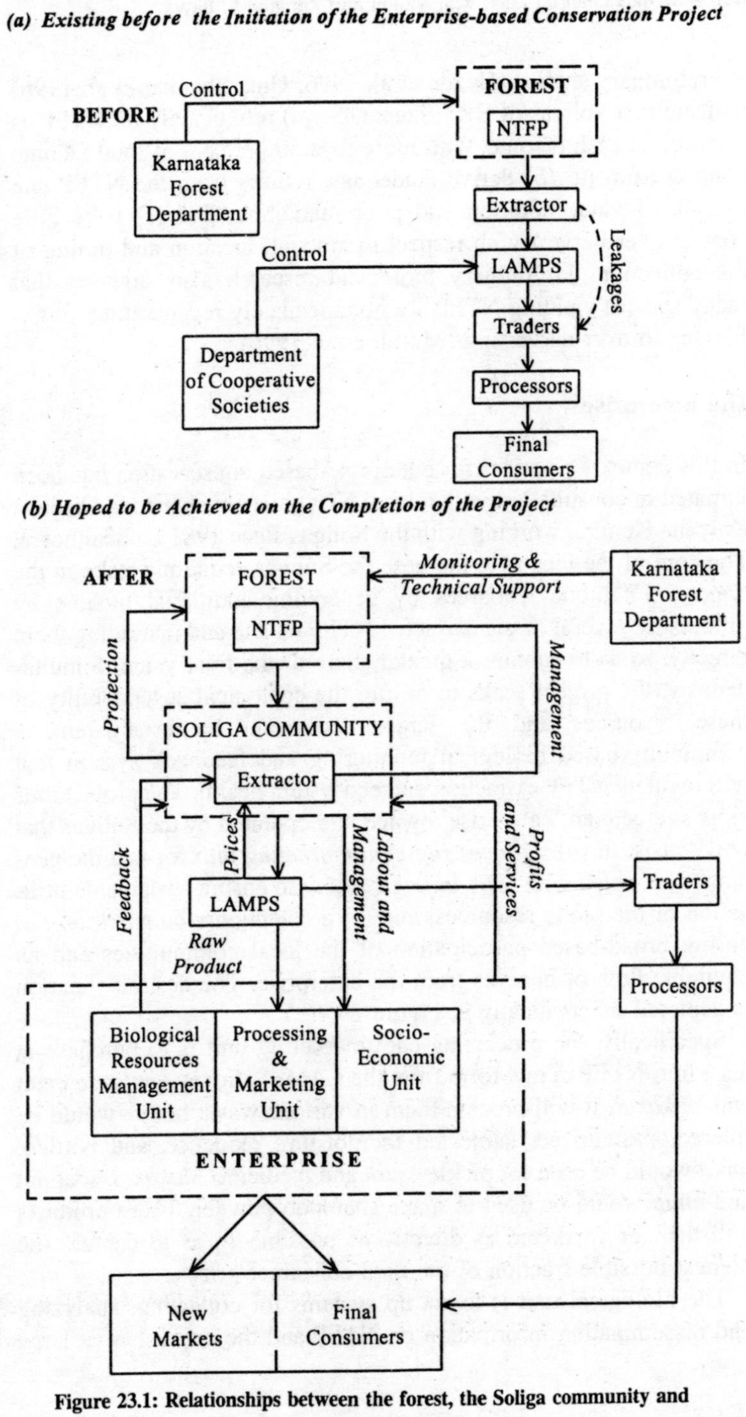

(a) Existing before the Initiation of the Enterprise-based Conservation Project
BEFORE
Control
FOREST
NTFP
Karnataka
Forest
Department
Extractor
Leakages
Control
LAMPS
Traders
Department
of Cooperative
Societies
Processors
Final
Consumers
(b) Hoped to be Achieved on the Completion of the Project
AFTER
FOREST
Monitoring &
Technical Support
Karnataka
Forest
Department
NTFP
Protection
Management
SOLIGA COMMUNITY
Extractor
Prices
Labour and
Management
Profits
and Services
Feedback
LAMPS
Traders
Raw
Product
Processors
Biological
Resources
Management
Unit
Processing
&
Marketing
Unit
Socio-
Economic
Unit
ENTERPRISE
New
Markets
Final
Consumers

Figure 23.1: Relationships between the forest, the Soliga community and the larger society

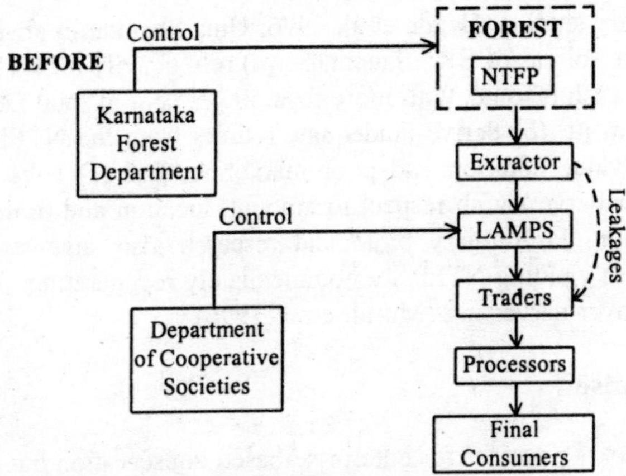

Preliminary studies (Hegde et al. 1996; Uma Shankar et al. 1996) indicate that Soligas of BRT Sanctuary: (*a*) rely heavily on NTFP as a source of cash income, with more than 50 per cent of total income coming from it;[2] (*b*) derive inadequate returns from the NTFP due to lack of value addition and poor marketing; and (*c*) have little control over harvest with respect to amount, location and timing of the collection. Preliminary biological research also suggests that many species yielding NTFP are not adequately regenerating, *possibly* due to over-harvesting (Murali et al. 1996).

The enterprise

In this context, a project for enterprise-based conservation has been initiated in consultation with a local NGO, the Vivekananda Girijana Kalyana Kendra, working with the Soligas since 1981 on healthcare. The core of the idea is to increase the Soligas economic stake in the Sanctuary's biotic resources by generating additional income by processing several of the extracted NTFP on site and marketing them directly, so as to capture a greater share of the final value. Simultaneously, the project seeks to ensure the ecological sustainability of these resources and the larger ecosystem by establishing a community-based biological monitoring and feedback system that will regulate NTFP extraction and ecosystem health. The project thus aims to create an 'enterprise' owned and operated by the Soligas that will consist of (*a*) a *processing-cum-marketing unit* for income generation, coupled with (*b*) a *biological unit* to ensure sustainable utilization of the biotic resources, and (*c*) a *community outreach unit* to ensure broad-based participation of the local communities and an equitable flow of benefits from the enterprise. The desired situation is depicted schematically in Figure 23.1(b).

Specifically, the processing-cum-marketing unit is to purchase at least four NTFP in raw form from the LAMPS: honey, *amla*, soapnut and *shikakai*. It will process them in various ways: honey would be filtered, pasteurized, subjected to moisture reduction and bottled; *amla* would be used for pickles, jam and medicine; *shikakai*, soapnut and *amla* would be used to make shampoo powder. These products will then be marketed as directly as possible so as to capture the highest possible fraction of the final consumer prices.

The biological unit is to set up systems for collecting, analyzing and disseminating information on NTFP and the ecosystem at large and for generating recommendations for modifying harvesting practices and other resource management measures. This will be done at two complementary levels: the level of the community (necessarily simpler and rule-of-thumb) and the level of the enterprise (more sophisticated, by scientifically trained staff).

The original objectives of the community outreach unit of the enterprise were to ensure participation in, training for, and ultimate handover of the food processing unit (FPU) to the Soligas, and to facilitate the setting up of community-based biological monitoring. Subsequently, it became clear that the smooth functioning of the processing units and the passing on of enterprise profits to collectors through better prices for the raw NTFP depended critically upon reforming the functioning of the local LAMPS. The LAMPS is also the key body through which forest management efforts have to be channeled, since it has the NTFP collection rights from the FD. LAMPS reform has therefore been added to the list of objectives of the community outreach component.

In addition to setting up this broadly defined community-based enterprise, the project includes substantial online applied research-cum-monitoring by researchers. The idea here is that many questions relating to the specific resource management strategies to be adopted by the enterprise, or the approach and institutional structure to be adopted for community involvement in the enterprise, will require an understanding of the interactions between the local community (and other actors) on the one hand and the NTFP resources and broader ecosystem on the other. Developing such an understanding and monitoring outcomes (covering biological and socio-economic aspects) so as to provide ideas and feedback to the enterprise is the objective of this research-cum-monitoring activity.

To summarize, the project is an *integrated* plan for CBC that tries to mesh increases in *incentives* with increases in *responsibilities* and in *capacities*. It increases the economic stake of the Soligas through better NTFP prices, increased employment, and profits. It simultaneously increases their responsibilities by requiring them to monitor NTFP resource sustainability and wider ecosystem effects, and also to ensure fairness in distribution of benefits within their community. And it seeks to achieve this by increasing the capacities of the Soligas for financially sound and profitable processing and marketing, for socially acceptable management of the enterprise, and for biologically sustainable management of the resource. The project is initially focused on the part of the Sanctuary that falls in Yalandur taluka (sub-district), a forest area of about 110 sq. km containing nine Soliga settlements with a population of about 1,400.

Activities to date

The project was initiated in March 1995, and is currently funded for three years. The progress achieved so far is described briefly below.

Processing and marketing for income generation

The processing-cum-marketing activity consists primarily of two units: a Honey Processing Unit (HPU) and a Food Processing Unit (FPU).[3] The project had also planned to increase Soliga incomes by initiating apiculture with another species of honey bee (*Apis cerrana*). However, a nationwide epidemic of sac brood disease on this species has prevented this activity from taking off. These units were set up in the premises of and with infrastructural support from the local NGO. The main features of and physical targets achieved by each value-adding activity/unit are described below, followed by a brief description of the marketing strategies adopted, the net profits generated so far, and the level of Soliga staff training achieved. The funding from BCN has covered all capital costs, all operational costs for the first two years, and salaries of all specially hired professionals for three years.

The HPU is designed to process honey, currently collected primarily from wild rock bees. Begun in 1995, the unit has the capacity to process 30 tonnes of honey per year. The total revenue generated in 1996 from honey was Rs 700,000; the profit margin is approximately Rs 20,000 per tonne.

The food processing plant is intended to prepare pickles, jam, other food products and also shampoo powder. The principal NTFP being used is *amla* fruit for the food products and *shikakai* (with *amla*) for shampoo powder. In 1996, 2,000 kg of *amla* was processed, about half for pickle and half for shampoo powder. In 1997, the target is to process about eight–nine tonnes of *amla* fruit to generate a revenue of about Rs 450,000 and a profit of about Rs 60,000 from the various products.

To facilitate marketing, necessary regulatory approvals have been obtained and trademarks registered. In particular, the HPU has obtained

'Agmark' certification of quality from the Indian Standards Institution. The FPU has received FPO certification (Food Products Order from the Central FPO Licencing Authority). Both units market their products under the registered trademark 'Prakruti' (meaning 'nature'). In marketing these products, a conscious strategy of product and channel diversification has been followed. Honey is currently being marketed through the state-owned Khadi and Village Industries Commission, and through wholesale and retail outlets in Bangalore. The same outlets will be used for FPU products. A retail outlet has also been opened near the BR Temple, which will sell all these products.

Over the period March 1995–December 1996, the HPU and FPU together made a net profit of about Rs 140,000 and generated about 2,400 person-days of wage-labour employment for the Soligas. Skill improvement has been significant on the production side: it is estimated that the production activities in the HPU will be entirely Soliga-managed within a few more months.

The processing-cum-marketing activities are managed by a professional manager, and the local NGO has provided accounting and secretarial support. Initially, a Management Committee comprising six–seven Soliga representatives, an NGO representative and the authors of this paper was formed to oversee the operations of the enterprise and the project as a whole, with the Soliga representatives being nominated by the NGO. This committee, however, failed to function: the Soliga members did not show up or participate, the nominees kept on changing, the meetings were called infrequently, discussions were repetitive and superficial. The exact powers and responsibilities of this committee were not clear; *de facto*, all operational decisions about the processing-cum-marketing activity were taken by the NGO.

The causes of this lack of community involvement and our efforts to regenerate community interest are described later below. Suffice to say that it has now been decided that the Soligas will set up a separate registered society to which the NGO will formally hand over all the assets of the enterprise by 1 April 1997, and the Executive Committee of this society will manage the enterprise, with advice from us and the NGO, and with professional staff wherever required.

Enterprise-based biological resource monitoring

The basic motivation for systematic biological monitoring is twofold. First, the enterprise itself must have a way of knowing how its activities are affecting resource sustainability and overall biodiversity, so that it may take appropriate management measures. Second, if the concept of CBC in general and enterprise-based or incentive-based conservation in particular is to be shown to be viable (so as to replicate it on a larger scale), there must be a demonstrable improvement in the resource and ecosystem condition following the setting up of the enterprise. While the first clearly needs to be community-based, the second may perhaps also require some outsider participation to ensure reliability and rigor.

The project visualized that this enterprise-based biological monitoring might be in two forms: a simpler, community-based monitoring system, and a more scientific system using ecologists employed by the enterprise. Till recently, the latter was being conducted by the project's biological research team working somewhat independently of the enterprise and the Soligas, although on topics related to the enterprise's activities. Starting late 1996, one of the biologists in this team began to work directly with the enterprise.

Prior to initiating efforts for community-based monitoring, the project conducted a rapid assessment of traditional knowledge and attitudes to contextualize our efforts. This assessment revealed that while the Soligas' knowledge is very substantial and detailed, certain seasonal information (such as the distribution of *amla* fruit) is unevenly distributed across the community, and also that traditional knowledge is probably not sufficient to ensure sustainability in the face of the recently commenced intensive (commercial) NTFP extraction. It also revealed the limitations of our own 'rigorous' biological monitoring programme. The Soligas repeatedly pointed out that fire, now suppressed vigorously by the FD as a part of its sanctuary management plan, was actually an integral part of the ecosystem and lack of fire may be partly responsible for the lack of regeneration of a number of the species. Our original studies had only focused on extraction pressures, missing out this vital factor.

Progress towards setting up the community-based resource monitoring system, however, has been much slower than desired. So far, a few Soligas have been trained in systematic monitoring of resource extraction and in NTFP availability estimation; community interest and participation has, however, been slow in coming. In November 1996, the project got key persons (collectors, LAMPS agents and directors) in the community to estimate the size of the *amla* crop in their traditional manner, and found it to be similar to the estimates arrived at by the project through more rigorous and systematic methods. In December 1996, a training programme on participatory resource mapping was conducted with the help of a resource person from Centre for Earth Science Studies, Trivandrum. Using this, a comprehensive programme of participatory monitoring of *amla* harvest and regeneration was initiated in January 1997, which included pre-harvest discussions, online monitoring of harvest percentage, presence of parasites and seedlings, and post-harvest feedback sessions on all harvest days. The response obtained was encouraging; similar efforts are planned for other products in the future.

The difficulties faced in devising and institutionalizing a community-based resource monitoring system are myriad. One may be that our 'scientific' training makes it difficult for us to conceptualize a socially acceptable monitoring system. And there are hardly any other sites where such systems have been set up that can serve as a guide. But the major barriers to setting up such a monitoring system are the facts that the community today has no incentive to get involved in enterprise activities (as even at the end of the second year it has received very few tangible benefits), that it is yet to see any tangible effects of its own recently intensified NTFP extraction, and that even when it does wish to act on the basis of new information it knows that it has very little say in forest management.

With the transfer of the enterprise to their hands, we hope that the involvement will increase. The project is also trying to set up some specific incentives for the NTFP collectors in the form of bonus payments from enterprise profits, and experimenting with different techniques for making the monitoring simple yet useful. The results of the 'scientific' research are being taken back to the community to show that there is in fact some cause for concern over certain aspects of the forest ecosystem, such as the regeneration of NTFP species. We are also trying to emphasize that systematic monitoring can be used as a tool or argument for gaining greater control over the resource. Simultaneously, efforts to lobby the FD to grant greater control to the local community also continue.

Community outreach: Involvement, empowerment, and benefit distribution

As described earlier, the project was conceived in consultation with an NGO that has been working for Soliga development for the past

15 years and most of whose staff and board members are Soligas themselves. The enterprise activities have been based physically and operationally in this NGO, which had already initiated a number of vocational training programmes and even some small-scale NTFP processing activities with the Soligas before our project began.

Thus, in one sense the Soliga community has been aware of and involved in the enterprise from day one. On the other hand, the enterprise is still quite far away from its goal of true community involvement, i.e., making the enterprise activities entirely managed, controlled and owned by the Soligas. Soliga response to all outreach efforts was quite lukewarm for the first year-and-a-half. The meetings of the Soliga Managing Committee set up for overseeing all the enterprise activities initially evoked little response in the community.

Subsequent intensive and candid discussions with the community and introspection amongst ourselves indicated two major problems. First, in spite of us having a local, well-established NGO as our collaborator, the Soliga community was not at all well-informed about the basic objectives of the project and their role in it. The Soligas saw the project as just another activity run by the local NGO in which they would get some wage labour, not as an enterprise that they would jointly manage and eventually own. Indeed, the NGO itself was unclear about what 'community-based enterprise' really meant. It became apparent that the NGO was overloaded with many other programmes and projects, and that separate community outreach efforts needed to be initiated for spreading the message and building Soliga capacity for CBC.

Second, these consultations also reinforced the feeling that a centralized, capital-intensive, and technologically sophisticated approach to processing was not generating sufficient interest among the dispersed, seasonally employed, largely illiterate and infrastructurally constrained Soliga community. The people repeatedly expressed a need for processing activities that could be taken up at the *podu* (hamlet) or even household level.

To tackle the first problem, the project hired a social worker and began an intensive community outreach campaign starting September 1996. In January 1997, a public meeting was held in which the local NGO promised to hand over the units along with profits and working capital to the Soligas. The Soligas, with the help of the community outreach unit, are now in the process of forming and registering a society that will operate on the lines of a cooperative to take over and manage the enterprise.

The second problem cannot be solved immediately or directly, since the investments in the processing units have already been made. The project has therefore tried to broadbase the benefits flowing from the enterprise by including LAMPS reform as one of its activities, and will also attempt to pass on some of the profit margins in the enterprise to the collectors through bonus payments on the NTFP they contribute. Finally, possibilities for decentralized processing of some other NTFP are being explored.

Efforts at reforming the functioning of the LAMPS have evoked a somewhat more enthusiastic response from the community. These efforts were initiated at two levels: reforming the local LAMPS, and changing state-wide policies towards LAMPS in general. The first level directly involved the BRT Soliga community, while the second level targeted the Karnataka-wide tribal community. At the local level, a greater awareness has been generated in the community about the malpractices occurring in the LAMPS, demands are being made by the Soliga members to revise the pricing system so as to reduce LAMPS' and agents' margins, and some improvements in the NTFP tendering/auctioning system have been attempted. At the state level, significant momentum has been generated amongst tribal organizations and tribal development NGOs to push for LAMPS policy reform, resulting in the drafting of a detailed action plan that is being finalized and submitted to the government (Lele et al. 1996).

Applied research-cum-monitoring

This is an interdisciplinary effort involving biologists and social scientists. Its objectives are to generate rigorous information on the overall vegetative landscape, the distribution, regeneration and productivity of NTFP species in particular, the effects of current extraction patterns and other factors (weeds, fire) on these variables, and the socio-economic factors influencing Soliga extraction practices and their ability to derive income from the forest. The results obtained so far are described very briefly below; for further details, including methodology, see Bawa et al. (1997).

Vegetative sampling covering the entire 540 sq. km Sanctuary area, now being incorporated into a Geographical Information System (GIS) and combined with satellite imagery, has produced a detailed picture of the overall vegetation in the Sanctuary. It indicates that the distribution of NTFP species is very uneven, and that exotic weeds (*Lantana* spp. and *Eupatorium odoratum*) are widespread. Regeneration studies suggest that some of the NTFP species (such as *amla* and *aralekai*) are not producing adequate numbers of seedlings. There is evidence from controlled harvest experiments that, in the case of *amla*, current levels of fruit harvest could be partly responsible for this. However, further fieldwork suggests that weeds and fire may also have an adverse effect on regeneration, and studies are being initiated to thoroughly understand these effects. The life of *amla* trees may also be affected negatively by a semi-parasitic mistletoe vine, and Soliga harvesters have begun pruning these parasites to ensure future productivity. Studies have also been carried out for some medicinal plant species that could be used for medicinal products in the future, and they indicate significant variation in the requirements for sustainability across species (Murali 1997).

Results of monitoring the community's participation and intensity of NTFP collection as a whole and harvest of *amla* (supposedly the most popular NTFP) in particular are as follows. Soliga households are differentiated into traditional ('hard-core' or 'full-time') NTFP collectors who collect all products (15–25 per cent of all households), marginal or 'part-time' collectors who only get involved in the relatively unskilled and lucrative collection of fresh *amla* fruits (40–50 per cent), and those who are not involved in commercial NTFP collection at all (35 per cent), the last fraction being much higher than anticipated. The project also found that non-participation in *amla* harvest is significantly correlated with household member(s) being engaged in salaried jobs or steady wage work, and that the biggest beneficiaries from NTFP harvest are the tribal agents appointed by the LAMPS who earn commission on each kilogram harvested, while also acting as moneylenders.

A detailed study of the extraction of wild honey (the NTFP with biggest margins for the processing unit) showed that timing of extraction significantly affects productivity (kilogram of honey per comb) and possibly sustainability (through its impact of larval loss). The timing of extraction is clearly affected by the nature of tenure on the honey resource: open-access trees/cliffs get harvested earlier than is optimal. The establishment of tenure itself appears to be a complex social process, in which the emerging tribal elite (LAMPS agents) play key roles. These findings are being incorporated in developing a detailed plan for community-based honey extraction, training, and experimentation during the next honey season.

Implications for CBC

While our project is as yet incomplete, we can offer some insights based on our experience so far that could be useful when devising other CBC programmes. We have attempted to organize these insights broadly along the lines laid out in the overview paper of Kothari et al. (in this book).

Identifying the 'local community'

Identifying who represents the local community and getting their involvement is always the biggest challenge facing CBC, or for that matter any development activity. Outsiders are therefore tempted to take the help of local NGOs in such activities. Indeed, 'local NGO involvement' has been made a prerequisite for development activities by most funding agencies. However, one must be careful not to confuse 'local NGO' with 'local community'. Rural development NGOs are often set up by outsiders, and they may not necessarily represent the entire community. Further, they may not even facilitate community involvement: the NGO may be more concerned about self-development and can become an obstacle to genuine community empowerment. There is thus no simple formula for properly identifying who speaks for the local community; diverse approaches have to be tried and evaluated in a spirit of genuine self-reflection.

Another oversimplification occurs when we use the term '*the* local community': it suggests that there is one such single, homogeneous entity. This assumption is particularly easily adopted in tribal areas, because tribal communities are known to have strong communal traditions and cultures. Our experience shows that even tribal communities are no longer homogeneous. Both specialization and differentiation are occurring rapidly in these communities. Specialization, wherein some households move away from forest-based activities (typically into non-forest wage labour and sometimes into salaried jobs or trades) while some others remain 'hard-core' NTFP collectors, is perhaps the inevitable consequence of their contact with the modern economy. Simultaneously, tribal societies have drifted away (or been dragged away) from their communal and relatively egalitarian social organization to situations where individuals in the community (like some of the tribal LAMPS agents) become exploiters (usually with some external support) and an elite is formed. Another 'tribalism', that women have a more or less equal status as men, is again becoming a thing of the past as tribal societies 'learn' from the gender-biased dominant communities around them.

Any interventions, including CBC programmes, will at the very least need to be sensitive to such variations in interest and differences in power. The interventionists will have to walk the tightrope between ensuring participation of the genuinely needy and forest-dependent sections and the ambitions (and even abilities) of the elite who will see these programmes as fresh avenues for self-gain. Extra efforts and specific strategies must be devoted to ensuring that the various interest groups within the community are identified and the opinions of the weaker ones are sought.

Form, magnitude and mode of benefits

The fundamental assumption of the BCN approach is that an economic incentive is necessary to get the community involved in biodiversity conservation, and that one way to increase this incentive is through value-added processing using capital investments and professional expertise. Our experience so far shows that the form of the incentive may be more than just economic, that the magnitude of economic incentives required for participation is hard to predict, and that economic benefits may often be more easily or effectively generated through alterations of property rights than through conventional modes of 'enterprise' or 'entrepreneurship' development.

While economic incentives are important, the legitimacy and political empowerment that may result from involvement in CBC can be an added incentive and often a more effective tool for mobilization. This seems to be particularly true in the context of monitoring the community's own resource use: the argument that 'this monitoring is necessary to show outsiders, including the Forest Department, that we [the Soligas] do not over-harvest the resource and hence to strengthen our demand for forest rights' was found to be critical in generating Soliga participation in resource monitoring.

We also found that there is palpable tension between two distinct schools of development thinking: the school believing in a capital-intensive, centralized, highly mechanized, professionally managed format (so as to ensure competitiveness in today's 'global economy'), and the school which believes in a low-cost, decentralized, appropriate technology-based, and locally managed format (so as to ensure broad-based participation, genuine empowerment and opportunity for social adaptation). The latter format may yield slower and less spectacular results than the former, but it is likely to be more socially sustainable. However, funding agencies by their very nature are tempted to go for the former approach, both for its speed, its visibility, and the compulsion to throw money at any problem. CBC proponents must resist this temptation and choose a middle path depending upon local needs and capabilities.

Indeed, capital investment or technological intervention may be less important than (and even fruitless in the absence of) simple (but politically difficult) changes in the property rights arrangement. For instance, the fundamental problem with LAMPS today is not that the tribals lack adequate capital or marketing know-how or managerial expertise (as government reports claim), but that they lack clear and secure rights of access to NTFP and genuine control over the day-to-day operations of their cooperative (Lele and Rao 1996). Reforming these arrangements (so as to narrow the large gap between prices given to individual NTFP collectors by the LAMPS and prices obtained by the LAMPS subsequently when auctioning the products) would require no capital but could yield far more per capita benefits to the tribals than capital- and technology-intensive processing. In fact, the latter gains are dependent upon the former changes.

Will, however, such locally-generated benefits be *generally* enough to ensure people's participation in conservation? This is a much-debated question in the literature on conservation policy and CBC. For instance, Gadgil and Rao (1994) have argued that external subsidies will usually be necessary to ensure local compliance and support for conservation practices. While our project is still incomplete, preliminary indications suggest that the local community should be able to generate sufficient economic income and non-economic benefits from the local biodiversity through both controlled NTFP harvests and through even less disturbing uses such as eco-tourism to obviate the need for external subsidies (the mechanism of which may be suspect and the very act perhaps degrading). Clearly, much further work on assessing the potential of various options is necessary for designing CBC programmes and policies.

On the other hand, we would also caution against the simplistic belief that economic incentives are 'sufficient' to ensure conservation. In fact, standard economic theory predicts, all other things remaining constant, increased extraction when product prices increase. If, for instance, in the existing situation the Soligas obtain significantly higher prices for a product, they will most probably extract more of that product without thinking about its future sustainability. Often, the major reason for this behaviour is again insecurity and incompleteness of tenure. Tenure must include not only long-term rights of access, but also an enforceable right to exclude others, and the right to 'manage' the forest resource as a whole.

Role of indigenous and modern knowledge systems

The breadth and depth of local knowledge about local ecosystems is undoubtedly very substantial. This is particularly so in tribal communities, such as the Soligas. Legitimizing and nurturing this knowledge is an important task in CBC-type programmes.

At the same time, we would like to caution against romanticization of traditional knowledge to the point where one allows no role for modern methods of data collection or interpretation. It must be remembered that traditional knowledge has evolved over generations, typically under relatively static and low-intensity regimes of human–nature interactions. It is therefore unlikely that traditional knowledge will be adequate to understand and anticipate the full impact of the rapidly expanding forms and intensities of these interactions, such as commercial-scale harvests or air/water pollution by tourism. More than simple documentation of traditional knowledge, what is required is a way of involving those who serve as repositories of this knowledge in the day-to-day management under CBC programmes, such that they can begin to adapt their knowledge to new situations. Simultaneously, more formal 'scientific' monitoring will have an important role to play, especially in the initial stages. The challenge for the scientists will be how to share their methods and models with the local community in a manner that allows local perceptions to be incorporated and tested by local communities themselves.

Institutional issues

While the project is yet to reach the stage where institutional issues come to the fore, we anticipate the following main issues to come up: how the balance between individual and community benefits (or active participants versus non-participants or collectors versus non-collectors) is struck; and what will be the organization interfaces with the main governmental agency (the FD in this context) and with the local political bodies (like *panchayats*).

Legal and policy issues

We need not elaborate here upon all the aspects of Indian laws (Wild Life Act, Forest Act, etc.) that we found detrimental to our attempt at CBC. Suffice to say that in the specific context of NTFP-based enterprises, the biggest hurdle will be that the Wild Life Act proscribes such extraction from National Parks; its amendment is therefore a prerequisite to taking up such activities in these areas.

Our experience also shows that between the Act and the field situation are innumerable administrative policies and procedures that serve to effectively marginalize local communities even when the law gives them a role. A classic example is that of the LAMPS, which do not always get the NTFP extraction lease even though there is a Government Order in Karnataka requiring that the lease be preferentially given to local LAMPS. Thus, CBC programmes will have to deal with many of these written and unwritten rules of the game that politicians and bureaucrats play.

NOTES

1. The average size of a Soliga landholding is 0.6 hectare, but approximately 30 per cent of the households have no access to cultivable land.
2. Other major sources of livelihood include subsistence agriculture and wage labour on non-Soliga orchards, coffee plantations, and FD works.
3. In addition, we have been involved in helping the local NGO manage another unit—a Herbal Medicinal Plant Unit (HMPU)—that it had set up with funding from the Foundation for Revitalization of Local Health Traditions (FRLHT) with the same concept in mind: increasing tribal incomes through processing and marketing of ayurvedic medicines using medicinal NTFP from the Sanctuary. Unfortunately, restrictions on the extraction of NTFP prevent Soligas deriving any additional income from medicinal NTFP harvest, while the location of the unit away from the Sanctuary prevents any Soliga participation in processing and management. Hence the activities of the HMPU are not described herein.

REFERENCES

Bawa, K.S., S. Lele, K.S. Murali and **B. Ganesan.** 1997. *Management of Tropical Forests for Extraction of Non-Timber Forest Products*. Technical Report. Biodiversity Support Program, Washington, D.C.

Gadgil, M. and **P.R.S. Rao.** 1994. A System of Positive Incentives to Conserve Biodiversity. *Economic and Political Weekly* 29 (32): 2103–107.

Hegde, R., S. Suryaprakash, L. Achot and **K.S. Bawa.** 1996. Extraction of Non-Timber Forest Products in the Forests of Biligiri Rangaswamy Hills, India–1. Contribution to Rural Income. *Economic Botany* 50 (3): 243–51.

Lele, S. and **R.J. Rao.** 1996. Whose Cooperatives and Whose Produce? The Case of LAMPS in Karnataka, in R. Rajagopalan (ed.). *Rediscovering Cooperation.* Institute of Rural Management, Anand, Gujarat.

Lele, S., R.J. Rao, Nanjundaiah and **V. Muthaiah.** 1996. Re-lighting LAMPS: A Draft Action Plan for Revitalizing the Tribal Cooperatives in Karnataka, in R. Rajagopalan (ed.). *Rediscovering Cooperation.* Institute of Rural Management, Anand, Gujarat.

Murali, K.S. 1997. Use of Medicinal Plants in a Sustainable Way—Assessment of Impact Due to Extraction. Final report submitted to the Foundation for Revitalization of Local Health Traditions. Tata Energy Research Institute, Bangalore.

Murali, K.S., R. Uma Shankar, K.N. Ganeshaiah and **K.S. Bawa.** 1996. Extraction of Non-Timber Forest Products in the Forests of Biligiri Rangaswamy Hills, India. 2. Impact of NTFP extraction on regeneration, population structure, and species composition. *Economic Botany*, 50 (3): 252–69.

Uma Shankar, K.S. Murali, R. Uma Shankar, K.N. Ganeshaiah and **K.S. Bawa.** 1996. Extraction of Non-Timber Forest Products in the Forests of Biligiri Rangaswamy Hills, India. 3. Productivity, Extraction and Prospects of Sustainable Harvest of Amla *Phyllanthus emblica* (Euphorbiaceae). *Economic Botany* 50 (3): 270–79.

24

Tribals and conservation: Chinnar Sanctuary, India*

U.M. Chandrashekara and S. Sankar

Background

Location

The Chinnar Wildlife Sanctuary (CWS; see Map 24.1) with an area of 90.44 sq. km is located in the Southwestern Ghats of Kerala, lying between 10°15′N and 10°21′N latitudes and 77°5′E and 77°6′E longitudes. It is bounded on the east by the interstate boundary between Kerala and Tamil Nadu, on the west by the Marayoor Forest Range and Eravikulam National Park, on the north by the Chinnar river which also happens to be the Kerala–Tamil Nadu boundary, and on the south by the Marayoor Forest Range.

Geographical features

The area is characterized by an undulating, rugged and rocky terrain with altitudes ranging from 500 to 2,200 msl. Geologically, the valley is comprised of gneissic metamorphic rocks from the Archean shield. The valley is deeply eroded by the rivers Chinnar and Pambar, and their tributaries. Weathering is very intense as the predominant rock type is gneissic. Valley fills comprise mainly unconsolidated gravels and are of recent origin.

* The authors are grateful to Dr K.S.S. Nair, Director of Kerala Forest Research Institute, and Professor P.S. Ramakrishnan and Dr K.G. Saxena of Jawaharlal Nehru University, New Delhi for their encouragements. This paper is based on work carried out by the authors for a research project on biodiversity conservation within the context of traditional knowledge and ecosystem rehabilitation in Chinnar Wildlife Sanctuary. The financial support by Mac-Arthur Foundation and UNESCO for the project is acknowledged. The authors also extend their thanks to other investigators of the project.

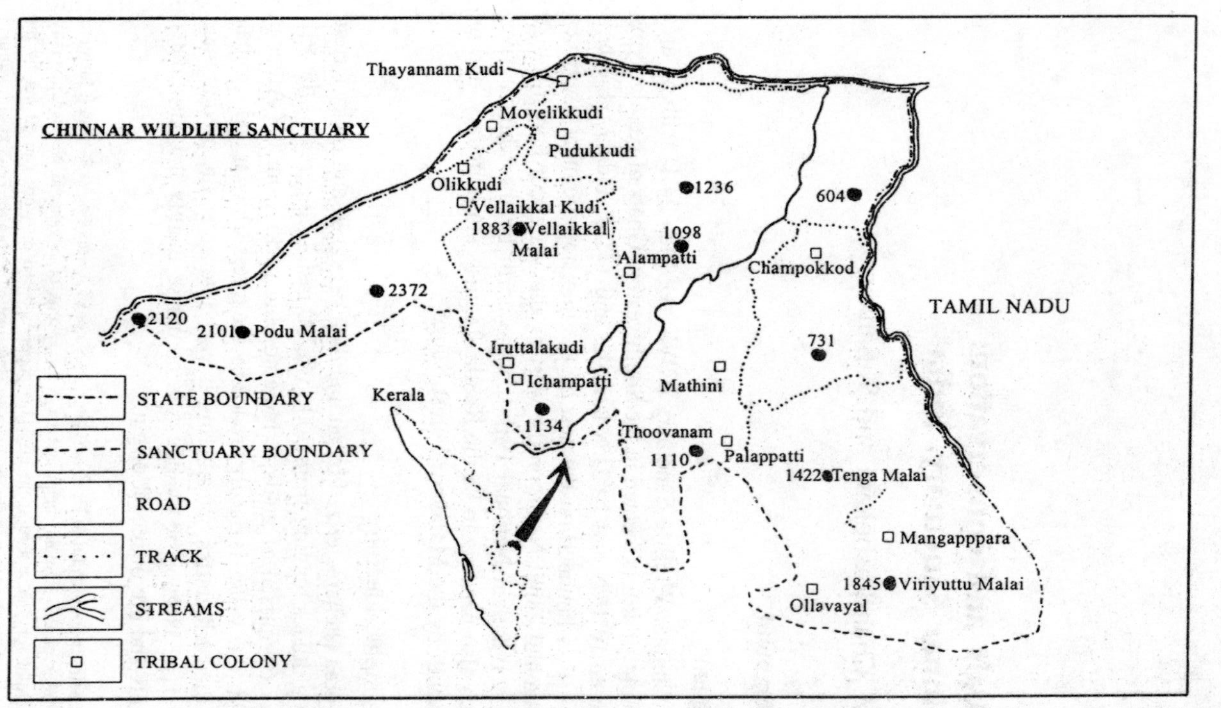

Map 24.1: Location map of Chinnar Wildlife Sanctuary

While the soils of scrub jungle contain more gravel content (12 to 43 per cent) those of Shola forest possess less (zero to 7 per cent). The soils are sandy to sandy loam in texture. The soil reaction varies from slightly alkaline (pH 8.2) to strongly acid (pH 4.2) depending on the vegetation type. The organic carbon content is very low in the scrub and riparian forest (0.49 to 1.55 per cent), medium in the dry deciduous forest (0.76 to 2.09 per cent) and very high in the Shola forest (4.55 to 14.03 per cent).

The area falls in the rain shadow region and the average annual rainfall varies from 500 to 800 mm. Rainfall is mostly concentrated in the months of November and December, with 40 to 44 rainy days in a year. About five–six months are dry. The mean temperature of the coldest month ranges between 16°C and 23°C, and the minimum temperature is below 15°C.

Vegetation and land cover

The vegetation of the area primarily belongs to the *Terminalia-Anogeissus latifolia-Tectona grandis* series (Puri et al. 1983). Pristine patches of this vegetation are confined to small pockets in inaccessible localities, with the rest of the forest tract being degraded. A major part of vegetation in the Sanctuary belongs to the secondary seral stages of various gradations. According to S.C. Nair (1991) 11 vegetation types (using the classification developed by Champion and Seth 1968) can be identified within the Sanctuary:

- Moist teak-bearing forest 3B/C1
- Southern secondary moist mixed deciduous forest 3B/C2/2S2
- Riparian fringing forest 4E/RS1
- Dry teak forest 4A/C1
- Dry deciduous scrub 5A/DS1
- Dry bamboo brakes 5A/2S1
- Secondary dry deciduous forest 5A/C3
- Southern subtropical hill forests 8A/C1
- Southern montane wet temperate forest 11A/C1
- Southern montane wet scrub 11A/DS1
- Southern montane wet grassland 11A/DS2

Apart from biotic pressures, invasion of weed species, low rainfall and prolonged dry periods have resulted in shaping the present vegetation.

The land cover map of Chinnar prepared from the satellite imagery of 1993 indicates that there are 11 land cover categories in which eight vegetation categories can be differentiated: dry deciduous forests (28 per cent), scrub forests (20 per cent), exposed rock (16 per cent), barren land (11 per cent), degraded forest (8 per cent), moist deciduous forest (7 per cent), agricultural land (6 per cent), riparian forest (1 per cent), forest plantation (1 per cent), grassland (1 per cent) and *shola* forest (0.2 per cent).

Fauna

CWS harbours a wide range of herbivores and carnivores, including the Elephant (*Elephas maximus*), Sambar (*Cervus unicolor*), and Barking deer (*Muntiacius muntjack*), Wild boar (*Sus scrofa*), Bonnet macaque (*Macacca radiata*), Nilgiri langur (*Presbytes johnii*), Wild dog (*Caun alpinus*), Small Indian civet (*Viverricula indica*), Tiger (*Panthera tigris*), Leopard (*Panthera pardus*), Malabar giant squirrel (*Ratufa indica*), Grizzled giant squirrel (*Ratufa macroura*) and Gaur (*Bos gaurus*). The latter four are recorded as threatened species.

Socio-economic profile

Tribal communities

In the CWS, there are 11 settlements belonging to the tribal groups Muduvas and Hill Pulayas (Table 24.1). Muduvas follow a traditional way of life and appear to have a wealth of traditional knowledge in agriculture and medicine. The Hill Pulayas, on the other hand, are not original tribals of this Sanctuary. They migrated from Tamil Nadu and engage themselves mainly as agricultural and non-agricultural labourers for the Muduvas, and for other farmers residing outside the Sanctuary.

Existence of a social council in each settlement, especially in those of Muduvas is still common. This social council, created to discuss and solve problems faced by its members, is headed by the *moopan*. The *moopan*, who is therefore, the socio-political and religious leader of the hamlet, presides over the meeting of the council. However, in recent days, the position of *moopan* in the community has been eroded due to the influence of outside socio-political forces.

Cropping patterns

Villagers primarily carry out mixed cropping of staple food crops, and lemongrass (*Cymbopogon flexuosus*) cultivation. Staple food

Table 24.1: Population size and land area in different settlements in Chinnar Wildlife Sanctuary

Settlement Name	*No. of Houses*	*Population (Approx.)*	*Extent of Area Allotted**
Hill Pulayas			
1. Champakkad	27	200	120 acres
2. Palapetti	30	120	80 acres
3. Itchampetti	35	175	100 acres
4. Alampetti	47	190	180 acres
Muduvas			
1. Thaiannankudi	11	70	70 acres
2. Iruttalkudi	26	150	90 acres
3. Pudukudi	14	60	(not known)
4. Olikudi	17	100	(not known)
5. Mavelkudi	15	65	(not known)
6. Mangapara	14	67	(not known)
7. Ollavayal	16	68	(not known)

* According to informants of each settlement.

crops such as finger millet (*Eleusine coracana*), maize (*Zea mays*), sweet potato (*Ipomoea batatas*), amaranth (*Amaranthus* spp.), beans (*Dolicos lablab*), mustard (*Brassica camprestris*), pigeon pea (*Cajanus cajan*) etc., are being cultivated, following enforced sedentary form of slash and burn agriculture system with short (three years) fallow period (see later sections). A census survey in nine settlements in the Sanctuary made during the period 1995–96 indicated that about 1 per cent (86.7 hectares) of the total area of the Sanctuary is under cultivation of which about 0.45 per cent (40.5 hectares) is under mixed food crop cultivation while about 0.52 per cent (46.7 hectares) is under lemongrass cultivation. (The seeming discrepancy between this figure of 1 per cent and the satellite-based land use figure of 6 per cent under agriculture given earlier, is explained by the fact that the latter is the total land used for cultivation during the full cycle of shifting cultivation whereas the former is the land used in any given year, the rest lying fallow.) The size of the total farm holding per family is not more than three hectares in any given settlement. This is because of factors such as how much time a man can spend for clearing the land without sacrificing other economic activities such as cattle rearing and wage labour inside and outside the Sanctuary, and also how large an area the home labour will be able to manage. Although there is no limitation on how large a field a family can farm, there is a limit on how large a field a family can manage.

Official efforts at conservation

History

Before the declaration of Chinnar Wildlife Sanctuary the area was known as Chinnar Reserved Forest. Selection felling operations were carried out by the FD in 1979 in some localities. The area, further degraded by grass invasion and fire, has been reduced to discontinuous thorny thickets and pseudo-steppes. In order to repair the damages caused by selection felling to the forest ecosystem, attempts have been made to raise plantations of a few exotic and indigenous species of trees. Although most of these plantations are small experimental plots of a few hectares, the activity of preparing the land for plantations has resulted in opening up of canopy and weed invasion. At the same time, most of these plantations have failed and resulted in open scrub jungle (Ramesan, 1990a). Though the area receives low annual rainfall, the sporadic but heavy downpours during the NE monsoon cause heavy runoff in canals and small streams. The erosion of the surface soil has been a serious threat to soil already exposed by fire. This is indicated by a high gravel content of the soil in most parts of the forest area. Apart from this, anthropogenic pressures mainly from outside the Sanctuary are in the form of road traffic, firewood collection, grazing, agricultural activities and fire. Taking into account the wide spectrum of biological diversity in the region, and the magnitude of the threats, the Government of Kerala declared the area as Chinnar Wildlife Sanctuary in 1984. Now this Sanctuary is under the control of the Wildlife Warden of the Idukki–Eravikulam Wildlife Division.

According to sources from the FD their efforts are giving promising results in terms of wildlife conservation and ecosystem rehabilitation. They claim that after the declaration of the area as a Sanctuary, smuggling, poaching, incidence of fire, and illegal harvesting of natural resources such as firewood and non-wood forest produce have declined drastically. Similarly, the nature education programme being conducted by the Sanctuary managers for the people in and around the forest is helping to convey the importance of nature and environmental conservation, and offering opportunity to enlist the people's support in implementing policies and programmes.

Drawbacks

Mounting evidence suggests that environmental policies and developmental interventions themselves have, at times, been the contributing factors for poor results in conservation and ecosystem rehabilitation efforts. Some of these are: (*a*) introduction of inappropriate planning and implementation, and an overly optimistic drive of planners to enforce totally different types of land use systems; and (*b*) impacts of insecurity among the local communities regarding the land and resource tenure. These and other reasons have led to serious people–Sanctuary conflicts.

The establishment of settlements for tribals and forcing them to practice a near-sedentary form of slash and burn agriculture with short fallow periods is one such management practice which is adversely affecting the conservation in the Sanctuary. In earlier days, cultivators were using sloping lands with fertile soil interspersed with rocky landscapes situated at higher elevation to cultivate mixed food crops. This is because these locations are suitable to crops since mist provides the necessary moisture and crop raiding by animals such as elephants is minimum. At the same time, it may be pointed out here that even now, in any given year, area under food crop cultivation by all settlements is not more than 0.47 per cent of the total area of the Sanctuary. However, the Department feels that the traditional agricultural system of tribals is one of the major causes for ecosystem degradation. Therefore, the Department has forced the tribals to adopt a sedentary form of cultivation. This kind of cultivation, however, leads to the use of rather unsuitable lands for cultivation of mixed crops in most of the settlements, as there is no option for selecting suitable lands. Therefore, as the informants of each settlement put it, the number of families who cultivate food crops and the area under such cultivation is declining drastically.

Consequent to this is the introduction of lemongrass cultivation by the tribals in the Sanctuary. The diffusion of lemongrass leads to both positive and negative impacts on the ecosystem and on the farmers:

Advantages:
(To farmers)

- Easy crop to manage
- Suitable crop for drought condition

- Crop damage by animals (elephants) is minimum
- Income-generating profitable crop
- Labour intensive
- Can be used as guard plant of food crops against herbivores
- Controls insect pests to a certain extent

(To forest)

- Prevents and/or does not cause soil erosion
- Rocky waste lands can also be utilized to grow this crop
- Debris of the lemongrass is not being removed from forest area. Thus, it will be recycled in the system itself

Disadvantages:

(To farmers)

- Does not have any food value
- The waste grass (extracted) is not being used
- More fuelwood is required for distillation
- Price of the oil is determined by merchants, not by farmers
- There are chances of exploitation of farmers by merchants

(To forest)

- Leads to damage and use of fuelwood trees of the forest
- Causes water pollution

It is estimated that every year 11,000 kilograms of fuelwood is being used for distilling oil from lemongrass cut from 1 hectare land. The number of mature trees, stumps, tree saplings and seedlings in a 1 hectare lemongrass farm was 84, 12.3, 11.7 and 37.5, respectively. This indicates that the regeneration of trees in and around lemongrass farms is poor, and at the same time whatever trees do grow are severely damaged (KFRI 1996).

Non-timber forest produce (NTFP) has been playing a very important role in the tribal economy. The rights of tribals to NTFP for bona fide domestic consumption and for sale are recognized in these Reserved Forests (Sreekumar and Aravindakshan 1986). Some of the NTFP which are being collected by tribals are *Mangifera indica* (mango), *Terminalia chebula* (Chebulic myrobalon), *Emblica officinalis* (Gooseberry), *Acacia concinna* (*Cheevakka*), *Acacia intsia* (*Inja*), *Phoenix loureii* (*Icham pullum*), *Terminalia bellirica* (Belleric myrobalon), *Sapindus emarginata* (soap nut), *Garcinia gummi-gutta* (*Kodam puli*), honey and wax. However, due to the non-availability of such produce near their new settlements and the risk involved in collection, the tribals are least interested in gathering and selling NTFP. A tribal cooperative society situated in the nearby town procures the NTFP from the tribals. The establishment of tribal cooperative societies helped partly in eliminating private traders and partly in providing a fair price to the gatherers. According to the local people, the procurement prices of the products are determined by the government and revised during certain years. But, in general, the price increase declared by the government is only marginal and significantly lower than the open market price. Though the societies could succeed in eliminating private traders, they have failed to give a fair price to the tribals considering the factors such as risk and time involved in the collection and distance to be travelled. Therefore, NTFP collection is now not prevailing as a dominant sub-system of the land use in the tribal community. In the meantime, instead of developing new and efficient methods for fixing the price of the NTFP, the Department has assigned the task of collection and selling of these products to contractors who in general belong to areas outside the Sanctuary. At present, with the large appreciation in value and also importance of some items of the minor products in the state economy, the collection operations have been commercialized. It is also important to mention here that at present, tribals are working on wage basis in collecting forest products, and they are deprived of their rights to collect, use and sell these products. This will inevitably lead to some friction between the forest dwellers and managers.

An increasing number of fire incidents in the Sanctuary in recent years can be attributed to the insecurity in land tenure and job opportunity for the forest people. Not only the management plan of this Sanctuary (Ramesan 1990a) but also of other Sanctuaries and Wildlife Reserves (Gopinathan 1990; M.S. Nair 1990; Ramesan 1990b; 1990c), where tribals are a part of the ecosystem, have stressed that biodiversity conservation and wildlife management are possible only if the tribals are relocated outside the Sanctuary. This kind of attitude from the forest managers is responsible for the unrest in tribal minds and a feeling of alienation from the forests of the Sanctuary. This manifests as forest fires, the worst threat to forests and biodiversity. Similarly, the policy of cutting down the number of fire watchers by the Department means unemployment for a large number of tribals. Therefore, a widespread protest against such a policy has also to be seen.

Even for the conservation of genetic wealth of crops prevailing in the Sanctuary, enforced sedentary form of cultivation is not a suitable land use system. For example, Meenthani kepa, Poovan kepa, Akkini kepa, Elam kepa, Poothalashi, Kuruvi, Thotath mutti, Cheriya kepa, Karim kanni and Matti kepa are the 10 varieties of finger millets which have been recorded from the tribal settlements of the Sanctuary (KFRI 1996). Traditional cultivators maintained these varieties of finger millets because they could stagger the planting depending on the conditions under which a given variety will do best. Enforced sedentary form of cultivation by providing only limited kinds of microsites can ultimately cause genetic erosion of the crops.

Conservation and development efforts of other agencies

Apart from the FD, other government departments are also implementing various schemes within the Sanctuary. For example, in the Champakkad Settlement, the Water Authority has installed a diesel pump-set to irrigate dry lands and to provide drinking water to the settlers. However, the water supply did not last long as the local people were not in a position to run and maintain pump-sets. The Department of Animal Husbandry plans to assist tribals to rear cattle and cultivate grass for fodder so that the pressure on forest land could be reduced while the economy of the tribals is improved. However, like many other projects, this project may also work well in the initial stages due to external financial support. Once the input in terms of money, fertilizers, water, etc. is stopped, fodder cultivation is likely to be discontinued and pressure on the forest for fodder requirement would increase manifold. It is also necessary to mention here that most of these schemes are being implemented by other Departments without informing the Sanctuary authorities.

Need for community-based conservation approach

Based on the above mentioned observations, it is clear that ineffectiveness of conservation programmes taken up by the government departments, is due to policies and methods and lack of understanding of the socio-cultural and economic conditions existing at the landscape level. In this context, it is necessary to view tribals and their land use systems, and forestry and biodiversity, as part of a single ecosystem.

In the past several years, suggestions have been made to consider community-based systems for effective resource conservation and management (Lawry 1990; Wells and Brandon 1992). It seems that the community-based approach is suitable even for the conservation of biodiversity in Chinnar Wildlife Sanctuary. One reason for this is that the local community has the potential to provide protection to the forests, in the form of thousands of guardians. However, this is possible only when security of land tenure is provided and adequate opportunities for earning livelihood are provided to the people. This will also stop unsustainable resource use. Therefore, policies such as displacement of tribals from the Sanctuary and forcing people to adopt sedentary form of agriculture should not be undertaken.

Muduvas and Hill Pulayas of the Sanctuary can be described as multi-user strategists. They combine agro-pastoralism with periodic wage labour and NTFP collection. They have extensive environmental knowledge, are widely spread in small groups, and have relatively simple but efficient technology. Long-term sustainability of the Sanctuary may be ensured by encouraging them to continue to use social mechanisms to resolve community conflicts including bringing pressure on individuals involved in over-exploiting resources. A strong community organization could also serve to control destructive resource exploitation by outsiders.

Agricultural systems are often repositories of unique genetic variability. As mentioned above, there are several varieties of finger millets grown by tribals of the Sanctuary. Similarly, one can also find technological diversity to deal with this 'cultural biodiversity', the natural vegetation and also the local soil moisture content. Community-based efforts will provide an opportunity to recognize and support indigenous adaptive technologies as well as help search for new or improved appropriate technologies for the ecosystem regeneration and biodiversity conservation.

Traditional farming systems use local products and local techniques, have roots in the past and have evolved to their present state as a result of the interaction with cultural and environmental conditions of the region (Gleissman 1985). But, the local adaptation does not make the farmer non-innovative and tied to unchanging methods (Chandrashekara 1986). Reliance on local materials, energy sources and technical knowledge does not imply a lack of willingness to try something new (Padoch and de Jong 1987). Certainly, no traditional agricultural community is today doing precisely what it was doing a generation ago. This is true in case of the Chinnar Wildlife Sanctuary also. The introduction of lemongrass reveals the farmers' willingness to innovate and experiment. The diffusion of this new crop allowed farmers to become more productive and economically efficient. It may also be pointed out here that from among different sub-systems of the land use in the Sanctuary, a larger emphasis is placed on lemongrass cultivation. Because of this it appears that there is less interest among farmers in other sub-systems such as hunting and gathering. This sub-system is responsible for degradation of forests due to extraction and use of fuelwood. Proper strategies such as developing agro-forestry based on a mix of lemongrass, fuel, and multipurpose plants, with the involvement of the community, can make the lemongrass farming system ecologically viable and sustainable in this area. At the same time, this participatory effort will also provide an opportunity to the community to understand and analyze the strengths and weakness of lemongrass cultivation, especially in relation to the conservation values of the Sanctuary. To build up the trust between the people and the managers of the protected area (PA), it is important that a will is expressed by the PA managers to protect the forest people and their habitats from being swallowed up for the profit of others.

Discussion with the members of different settlements in the CWS indicated that more stress should be given to conservation and sustainable utilization of natural resources. Local people argue that they must have secure access to resources, including land, labour, capital and development-related information in order to survive (KFRI 1996). They must also be given an opportunity to participate in determining the kinds of projects to be implemented in the area. People can no longer be considered, as mere 'target groups' in the resource management process. Yet, it is common for the implementing agencies to view people as 'problems' and regard themselves as embodying the 'solutions'. Such paternalistic and technocratic attitudes have not worked in the past and are doomed to fail in the future also. Community concerns and involvement should be seen as crucial to biodiversity and habitat conservation.

Proposals for community-based conservation efforts

In the past, in each settlement in the PA, traditional social councils with full social powers were active as governing institutions. Traditional practices, indigenous knowledge and strategies regarding conservation and sustainable management of natural resources were supported and controlled by these social councils. These were highly sophisticated systems and were adapted to local situations. However, many of these have been lost or broken down in recent times, due to mounting pressures from outside. Revival and/or strengthening of these traditional social councils should be the basis for CBC.

At the same time, the local communities should have the right to use the resources, to determine the mode of usage, and to determine the distribution of benefits accruing (Murphee 1993). Indigenous conservation strategies evolved, modified and adopted by forest dwellers are often sustainable and well suited to the complex local ecological, socio-cultural and economic conditions. However, these strategies are not easily accessible to forest policy-makers and scientists since they are not recorded in the written form. Future efforts on CBC should consider and incorporate these forms of threatened knowledge. To achieve the above, identifying indigenous resource management practices, harmonizing indigenous and modern technologies, conducting research with the local communities, while concentrating mainly on finding solutions to conservation and sustainable management of natural resources, are important. Subsequently, the policies should be formulated such that they ensure genuine participation by all sectors of the communities in micro-level planning and implementation in their area.

REFERENCES

Champion, H.G. and **S.K. Seth.** 1968. *A Revised Survey of the Forest Types of India.* Government of India, Dehradun.

Chandrashekara, U.M. 1986. Strengths and Weaknesses of Traditional Systems of Bamboo Cultivation in Rural Kerala. *Agroforestry Forum* 7: 21–23.

Gleissman, S.R. 1985. Economic and Ecological Factors in Designing and Managing Sustainable Agro-ecosystems, in T.C. Edens, C. Fridgen and S.L. Battenfield (eds). 1985. *Sustainable Agriculture and Integrated Farming Systems.* Michigan State University Press, Michigan.

Gopinathan. 1990. *The First Management Plan for Wyanad Wildlife Sanctuary: 1990–91 to 1999–2000.* Kerala Forest Department, Trivandrum.

KFRI. 1996. Biodiversity Conservation within the Context of Traditional Knowledge and Ecosystem Rehabilitation in Chinnar Wildlife Sanctuary. Interim Report of MacArthur-UNESCO funded Project. Kerala Forest Research Institute, Peechi, Kerala. (mimeo).

Lawry, S. 1990. Tenure Policy Toward Common Property Natural Resources in Sub-Saharan Africa. *Natural Resources Journal* 30: 403–22.

Murphee, M.W. 1993. *Communities As Resource Management Institutions*. Sustainable Agriculture Gatekeeper Series No. 36. International Institute for Environment and Development, London.

Nair, M.S. 1990. *Management Plan for Periyar Tiger Reserve*. Kerala Forest Department, Trivandrum.

Nair, S.C. 1991. *The Southern Western Ghats: A Biodiversity Conservation Plan*. Indian National Trust for Art and Cultural Heritage, New Delhi.

Padoch, C. and **W. de Jong.** 1987. Traditional Agroforestry Practices of Native and Ridereno Farmers in the Lowland Peruvian Amazon, in H.L. Gholz (ed.). 1987. *Agroforestry: Realities, Possibilities and Potentials*. Martinus Nijhoff, Dordrecht.

Puri, G.S., V.M. Meher-Homji, R.K. Gupta and **S. Puri.** 1983. *Forest Ecology, Vol. 1*. Oxford & IBH Publishing Co., Calcutta.

Ramesan, R. 1990a. *The First Management Plan for Chinnar Wildlife Sanctuary: 1990–91 to 1999–2000*. Kerala Forest Department, Trivandrum.

———. 1990b. *The First Management Plan for Idukki Wildlife Sanctuary: 1990–91 to 1999–2000*. Kerala Forest Department, Trivandrum.

———. 1990c. *The First Management Plan for Thattekad Bird Sanctuary: 1990–91 to 1999–2000*. Kerala Forest Department, Trivandrum.

Sreekumar, K. and **M. Aravindakshan.** 1986. Creation of a Life Supporting System for Tribals: A Key Element in the Eco-development of Western Ghats, in K.S.S. Nair, R. Gnanaharan and S. Kedarnath (eds). 1986. *Eco-development of Western Ghats*. Kerala Forest Research Institute, Peechi, Kerala.

Wells, M. and **K. Brandon.** 1992. *People and Parks: Linking Protected Area with Local Communities*. World Bank, WWF and USAID, Washington, D.C.

25

Communities protecting coastal resources: Rekawa Lagoon, Sri Lanka*

S.U.K. Ekaratne, John Davenport,
D. Lee and R.S. Walgama

Introduction

Sri Lanka is a relatively small tropical island of 65,610 sq. km having a coastline of 1,585 km (NARESA 1991) and situated off the southern tip of India. Thirty-two per cent of the present population of 18 million people are concentrated along the coastal areas of the island (CCD 1992), with increasing pressure being brought to bear on natural resources leading to their over-exploitation, degradation and alteration.

Among the coastal resources, brackish water habitats such as mangroves, lagoons and estuaries have become degraded and altered through conversion into ponds for the culture of the tiger shrimp or prawn, *Penaeus monodon* (the terms shrimp and prawn have been used interchangeably in this paper). The high investment return of this export-oriented industry has witnessed the entry of eager investors clamouring to enter this field of economic activity resulting in unbridled expansion. The unregulated expansion of prawn farming in brackish waters which is presently concentrated along the north-western coast has resulted in multiple problems, including the disruption of the socio-economic structure of local coastal communities, the large-scale destruction of brackish-water natural resources such as mangroves, lagoons, estuaries and the alteration of water quality

* Many persons at the project site, including Mssrs D.A.L. Nimal, Tissa Ariyaratne, Ranjith (the schoolteacher mentioned in the text) as well as many fishermen of Rekawa assisted us in carrying out our work. This assistance is gratefully acknowledged.

through unregulated discharge of prawn farm effluent into coastal habitats (Corea et al. 1995). The recent outbreak of white spot disease in farm-cultured tiger shrimp, that resulted in the closure of the majority of operational prawn farms, is a direct result of this deterioration of water quality, which is also the prime reason for depressed production that this industry is presently experiencing.

The depressed production of high value tiger shrimp from existing prawn farms, primarily due to poor management practices, has induced investors and planners to examine the availability of other suitable locations for the future expansion of this industry. The coastal habitats of the south have been identified as potential areas for extending tiger shrimp farming, with one such area being the lagoon in a village known as Rekawa. A quiet village of 5,373 inhabitants, it retains traditional and conservative lifestyles and value systems.

The Rekawa lagoon has a community of traditional fishermen who engage in seven-month prawn fishery using the *kraal* traps (17 per cent), cast nets (9 per cent), and gill nets (74 per cent), and with the aid of traditional non-mechanized boats. The *kraal* trap is a passive trap that is made from vertically positioned thin panels of bamboo lashed together to form a basket-like structure using coconut fibre rope. The intervening spaces in this basket-trap act as a mesh barrier, and trap the fish and shrimp that travel into it. Valve-like wings prevent the escape of the trapped animals and are collected by the fishermen at predetermined intervals. Predominantly two varieties of tiger shrimp (prawns) *P. indicus*, and (in lesser quantities) *P. monodon*, are harvested by the closely-knit conservative rural fishing community.

The fishing community has adopted several participatory measures to sustainably manage their aquatic resources against depletion and outside intervention. The successful adoption of management practices that are seen as benefiting the community has been made possible largely through the fisherfolk banding together to form a Lagoon Fishermen's Association. This Association now stores the prawn catch in a community freezer to sell it directly to consumers whereas it was traditionally sold to itinerant middlemen at the landing sites. Periodically, the community reviews the status of their lagoon prawn resources and adopts resource management practices through regulating fishing gear and fishing effort. A programme of stock enhancement of lagoon prawns was discussed with the fishing community. With their participation, the stock enhancement programme was started recently. The fishing community is earnestly learning the process in the hope that they would be able to practice it themselves as an environmentally and socially friendly alternative to polluting prawn farms and as a way of sustainably managing their own lagoon resources.

As an alternative to the potentially destructive practice of prawn farming, stock-enhancement of lagoon prawn was discussed as a part of the Special Area Management (SAM) Plan for the area (see later sections). The Enhancement of Rekawa Lagoonal Prawn Fishery Project, implemented by the Universities of Colombo and Millport, Scotland, was started in mid-1995. The aim of the project was to enhance the natural *P. monodon* population of the lagoon. The project was planned and implemented with the active participation of the fishing community. It was funded by the ODA (MRAG), UK. The present paper is an analysis of this stock-enhancement programme, looking at how such a programme can empower and induce the local stakeholder community to manage their natural resources to benefit them equitably with minimal ecological disturbance and least social disruption.

Area profile

Rekawa is a village located about 200 km south of the capital, Colombo, and situated along the south coast in Hambantota district in the Southern Province of Sri Lanka. The Hambantota district is the least developed district in Sri Lanka (Sullivan et al. 1995).

The majority of the villagers are self-employed. About half of the village community at Rekawa is engaged in agricultural pursuits (47 per cent) while over a quarter (28 per cent) is engaged in fisheries. Around 10.4 per cent of Rekawa inhabitants fish in the lagoon (Ranaweera Banda et al. 1994). The low-income status of the Rekawa villagers can be gauged from the high percentage of families (54 per cent of the 1,184 families) that are recipients of the government-assisted scheme through which low-income families receive financial assistance—the Janasaviya Programme. Housing-related facilities such as sanitation, pipe-borne water, electricity, etc., are available only to a minority of houses. About half the population (48 per cent), is in the 19–55-year age group. Thirty per cent is in the schoolgoing 5–18-year age group and 10 per cent of the villagers are above 55 years (Ranaweera Banda et al. 1994).

Due to the low-income status of many villagers, the women in the community engage in whatever activities are available for supplementing their meagre family incomes, particularly by utilizing easily available natural resources from the surrounding environment. Some of the readily available natural resources such as mangrove, fuelwood and coral lime are extracted and used or/and sold, even though their exploitation is illegal. For example, 8.5 per cent of the villagers extract coral lime from the sea (Ranaweera Banda et al. 1994), even though the mere possession of coral has been a punishable offence for over 20 years (Ekaratne 1990).

The Lagoon

The Rekawa Lagoon is a 250 hectare brackish waterbody that is fringed by 200 hectares of mangroves. It has a freshwater input through the 32 km long Kirama river and the 42 km long Urubokka river. Its connection with the sea is closed periodically by sand-bar formation. With the closure of the lagoon mouth, the water level rises until the threat of inundation of upland farmed land compels the villagers to breach the sand-bar manually, thus allowing the water level to subside.

The major commercially important occupation of the people in the Rekawa village is fishing. The lagoon waters are inhabited by both finfish and shellfish, based on which the lagoon has sustained a fishery from the time that records have existed. Several edible species of finfish have been recorded which are exploited mainly for subsistence fishery. The commercial fishery is mainly for shrimps where *P. indicus* forms the bulk of the harvest, with the more expensive tiger shrimp, *P. monodon*, making up a smaller amount of the catch (Ekaratne et al. 1996). Other lagoon resources comprise mangroves which cannot, however, be legally extracted.

Organization, benefit-sharing and prevention of misuse

Increased generation of income was, therefore, thought to be a measure that could divert the villagers away from over-exploitation of such natural resources. These were some of the considerations that led to the Rekawa area being included as a site for implementation of a Special Area Management (SAM) Plan drawn up and implemented by the Coastal Resources Management Project (CRMP) under USAID sponsorship.

The drawing up of the SAM Plan report had engaged the services of shrimp farm developers from the north-western prawn farm areas to identify suitable areas for prawn farming. These farmers were already eager to expand their farming operations to the south coast and even the final SAM report speaks somewhat favourably of introducing prawn farming to selected areas of the Rekawa lagoon. However, the villagers had strenuously voiced and demonstrated their protest against the introduction of such practices, in view of the environmental and socio-economic disturbances that the north-western areas of the island had experienced in the wake of prawn farm expansion. There was also the fear of effluents from prawn farming reducing natural productivity of the lagoon (RSAMCC 1996).

Early records of shrimp catches from the Rekawa lagoon had indicated the existence of tiger shrimp in the lagoon. With this background, several discussions were conducted with the villagers proposing stocking of the lagoon with tiger shrimp as an environmentally and socio-economically friendly alternative to intensive prawn farming. The harvest from the lagoon stocking exercise was to go to the fishermen who would manage the fishery through their Rekawa Lagoon Fishing Cooperative Society (RLFCS). It was thought that the increased income from the tiger shrimp harvests would remove the economic compulsion for fishermen's families to harvest natural resources such as coral lime and mangrove wood, so that while the prawns/shrimps stocked into the lagoon would involve stakeholder villagers in Community Based Resource Management (CBRM), it would also lead to the better management of the natural coastal resources of the Rekawa area.

The early part of the programme involved the appraisal of the prawn fisheries and water characteristics of the lagoon through fortnightly sampling. This data base was used to assess the numbers of tiger shrimp that the lagoon could sustain as well as the lagoon areas and times of the year that would be favourable for stocking without adversely affecting existing lagoon fisheries. Following such assessment, stocking of lagoon waters with tiger shrimp was carried out with the active involvement of the local village community where they were shown the procedures that were necessary for a successful restocking operation.

The major occupation of the villagers being lagoon fishing, they ascribe high importance to any activity related to the fishery resources of the lagoon. When the possibility of outside businessmen coming into Rekawa to open up prawn farms became a topic of discussion, the local fisherfolk were induced to band together, realizing that a united force could resist business pressure and interests more strongly than individual voices of local fisherfolk. Thus was formed the RLFCS.

The RLFCS was registered in 1995, the first such lagoon fishery society to be registered. The RLFCS participated in a National Aquatic Resources Agency (NARA) research exercise on shrimp fishery and ecology. This led to the formulation of management guidelines for the lagoon by the RLFCS. The society also takes responsibility of the management of the lagoon (RSAMCC 1996).

The RLFCS is a voluntary organization composed of the fishermen acting on an equitable basis. The membership has regular meetings at a community centre or in a local school where common issues and areas of conflict are discussed. Following such discussions, the membership will adopt a majority decision which becomes binding on all its members. Conflicts are therefore resolved by discussion and consensus, although there have been instances where some members have acted against the majority decision in briefly catching prawns over banned periods, until the community discovered this and adopted remedial measures.

The members elect, from among its membership, a President and a Secretary at its Annual General Meeting. Meetings are held regularly throughout the year and are chaired by the President of the RLFCS, or in his absence, a pro-tem Chairman is elected at that meeting. Outside people are invited to attend the meetings, specially when matters are to be discussed in areas where the membership lacks expertise, such as the first author being invited when shrimp restocking was being discussed. The local schoolmaster often serves as an advisor to the society membership and since he is a learned member of the village, fisherfolk often seek counsel from this experienced schoolmaster who has the welfare of the members at heart and exerts a productive influence on the community.

The structure of the RLFCS is thus simple, with a President and Secretary serving as its principal officers and being empowered to represent them at government or other outside meetings. This representation is carried out in a responsible manner as the issues are discussed and a majority of the decisions arrived at, prior to presenting views at other fora. Since all members are available at a local level, difficulty is not experienced at convening local meetings.

High participation and awareness levels may be due, in part, to the role of the 'catalyst' in the Rekawa lagoon area. This catalyst is one of nine young people from the community, trained and then posted as volunteer community organizers in January 1995. Their responsibilities are to work closely with village officials, community leaders and volunteer organizers to create awareness of environment and natural resources conservation, to help strengthen organizations and develop self-reliance (RSAMCC 1996). These catalysts were posted in the Rekawa area divisions and one each was appointed to work with the RLFCS and the Federation of Womens' Organizations.

Community mobilization in Rekawa has been a strength of the SAM process at Rekawa. The Rekawa Steering Committee, which influences resource management patterns in the area, arrives at decisions that are in consonance with the local communities' wishes (Lowry et al. 1997), since local communities make their voices heard as they are represented at the meetings. There is also a high level of local leadership. The Presidents of the RLFCS and other local bodies such as the Rekawa Development Foundation are able to represent community concerns effectively to government agencies.

Successes and opportunities

Monitoring studies of the fishery to assess the success of the restocking programme revealed that the catch of tiger shrimp before the restocking exercise amounted to only 2.5 per cent by weight, while after the restocking programme tiger shrimp catches increased to 100 per cent by weight, indicating that the restocking programme was successful.

The success of the villagers at warding off outside businessmen from starting prawn farms using the lagoon resources that were traditionally available to them is considered as a major benefit that arose from forming the RLFCS. They feel that environmental degradation from prawn farms would have affected lagoon productivity and, consequently, their fishery income. They also feel that, as a Society, they were able to prevent the social repercussions of changed lifestyles that high-income generating prawn farms bring to rural villages through the influx of paid workers and high technological inputs that would invariably have affected their quiet and traditional lifestyles. Another benefit that fisherfolk perceive as having flowed from the RLFCS is that they have become empowered by being able to send their elected representatives to meetings and fora to which they could not transmit their views earlier. They feel that their presence and rights are now being recognized by the state machinery and by officials outside the village. The prawn restocking programme is also viewed as a benefit stemming from forming the RLFCS, since they were able to express and discuss concerns, benefits, modalities and timing of prawn releases/captures through meetings convened by the Society.

The immediate day-to-day benefits of the RLFCS to its members arise through ensuring equitable benefits to its membership by a process of discussion on fishery harvests and marketing. When the fishing season commences, the members meet and discuss the number of gear types that would be allowed and the locations where these gears could be deployed. With the prawn restocking programme, members were able to realize and convince the fishermen that catching the prawns after they attained a size that fetched a higher market price was beneficial to all fisherfolk of the entire community, rather than individual benefit that would accrue to individuals by catching small-sized prawns from this 'common property resource'. With such a belief, the RLFCS was able to impose a ban on fishing over a period of time until the prawns grew to a readily marketable high-value size. In catching the prawns, members were able to allocate areas and gear types to the fishermen taking into consideration their traditional practices so that the high-value resource would not benefit only a few fishermen of the village. The RLFCS negotiated, and now owns, a freezer where the fishery catch is stored and then sold directly to the consumer and hoteliers, fetching better prices for the fisherfolk.

The sale price ranges from Rs 400 to 500 per kg bringing a landed value of around Rs 90,000 (production costs for the release were about Rs 85,000). Although the catch has tailed off lately, fishermen continue to catch a few specimens since these larger-sized prawns command a premium price. With these catches being accounted for, the cost of the restocking operation has been recovered so that the restocking along with community participation has proved a success. This cost-recovery calculation does not take into account the conservation of the environment that has taken place with the restocking programme that is reported here, as compared with intensive prawn farming practices that would have adversely affected the coastal habitats concerned.

Failures and constraints

Although the association is not empowered legally to impose punishments, social pressure and chastisement usually act as a sufficiently strong punitive measure against misuse of resources within the close-knit traditional communities as existing in Rekawa. Lacking legal teeth to adopt punitive measures can be viewed as one of the problem areas which needs to be examined. Even so, when individual fishermen adopt fishing practices that are viewed as unsustainable or as adversely affecting the harvests of a larger group of other fisherfolk, the association would inform the local village-level official and the police and would ensure that the officials take remedial action.

The community appreciates that the release exercise has brought them good returns through increased catches of *P. monodon*. Some of them have expressed a wish that another release exercise be carried out in the lagoon, indicating that socially acceptable and environmentally friendly prawn farming can be carried out with the participation of the stakeholder community.

The shrimp fishery in Rekawa lagoon resumed in late September/October for the 1996/97 season. This was after a community-enforced ban on fishing against the use of small-mesh gill nets (1.5 and 1.75 inches). The ban, which was enforced by the community as a safeguard against catching shrimp released by our project before they grew to a size that would command a high market price, was lifted 'officially' on 30 September (though a few fishermen had surreptitiously deployed nets of this size over the last week of September). During this enforced ban, the community allowed their fishermen to fish only with large-mesh nets (3.5 and 4 inches) so that the growing tiger shrimp would not be caught.

Conclusion

The success of the programme depended on the participatory measures that were collectively adopted and implemented by the community.

One of the first such facilitating measures was the formation of a RLFCS by members of the fishing community which has enabled them to manage their fishery resource sustainably. This was done through the community periodically reviewing the lagoon prawn resource and regulating fishing gear by regulating the mesh size, the effective size of the trapping mechanism of the *kraal* trap, boats, number of fishermen, outside fishing, etc. The RLFCS has become stronger and they now even store the prawn catch in a community freezer to sell it directly to consumers whereas it was traditionally sold at landing to itinerant middlemen. Through this practice fishermen also benefit by getting enhanced prices.

The programme of stock enhancement can therefore be viewed as an environmentally and socially friendly alternative to prawn farms provided that the community is actively participating and also learning the process of resource management.

The next step is to determine how this process can be further facilitated such as through institutionalizing tenurial rights, exclusion rights, etc., as well as giving the community a greater understanding of resource management.

REFERENCES

CCD. 1992. *Coastal 2000: Recommendation for a Resource Management Strategy for Sri Lanka's Coastal Region*. Vol. I and II. Coast Conservation Department, Sri Lanka.

Corea, A.S.L.E., J.M.P.K. Jayasinghe, S.U.K. Ekaratne and **R.W. Johnstone.** 1995. Environmental Impact of Prawn Farming on Dutch Canal: The Main Water Source for the Prawn Culture Industry in Sri Lanka. *Ambio* 24: 423–27.

Ekaratne, S.U.K. 1990. Man Induced Degradation of Coral Reefs in Sri Lanka. Proceedings of the Fifth MICE Symposium for Asia and the Pacific: Ecosystem and Environment of the Tidal Flat Coast Affected by Human Activities. Nanjing University Press, Nanjing.

Ekaratne, S.U.K., J.D. Davenport, D. Lee and **R.S. Walgama.** 1996. The Use of Stock Enhancement and Community Participation in Brackish-water Natural Resources Management in Sri Lanka. Regional Workshop on Management of Coastal Resources. Natural Resources Energy & Science Authority of Sri Lanka, October 1996.

Lowry, K., N. Pallewatta and **A.P. Dainis.** 1997. Special Area Management Plan at Hikkaduwa & Rekawa: A Preliminary Assessment. Coastal Resources Management Project, Sri Lanka.

NARESA. 1991. *Natural Resources of Sri Lanka, Conditions and Trends*. Natural Resources, Energy & Science Authority of Sri Lanka, Colombo.

Ranaweera Banda, R.M., A. Premaratne and **I. Ranasinghe.** 1994. People, Resources and Development Potentials in the Rekawa SAMP Area. Report to CCD and CRMP, Colombo.

RSAMCC. 1996. Special Area Management Plan for Rekawa Lagoon, Sri Lanka. Coastal Resources Management Project of the Natural Resource and Environment Policy Project, USAID, Sri Lanka.

Sullivan, K., L. de Silva, A.T. White and **M. Wijeratne** (eds). 1995. *Environmental Guidelines for Coastal Tourism Development in Sri Lanka.* Coast Resources Management Project and Coast Conservation Department, Colombo.

Glossary

anganwadi	Creche
adhikar patra	Letter of authorization/permit
adhyaksha	Head
adivasi	Tribal
akkini kepa	Variety of finger millet
akhand shikar/sendra	Mass hunt
andolan	Mass movement
ang	Chief
anupa	Creatures living in marshy habitats
ayurved	Indian traditional system of medicine
Azadi Bachao Andolan	A nationwide federation of organizations opposing the entry of multinational corporations and consumerism
bajra	Pearl millet
Baragaon ki Panchayat	*Panchayat* of 12 villages
Baragaon ki Samiti	Body of 12 villages
beedi	Hand-rolled country-made cigarette
bhaisaj kalpana	Pharmacy
Bhesaj Sangh	A medicinal plants cooperative union
Brahmin	Higher caste in the Hindu caste system
bhusaya	Creatures who live underground
Charak Samhita	Classical Indian text on medicine
chas	River island
chaurasi	Forest watcher
cheek ki mazdoori	The selling of *cheek* (the gum exuded by *Boswellia serrata*) to local traders after its collection from the forest
chena	Shifting cultivation
cheriya kepa	A variety of finger millet
chhada	A patch of forest earmarked to fulfill local resource requirements
chipko	Hug/stick
chitadar	Forest watchers
elam kepa	A variety of finger millet
dalit	oppressed classes

dev bani/oran	Forest patches dedicated to the gods or sacred groves attributed to the common deity of the village
dharadi partha	A traditional system of forest conservation in which different castes and tribes were prohibited from cutting trees for use as fuelwood, and each different *gotra* protected a particular species
Digambar Jains	A Jain religious sect
dravya guna shastra	A system of indigenous knowledge
Ganga *uthana*	Oath of honesty taken in the name of Hindu Goddess Ganga
gayana	Bhilala (a tribe) myth of creation
Gayatri *pariwar*	A Hindu religious sect
ghee	Saturated fats obtained from milk
gochar	Grazing land
gotra	Clan
gram sabha	Village assembly including all adults of a village or a hamlet
Gram Vikas Kendra	Village Development Centre
Gram Vikas Mandal	Village Development Association
guzara forest	Public wastelands assigned for local resource use, formerly under the jurisdiction of district administration, later under FD
Harijan	Literally—'people of God': traditionally outcastes/untouchables in the Indian caste system
hujra	Gathering place maintained by the village elite in parts of the North-West Frontier Province
jagirdar	Landlord
jalaja	Creatures living in water, e.g., fish
jalachara	Creatures who require aquatic habitats, e.g., swan
jirga	Council of village elders
johad	Water tank
jowar	Sorghum
jhum	Shifting cultivation
Kabir *panth*	A religious sect following Saint Kabir
kakad bani	Forest on the common geographical boundary of two-three villages
kamla	Infective hepatitis
karim kanni	A variety of finger millet
kavu/kenkri	Forest groves of arid regions of Rajasthan
khadi	Hand-woven cloth
khot ail	Nomadic herding camp

kraal	Passive fishing traps made of bamboo strips
Krsi-sastra	A system of knowledge pertaining to agriculture
Kulhari Bandh Panchayat	A *panchayat* that prevents the carrying of axes into the forests
kurulu paluwa	Bird's area
kuruvi	A variety of finger millet
laah	A labour-sharing custom
lhakhang	Altar
lok samiti	People's committee
maharaja	Emperor
mahila mandal	Women's association
Mahila Mangal Dal	Village Women's Organization
mahila samiti	Women's committee
mahila sangathan	Women's organization
Mahila Vikas Mandal	Women's Development Association
mahua	*Madhuca indica* and also the liquor brewed from its flowers
mana-pathi	The system of paying a fixed amount of paddy or other foodgrains for services rendered
mandala	Altar
manduwa	A variety of finger millet
matti kepa	A variety of finger millet
mauza	Territorial boundary
mazdoori	Wage labour
meenthani	A variety of finger millet
moopan	Leader of the hamlet
mukhiya	Local headman
nakedar	Deputy ranger or checkpost guard
nakhdul	Tubers of a plant
Nari Bikas Sangh	Women's Welfare Organization
Nay Sol	Buddhist sacred text from Sikkim
nevad lands	'New field', now also connotes fields on encroached forest land
nighantus	Lexicons
nirvana	Enlightenment with freedom from the bondage of rebirth
nistar	Usufruct rights
oraan	Forest grove of arid regions of Rajasthan
panch	Members of the *panchayat*
panchayat	Institution of local government at the village level
panchayati raj	System of local-level governance
panch patel	Handful of village elders

Pang-Lhabsol	A Buddhist festival
pannia	Bread cooked between leaves
parikrama	Circumambulation
parishad	Assembly
patel	Headman/person of prominence
pavitravan	Sacred grove
pilar kaan	Large sacred grove
podu	Hamlet
pooran kepa	A variety of finger millet
poothlashi	A variety of finger millet
pradeshiya sabha	A locally elected body
pradhan	Head
prakruti	Nature
prasaha	Creatures who pounce on their prey, e.g., lion.
pratuda	Birds that break food with their beak before eating, e.g., cuckoo
pushya nakshatra	One of the 27 *nakshatras* (position of stars) according to the Hindu calender
rajah	Native prince
rajakariya	A system of traditional service tenure
rakhat bani	Forest area that belonged to a single village independently
Ramanand *panth*	A religious sect following Saint Ramanand
Ramayana	Indian mythological text
rundh/balbani	Forest area protected by the king/jagirdar/ zamindar for the purpose of hunting
sabha	Council/meeting
sakshan	A forum of people's institutions
samiti	Committee
sarpa kavu	Snake God
sarpanch	Head of the village council/panchayat
satyagraha	Peaceful resistance or struggle
shamlet dehs	Forest grove of arid region of Rajasthan
shastra	Science, system of knowledge
shamilat	Common property
shikargah	Hunting reserve
shingo naua	Forest guard
shola	High altitude stunted forest
shramadan	Free labour
Siddha	Traditional system of Siddhi medicine
sthalaja	Terrestrial animals, e.g., deer
stupa	Buddhist monument
Swa-Sahyog sanstha	Self-help group/organization

talai/talav	Pond
taluka/talika/tahsil	Administrative division below that of a district
talukdar	Local headman
tarai	Grasslands and *sal* forests between the Himalayas and the Indo-Gangetic Plain
ter	Treasure
thain	A traditional institute for conflict resolution
thotath mutti	A variety of finger millet
tola	Hamlet
tum badloge, yug badlega	If you change, the age will change
Unani	Traditional system of Egyptian medicine
Van Gujjar	Nomadic pastoralists of north India
vani	Forest groves of arid regions of Rajasthan
Van Panchayat	Village Forest Management Committee
Van Samiti	Village Forest Committee
Van Suraksha Samiti	Village Forest Protection Committee
vara	Rotational
Vedic	Belonging to the historical period during which the Vedas (a collection of Sanskrit hymns) were compiled
visha vaidya	Traditional healer for poison cases
vishkira	Creatures that pick food by searching with beaks, e.g., hen
Vrksh-ayurveda	Indian plant science
Yuvak Mangal Dal	Village Youth Organization
zamindar	landlord

Notes on contributors

R.V. Anuradha is a graduate of the National Law School, Bangalore (southern India) and has worked on environmental law issues for some years. She has been a recipient of the Darwin Fellowship at the School of Oriental and African Studies, London. In particular, she has focused on the implications of the Convention on Biological Diversity for India, especially its tribal and other local communities. As a member of Kalpavriksh—Environmental Action Group, she recently undertook a study of a benefit-sharing arrangement between a formal research agency, a herbal drug company and a tribal group in Kerala. Ms Anuradha is currently working with the law firm Amarchand, Mangaldass and Shroff and Co., New Delhi.

Narayan Banerjee is the Deputy Director of the Centre for Women's Development Studies (CWDS), New Delhi, which has been providing facilitation support to Nari Bikas Sangh (NBS), a grassroots federation of peasant women's *samitis* in Bankura district of West Bengal. NBS and CWDS were the first organizations to demand that women be made members of the Forest Protection Committees in West Bengal. Mr Banerjee continues to provide NBS regular support in its lobbying and advocacy work to make the state government's JFM policy more sensitive to poor women's needs and priorities.

Amita Baviskar has done her doctorate on the Narmada Valley, examining the relationship between tribals, their environment, and mass movements. While there, she was deeply involved with tribal struggles as part of the Khedut Mazdoor Chetna Sangath and the Narmada Bachao Andolan. She has also been a member of Kalpavriksh—Environmental Action Group. She is currently Lecturer at the Sociology Department of the Delhi School of Economics, New Delhi.

Kamaljit S. Bawa is a distinguished professor of biology at the University of Massachusetts at Boston. He has worked on conservation and the environment for the last 30 years in central America and South Asia. He has authored or co-authored more than 130 papers in professional journals and edited several monographs or books. His research is supported by the National Science Foundation, the MacArthur Foundation, the World Wildlife Fund, the International Plant Genetic Resources Institute and the National Institutes of Health. He serves on the editorial boards of several journals and is or has been a member of several national and international advisory panels.

Seema Bhatt worked with VIKSAT in Ahmedabad on one of the first attempts at bringing participatory management into protected areas in India. She handled several activities at the Environmental Services Group of the World Wide Fund for Nature—India, and subsequently at the headquarters of WWF–I in Delhi. She is currently with the Biodiversity Conservation Network, involved with promoting enterprise-based biodiversity conservation at various sites in India and other parts of Asia.

U.M. Chandrashekara is with the Kerala Forest Research Institute, Peechi, Kerala, and has worked on several forest-related issues in Chinnar Sanctuary and other areas.

Mitali Chatterjee is associated with IBRAD, a Calcutta-based NGO which has done extensive training of forest officers of several states on participatory forest management. She has also organized several studies and field workshops for enhancing women's participation in Joint Forest Management in West Bengal, eastern India.

D. Choudhury was in charge of the G.B. Pant Institute of Himalayan Environment and Development, North-east India unit, at Dimapur, Nagaland. He is currently at the University of Silchar, Assam.

K. Christopher has a training in law, and has been working on legal issues relating to environment. He conducted fieldwork in the Dalma Sanctuary, Bihar, east India, as part of a project on Participatory Management of Protected Areas at the Indian Institute of Public Administration. He is currently practising law in New Delhi.

Priya Das is an anthropology student, currently pursuing her studies in Canada. She has worked on aspects of traditional science and technology, and at the Indian Institute of Public Administration, New Delhi, on joint management of protected areas. For this and her anthropological work on traditional resource use practices, she has spent considerable time with communities inside Kailadevi Sanctuary, Rajasthan, western India.

Soumyadeep Datta is the founder member and Director of 'Nature's Beckon', an environment action group which played a crucial role in the conservation of Chakrashila forests of Assam. They were also instrumental in getting it declared a Wildlife Sanctuary.

John Davenport is Professor of Marine Biology at the University of London, stationed in Millport in Scotland as the Director of the University Marine Biological Station. He has extensive experience in marine biological research and is particularly interested in the physiology of estuarine and marine fauna.

Philip J. DeCosse is a resource economist at International Resources Group (IRG), Ltd., in Washington, D.C., where he provides consulting and analytical services to a variety of public and private sector clients. His analytical work has focused on resource economics and policy, monitoring and evaluation, facilitation of farmer input into The Gambia's national agricultural and natural resource policy statement, provision of technical support to Sri Lanka's Environmental Impact Assessment process, and elaboration of a conceptual approach for a region-wide institutional impact monitoring framework for Central and Eastern Europe and the independent states of the former Soviet Union. He worked for 15 months in Sri Lanka in 1995/96. DeCosse holds a Master's degree in Agricultural Economics from the University of Wisconsin–Madison.

S.U.K. Ekaratne is a professor at the Department of Zoology in the University of Colombo, Sri Lanka. He is also presently functioning as the Director of the Staff Development Centre of the University. He is a trained aquatic ecologist with a postgraduate degree from the UK and is conducting research programmes in reef ecology, estuarine ecology, coastal biodiversity and how communities can be harnessed for environmental conservation and management.

Madhav Gadgil is currently Professor of Ecological Sciences at the Indian Institute of Science, Bangalore and Chair of the Scientific and Technical Advisory Panel of the Global Environment Facility. His research interests include Population Biology, Conservation Biology, Human Ecology and Ecological History. His interest in the study of sacred groves and other traditional practices of nature conservation dates from 1975 with 'Plea for Continued Conservation of Sacred Groves of India' published in the *Journal of Bombay Natural History Society*. He has co-authored *This Fissured Land* and *Ecology and Equity*.

Shyamala Hiremath is the Project Director of India Development Service (IDS), an NGO based in Dharwad, Karnataka, which has been associated with the development of Joint Forest Management in Karnataka from its inception. IDS is working with several Village Forest Committees in Uttara Kannara district of Karnataka, and has developed innovative approaches for empowering women to participate in resource planning and management at the village level. Ms Hiremath was recently a member of a team set up to design the second phase of the Western Ghats Forestry Project.

Avanti Jayatilake is qualified as a botanist and an ecologist specializing in agro-ecosystems. He received his university education in Sri Lanka and postgraduate education in Australia under a Colombo Plan Scholarship. His experience includes working in government environmental agencies at various capacities. He has also worked with an international donor agency as an Environmental Program Advisor. Among his major contributions to the environment field are the formulation of three National Environmental Action Plans in Sri Lanka during the period of 1990–98, designing of several environmental projects and soliciting donor assistance that includes projects of Wetland Conservation, Community Participation in Natural Resources Management and involving local authorities in environmental management decision-making. In addition to these, he has also worked as a visiting lecturer for the Environmental Management M.Sc. Course conducted by the University of Moratuwa.

Sherine S. Jayawickrama was with the Natural Resources and Environmental Policy Project of the United States Agency for International Development (USAID), in Colombo, Sri Lanka. She is currently pursuing higher studies in the USA.

Arvind Khare has worked for several years on issues of people's participation in natural resource management in India. He was Executive Director of the Society for Promotion of Wastelands Development, New Delhi, and subsequently with the World Wide Fund for Nature–India. He is currently with the VM Foundation, Trivandrum, and continues to be involved with several national and international efforts, including the Indian network on Joint Forest Management and the Asian Forestry Network.

G.M. Khattak is a trained forester, and has worked in forestry management, education and research in Pakistan. He has also served as the Vice Chancellor of a general and an Agricultural University and as Chairman, University Grants Commission. He is also a fellow of the Pakistan Academy of Sciences. In recent years, he has veered from conventional forestry to integrated development of renewable resources with the participation of rural communities. As Senior Advisor, IUCN, he is now assisting in the transformation of the NWFP Forest Department into an efficient, transparent and participatory organization, in the implementation of the Sarhad Provincial Conservation Strategy (SPCS).

Ashish Kothari is a Lecturer in Environmental Studies at the Indian Institute of Public Administration, New Delhi, and a founder member of the environmental action group, Kalpavriksh. His focus in the last few years has been on conservation of biodiversity and protected areas in relation to local communities. He has also been active in mass movements such as the Narmada Bachao Andolan. He served on the Government of India's Environmental Assessment Committee for River Valley Projects, and is a member of the IUCN Commission on Protected Areas and the Commission on Environmental Sustainability and Planning. He is currently based in Pune, undertaking action research on community participation in conservation in South Asia.

B.J. Krishnan, born in an indigenous community in the Nilgiri Hills of Western Ghats, is an advocate by profession. He has acted as a senior legal consultant on environmental laws to governments for many years. He founded the 'Save Nilgiris Society' in the early 1980s, and successfully campaigned against many environmentally harmful projects in the Nilgiris. His current focus is on the Convention on Biological Diversity (CBD) and related matters. At present he is Secretary of The Blue Mountains School, Ooty, an alternative learning centre for children based on the educational philosophy of J. Krishnamurti.

D. Lee functioned as an overseas Postgraduate Research Assistant on the project Enhancement of Rekawa Lagoonal Prawn Fishery Project of the Universities of Colombo and Millport, Scotland. He is experienced and interested in prawn culture, and has co-authored an authoritative book on the subject with Dr John Wickins.

Sharachchandra Lele works at the Tata Energy Research Institute, and Institute for Social & Economic Change, Bangalore. He has extensively studied issues of land tenure and resource management in parts of the Western and Eastern Ghats, south India, and is involved with the biological monitoring and other aspects of enterprise development work in the Biligiri Rangaswamy Temple Sanctuary of South India.

R.K. Maikhuri obtained his Ph.D. in Ecology from the North-Eastern University, Shillong, in 1989. Presently, he is working as a Scientist at the Garhwal Unit of G.B. Pant Institute of Himalayan Environment and Develpopment (GBPIHED). He is an active team member of one of the core programmes 'Sustainable Development of Rural Ecosystems' of the GBPIHED. He has more than 50 research articles published in various national and international journals to his credit. He has also co-authored one book on the management of biological resources in the Himalayas. His current research interest is Restoration Ecology, and Village Ecosystems Function.

K.S. Murali works at the Tata Energy Research Institute, Bangalore, southern India. He is involved with the biological monitoring and other aspects of enterprise development work in the Biligiri Rangaswamy Temple Sanctuary of South India.

D. Myagmarsuren is the Director, National Service for Protected Areas and Ecotourism, at the Ministry of Nature and Environment, Ulaanbaatar, Mongolia.

Sunil Nautiyal is a senior research fellow with the Garhwal Unit of G.B. Pant Institute of Himalayan Environment and Development (GBPIHED), currently working for his Ph.D. degree dealing with the structure and functioning of village ecosystems in the buffer zone of Nanda Devi Biosphere Reserve.

Nirmalie Pallewatta works with the NGO March for Education, based in the University of Colombo, Sri Lanka.

Neema Pathak has had training in Environmental Sciences from Pune, western India, and received a Smithsonian Scholarship for a Training in Wildlife Management in China. She is currently based in Pune, undertaking action research in community participation in conservation in South Asia, while specifically focusing on a case in Gadchiroli district in Maharashtra. She is the principle author of a forthcoming book *Directory of National Parks and Sanctuaries in Maharashtra*, which she prepared at the Indian Institute of Public Administration. She has worked on various issues related to protected areas and community-based conservation. She is also a member of Kalpavriksh, an environmental action group.

Michel Pimbert is an agricultural ecologist by training with special interests in the policy implications of biodiversity conservation and in supporting participatory approaches to natural resource management. Dr Pimbert is currently Visiting Fellow at the Institute of Development Studies (UK) and consults with the United Nations and non-governmental organizations on policy and technical dimensions of sustainable agriculture and natural resource management. In his previous association with the World Wide Fund For Nature—first as the Head of the Biodiversity, Protected Areas and Species Conservation Unit of WWF–International and then as Director of WWF–Switzerland—he actively sought to integrate social justice and human rights issues into conservation. His most recent co-edited book, *Social Change and Conservation*, was jointly published by UNRISD and Earthscan.

Jules Pretty is Director of the Centre for Environment and Society at the University of Essex, UK. The current activities of the Centre include river basin and estuary management, climate change, environmental security, agricultural and food systems and community-based management of natural resources. Jules Pretty was formerly at the International Institute for Environment and Development where he was Director of the Sustainable Agriculture Programme from 1989 to 1997. He is a founding member of the Agricultural Reform Group, a trustee for the Farmers World Network and has published widely on national and social capital development in rural communities. His most recent book, *The Living Land*, focuses on agriculture, conservation and food system issues in Europe.

G. Raju has a degree in Agricultural Engineering and a postgraduate diploma in rural management. He has worked with an NGO specializing in the area of community forest management. He has worked with People's Institutions and their federation right from their evolution and has created a People's Institution forum in the Gujarat state of India. For the past five years, he has devoted his attention to devolving powers to People's Institutions in the forestry sector. He is currently engaged in developing a medicinal plant-based enterprise of the community for the sharing of the benefits to the community.

Manju S. Raju has had a long field association with Joint Forest Management (JFM) in Gujarat, western India, initially as a staff member of VIKSAT and subsequently as a freelance consultant and researcher. She has completed a research study on the factors influencing women's participation in People's Institutions in Gujarat, West Bengal and Uttar Pradesh. She has also undertaken a study of JFM in Himachal Pradesh and has contributed to preparing a number of field manuals.

P.S. Ramakrishnan is Professor of Ecology at the Jawaharlal Nehru University, New Delhi. He is a biologist by training with special interests in linking ecological and social processes in approaches to resource management. During the period 1957–74, he worked on plant ecological topics related to population ecology and ecological genetics. From 1974 to 1992, he set up and directed an interdisciplinary research project on shifting cultivation and sustainable development in north-east India. In 1988–89, he was the founding director of the G.B. Pant Institute for Himalayan Environment and Development at Kosi (Almora). He is currently involved with a network of scientists working in upland environments of the Himalaya and the Western Ghats, and linked up to global programmes such as DIVERSITAS. He is also a member of the Indian National Science Academy and the Third World Academy of Sciences.

K.S. Rao is an ecologist by training and currently heading the core programme 'Sustainable Development of Rural Ecosystems' of the G.B. Pant Institute of Himalayan Environment and Development (GBPIHED). He has more than 60 research articles published in national and international journals to his credit. He has also co-authored two books on the ecology of the Himalayan region.

Lipika Ray is State Forest Service Officer at the West Bengal Forest Department. She was posted as the Women's Coordinator in the Department for about two years, during which time she took several initiatives to increase women's participation in Joint Forest Management in West Bengal, eastern India.

S. Sankar is a scientist at the Kerala Forest Research Institute, Peechi, Kerala, southern India.

Madhu Sarin has a long association with equitable and participatory natural resource management beginning in 1980 with a pilot project in Sukhomajri, Haryana, northern India. She has been actively involved with Joint Forest Management (JFM) from its inception, and has been coordinating a national group on gender and equity in JFM as an advisor to the Society for Promotion of Wastelands Development since 1994. Among other publications, she has written the book *Joint Forest Management: The Haryana Experience*, published by the Centre for Environment Education, Ahmedabad.

K.G. Saxena is presently a faculty member in the School of Environmental Science, Jawaharlal Nehru University (JNU), New Delhi. Before joining JNU, he was associated with the G.B. Pant Institute of Himalayan Environment and Development (GBPIHED), Kosi-Almora, as the Scientist-in-Charge, 'Sustainable Development of Rural Ecosystems' core programme. He has published more than 80 research papers/articles on the various facets of the ecology and environment of the Himalayas. He has also co-authored/edited five books related to environmental issues in the Himalayas.

R.L. Semwal obtained his Ph.D. from HNB Garhwal University, Srinagar in 1992. At present, he is working as a Research Associate in the 'Sustainable Development of Rural Ecosystems' core programme of G.B. Pant Institute of Himalayan Environment and Development (GBPIHED). So far, he has published more than 15 research papers/articles in various national and international journals of repute.

Darshan Shankar helped to set up the Academy of Development Science at Karjat, Maharashtra, western India, which helps farmers build sustainability and diversity into their production systems. He is currently head of the Foundation for Revitalisation of Local Health Traditions, Bangalore, where he is involved with campaigning, research, documentation and other activities related to traditional medicinal practices.

Bharat Shrestha started his career as a university lecturer in project management and development project analysis in 1981 in Kathmandu, Nepal. He joined Agricultural Projects Services Centre (APROSC) as a professional Development Economist in 1983 and started research studies concurrently with community-based implementation activities in environment, forestry, water supply and other resource management. His working experience in this field includes Nepal, northern India, Sri Lanka, northern Thailand and Bangladesh. Mr Shrestha has a number of independent research reports and papers to his credit undertaken on behalf of the Asian Development Bank, World Bank, UNRISD, FAO and others. He is the co-author of the book *The Social Dynamics of Deforestation: A Case Study from Nepal* published by Parthenon, England.

Rajendra Singh set up the NGO Tarun Bharat Sangh in the rural part of Alwar district, Rajasthan, western India, in the early 1980s, and has since then been involved with villagers in trying to revive their natural resource base. In particular, his work has focused on rebuilding water security in the area and organizing people's movements against destructive development projects. He is also very active in various Gandhian networks in India.

R.S. Walgama functioned as a Research Assistant in Sri Lanka on the project Enhancement of Rekawa Lagoonal Prawn Fishery Project of the Universities of Colombo and Millport, Scotland. He is carrying out postgraduate masters-level research under the supervision of Professor Ekaratne.

J. Wickramanayake works as Special Projects Manager with the Asia Foundation, Colombo, Sri Lanka.

Muhammed Zuhair started his career as an Assistant Research Officer (trainee) at the Environment Research Unit (ERU) of the then Ministry of Planning and Environment, Maldives. He was awarded a fellowship for his graduation in Environmental Biology and Religious Philosophy in UK. On returning, he joined the Ministry of Planning, Human Resources and

Environment as an Assistant Environment Analyst. His main work involves conservation of biodiversity, establishment of protected areas and protected species, as well as undertaking ecological research and managing the follow-up to the international Convention on Biological Diversity for his country.

About the editors

Ashish Kothari is Lecturer in Environmental Studies at the Indian Institute of Public Administration, New Delhi, and Co-coordinator of the South Asian Regional Survey of Community Involvement in Conservation. He is a founder-member of Kalpavriksh, an environmental action group, and is a concerned activist of movements such as the Narmada Bachao Andolan. He serves as a member of two commissions of the World Conservation Union (IUCN) and of several committees of the Ministry of Environment and Forests, Government of India. Besides having contributed over 120 articles to journals and newspapers, he has also published *People and Protected Areas: Towards Participatory Conservation in India*; *Understanding Biodiversity: Life, Equity, and Sustainability*; and *Building Bridges for Conservation: Towards Joint Management of Protected Areas in India.*

Neema Pathak has had training in Environmental Science, and in Wildlife Management. She was a research associate at the Indian Institute of Public Administration, New Delhi, and is presently Co-coordinator of the South Asian Regional Survey of Community Involvement in Conservation. She is also a member of Kalpavriksh and has compiled the *Directory of National Parks and Sanctuaries in Maharashtra* (in press) besides having contributed several articles to various magazines.

R.V. Anuradha works in a legal firm in New Delhi and specializes in human rights and environmental cases. She was a research associate at the Indian Institute of Public Administration, New Delhi and a Darwin Fellow at the School of Oriental and African Studies, London. She is a member of Kalpavriksh and has written articles for various journals and newspapers. In particular, she has focused on the implications of the Biodiversity Convention for India.

Bansuri Taneja was a research associate at the Indian Institute of Public Administration, New Delhi. A member of Kalpavriksh and currently studying geography in the USA, she was also associated with the Biodiversity Conservation Prioritization Project of the Biodiversity Support Programme and USAID.